DEVELOPMENTS
in
INDUSTRIAL
MICROBIOLOGY

A Publication of the Society for Industrial Microbiology

DEVELOPMENTS
in
INDUSTRIAL
MICROBIOLOGY

Volume 1

Proceedings of the
Sixteenth General Meeting
of the
Society for Industrial Microbiology
Held at State College, Pennsylvania,
August 30 - September 3, 1959

SPRINGER SCIENCE+BUSINESS MEDIA, LLC 1960

ISBN 978-1-4899-5075-8   ISBN 978-1-4899-5073-4 (eBook)
DOI 10.1007/978-1-4899-5073-4

LIBRARY OF CONGRESS CATALOG CARD NUMBER 60-13953

FOREWORD

This volume of the proceedings of the Society for Industrial Microbiology is a record of the symposia, special lectures, and contributed papers presented at the Sixteenth General Meeting of the Society for Industrial Microbiology. The meeting was held from August 30, through September 3, 1959, at State College, Pennsylvania in conjunction with the 1959 annual meeting of the American Institute of Biological Sciences.

This is the initial issue of an annual series. It is the desire of the Society and the members of its Publications Committee that the data, discussions, and general knowledge reported in this annual series will be widely disseminated and be of help to the workers in the field of industrial microbiology. The largest part of the current volume is concerned with data obtained in a field of applied microbiological research and it was for that purpose that the Society was formed ten years ago.

Many persons are responsible for the development of this book. First and foremost are the authors themselves. Their cooperation and personal efforts have made the difference between "a book" and "no book." Furthermore, the contributors to symposia and lectures of past annual meetings deserve a word of thanks for their interest and efforts in making this a flourishing society with popular and challenging programs.

Certain individuals must be singled out for a word of praise because of their continued work toward a better society. Among these are the 1959-1960 officers of the society: C. C. Yeager, President; C. L. Porter, Past President; A. E. Prince, Vice-President; A. L. Kaplan, Treasurer; B. M. Miller, Secretary; E. Mrak, E. A. Walker, H. B. Woodruff, and E. L. Dulaney, Directors; and Mr. Earl Coleman, President of Plenum Press, who has had the courage to underwrite this book financially.

Finally, we should like to thank the members of the Publications Committee of the Society. Without their help and most able assistance your editor could never have written all the letters and inquiries, gathered all the papers, figures, and tables; read, reread, edited, and put them in publishable form. These persons, Chester Koda, Saul Rich, Joseph Becker, and A. Katherine Miller, comprised a fine editorial staff.

Brinton M. Miller
Editor and Chairman of
Publications Committee, 1959-60
Society for Industrial Microbiology

TABLE OF CONTENTS

CONTRIBUTED PAPERS
Mary O'Hara, Presiding

PRESIDENTIAL ADDRESS

INTRODUCTION

Eugene L. Dulaney

Director, Society for Industrial Microbiology

The Society for Industrial Microbiology was organized on December 29, 1949, at a meeting held in the ballroom of the Hotel McAlpin in New York City. Between 250 and 300 scientists were present. Dr. Charles Thom was elected chairman for the year 1950. He appointed the following as members with him of the Committee of Organization:

Dr. M. M. Baldwin, Battelle Memorial Institute, Columbus, Ohio.

Dr. Walter N. Ezekiel, Head Mycologist, Bureau of Ordnance, Department of the Navy, Washington, D. C.

Dr. F. G. Walton Smith, Director, University of Miami, Marine Laboratory, Coral Gables, Florida.

Dr. E. A. Walker, Production and Marketing Administration, Department of Agriculture, Washington, D. C.

Dr. W. L. White, Curator, Farlow Herbarium, Harvard University, Cambridge, Massachusetts.

Dr. C. L. Porter, Department of Biological Sciences, Purdue University, West Lafayette, Indiana.

This committee organized the first annual meeting of the Society in conjunction with the meeting of the American Institute of Biological Sciences in Columbus, Ohio, September 11-13, 1950. The Committee of Organization had prepared a constitution that was presented and adopted at a business session during this annual meeting on September 13, 1950. The constitution has been ammended on two occasions, owing to growth and changes in the Society. At present an extensive revision of the constitution is under way, preparatory to expansion of our activities. The Society has recently been incorporated as a nonprofit organization in Washington, D. C.

The Society for Industrial Microbiology developed from the need for an organizational home for industrial microbiologists. Its goal is the advancement of microbiological sciences as they apply to industrial materials and processes. At the organizational meeting, thirty-three individuals signed membership cards. Since that time, membership has grown to approximately 600 annual members and corporate members. Included in the membership are mycologists, bacteriologists, protozoologists, virologists, plant pathologists, chemists, and others representing all areas of microbiology as well as aspects of plant and animal sciences not strictly microbiological.

The Society is a member of the American Institute of Biological Sciences and has held its annual meeting with the A. I. B. S. since its founding. It is affiliated, also, with the American Association for the Advancement of Science and has held meetings with the A. A. A. S. in Philadelphia, December 27-28, 1951; in St. Louis, December 26-31, 1952; and in Washington, D. C., December 29-30, 1958.

In the past few years, an increasing number of inquiries concerning the Society and its annual program have been received. This has prompted the publication of the present volume. It will be noted that symposia make up the dominant part of the program. This is due in part to a reluctance of the Society to schedule concurrent sessions that would be required by a large number of short contributed papers. Undoubtedly, this situation will change in the near future as the need for more and varied symposia increases. In addition, there is a growing need for original contributed papers to allow a greater participation of the increasing membership on the program.

The title of the Society might lead one to assume a more limited area of activity than that which it encompasses. Whereas the established areas of microbiology are being served continually by positions on the program, new areas of microbiology are being entered as the need arises. The symposium on space microbiology is an example of this. In addition, subjects peculiar to regions in which meetings are held will be treated. As an example, petroleum microbiology will be emphasized at the 1960 annual meeting at Stillwater, Oklahoma.

The present volume represents the program of the tenth annual meeting of the Society. Publication of the proceedings of future annual meetings is planned on a regular basis. In addition, monographs on selected microbial topics are contemplated, as is a Society journal.

TRAINING FOR CAREERS IN MICROBIOLOGY

Dr. C. L. Porter, presiding

Panelists: Dr. L. S. McClung
Dr. H. Boyd Woodruff
Dr. C. L. Porter

INTRODUCTION

C. L. Porter

Department of Biological Sciences, Purdue University

As the topic for this symposium is worded we should explore the broad field of possible microbiological careers, which would include teaching, academic laboratories, industry, and federal and state laboratories. I think another possibility should be mentioned, writing and public relations work. To date, there does not seem to be much demand or compensation for purely literary efforts in the microbiological field. Most microbiologists do publish reports of their investigations in technical journals, but I believe that another sort of writing is important and would be, in itself, a career. There is a real need for men thoroughly trained in microbiology and preferably with some practical industrial experience to interpret in language which can be understood by the layman, achievements in microbiology. There are two reasons why this kind of career has little popular appeal among scientists: 1) There is a tendency among scientists to regard those who write about research instead of engaging in research as "second-class" citizens. 2) Individuals who are scientifically trained become incapable of using a language which is understood by the man in the street.

There is a real need for this type of popularized science if microbiology or any other science is to receive the recognition among the masses of the people that it needs to gain financial support, to be trusted, and to become respectable.

To tell the story of a scientific discovery, avoiding the scientific jargon which appears in technical journals, to simplify it, to make it of absorbing interest, and to avoid exaggeration is an art which few can master.

Science Service is to be commended for its contribution to this particular art. In my opinion it could be improved and the coverage could be greater.

3

Two general phases of microbiological careers have been covered in this symposium. Dr. McClung discusses the subject from the academic point of view and Dr. Woodruff explains industrial possibilities and problems.

I do not wish to encroach upon the ideas that are discussed by either of these gentlemen, but there are certain peculiar aspects which I would like to mention.

Teaching and research in academic laboratories do not attract the salaries that are paid in industrial laboratories, but there are other compensations which must be considered. Salaries should be sufficient to provide for necessities, including savings with a margin left over for the so-called luxuries. In academic life you have certain built-in luxuries which provide the kind of happiness and contentment that no amount of money will purchase. This intangible factor resident on college campuses should not be overlooked by a young person planning a career.

Employment in our various and varied industrial laboratories offers opportunity for practical service with which our colleges cannot compete. Industrial employment in the past was routine to the point of deadly monotony. This situation is materially changed today. Many industrial plants are housed in buildings more modern than are usually found on college campuses. Some physical plants which I have visited are situated on grounds that more resemble a college campus than the grimy and weed-strewn grounds that formerly looked more like an old state prison than an industrial plant.

New discoveries in biology, physics, and chemistry lend excitement and glamour to microbiological research that was nonexistent only a few years ago. I cannot think of a greater challenge to the active mind today than industrial employment in microbiology. The old regime formerly imposed by industry, which was a form of paid slavery, is to a great extent now past history. Employees receive greater encouragement now to present papers at scientific meetings, and to publish in technical journals. A few companies are broadminded enough to permit research to be carried on which is not done strictly for potential profits.

Industry offers several types of employment: research, development, production, sales, and management. I have requested Dr. Woodruff to enlarge on these possibilities.

Dr. McClung speaks on careers from the academic point of view and the training he believes necessary for entry into this field.

TRAINING FOR A CAREER IN INDUSTRIAL MICROBIOLOGY: ACADEMIC VIEWPOINT

L. S. McClung

Department of Bacteriology, Indiana University, Bloomington, Indiana

The remarks which I make concerning the academic point of view for the training program for a career in industrial microbiology are based both on my experience as the Chairman of the Department of Bacteriology in a typical Midwestern state university and also on my prior experience as an industrial microbiologist.

THE IDEAL PROGRAM

Any program of training should present sound, basic courses in the fundamental sciences in the undergraduate curriculum as a background for (a) the graduate program or (b) industrial positions at the bachelor's level. It appears to me that there is little difference between the needs of these two groups and therefore the suggested courses which I will list later will probably be the same for both types of individuals.

While the standardization of curricula, to the point where some national committee or other group may "certify" a given program, may have considerable merit, there are also some disadvantages to these standardized programs. One of the chief obstacles to overcome, should such programs be considered desirable, will be the reluctance of college administrators to accept the will of an outside group in the determination of the courses to be included in any degree pattern. An informal inquiry from a committee of the Society of American Bacteriologists, several years ago, presented a tentative plan to the deans of a large number of colleges and universities. There was an immediate strong voice of disapproval of the principle of the certification of the program, even though the particular program suggested had much merit and may already have been in force in many schools.

DESIRABLE COURSES

Any attempt to compile a list of "required courses" is likely to be met with many arguments for and against certain specific courses. Perhaps as an alternate procedure we may compile a list of "desirable courses." In the list given below the approximate credit value is included, based on a program of 120 credits for the bachelor's degree.

1. General Biology (or Botany and Zoology) (5-10 cr.)
2. Genetics (3 cr.)
3. Chemistry (Introductory, Qualitative, Quantitative, and Organic I as minimum; Biochemistry desirable but difficult to schedule) (20 cr.)
4. Mathematics through Calculus (6 cr.)
5. Physics (10 cr.)

6. English Composition and Humanities and Social Sciences and perhaps a foreign language (6-18 cr.)
7. Microbiology
   a) General Bacteriology, Bacterial Physiology (10 cr.)
   b) Mycology and perhaps Algology, Protozoology and Virology (5 cr. +?)
   c) Industrial Microbiology (3-5 cr.)
   d) Immunology and Taxonomy (10 cr.)
   e) Literature (or Library Methods) (1 cr.)

Some comment on the above list may be in order. I believe that all micro-biologists should have a firm foundation in general biology and hence the inclusion of the usual introductory course and the very important fundamental course of genetics. Certainly, the chemistry training should include at least four semesters, and although biochemistry is desirable, we find it difficult to schedule in the undergraduate program. The mathematics courses would be desirable for those continuing the graduate program and as a foundation for physics, and, if not required, should be strongly recommended. Training in English composition is usually necessary at the college level and certainly the microbiologist should have a foundation in the humanities and social sciences. Foreign languages would be required for graduate programs but would be less necessary for the person starting employment following the bachelor's curriculum.

With regard to courses in microbiology, I would feel that in addition to an introductory bacteriology course and a course in microbial physiology, there should also be some introduction to mycology and perhaps to algology, protozo-ology, and virology. A course in industrial microbiology should prove of much value though it must be admitted that certain phases of the laboratory work in this area can usually not be successfully handled in such courses since many problems require the industrial laboratory and substitute experience even with small-scale pilot fermentors or other small-scale operations is not suitable. I would regard immunology as a sufficiently basic science to be included in the list of desirable courses and the training in taxonomy might be either with bacteria or with other forms. My reference to a literature (or library methods) course is meant to refer to a course in which the student will be introduced to the abstract journals, index sources, and the review journals, as well as the original publications. I would also like to see in such a course extensive use made of library facilities in the prep-aration of a written "semester paper" or other long project.

A.B. DEGREE VERSUS B.S. DEGREE PROGRAM

In most colleges the A.B. program pattern allows the student a fair degree of choice in elective courses and thus for the students who transfer to the program late in their academic career there is less difficulty in meeting specialized re-quirements. The B.S. pattern, on the other hand, commonly is a more rigid program and offers only limited elective work in addition to the specialized course listing.

GRADUATE PROGRAM

Naturally, I would list as required at the beginning of a graduate program any course or area omitted from the group listed above. Beyond this list the student would be expected to broaden and strengthen his course background as a preparation for research in a problem of a basic nature. I believe that there are many such problems which often are listed as "applied" but I see no reason to compromise quality in the research training of the student. Additional chemistry is almost a requirement for graduate training in microbiology and certainly all candidates should have had at least one basic course in biochemistry.

CONCLUSIONS

It seems to me that in the above program it would be well to reemphasize the desirability of including sound, basic courses in the traditional sciences, and the interdependence of these, and I see little reason to differentiate the training pattern for those who expect to go into different areas of industrial microbiology. The individual soundly trained can adapt to the position available. Indeed I am frequently told by present-day industrial microbiologists that the function of the university is to give only the basic training with the thought that the particular industrial position which the individual assumes will quickly provide the necessary industrial aspects.

I believe that industrial microbiology must be presented to the college student as a challenging, rewarding field, which indeed I believe it is, and with indications that there is an increasing trend in many companies to be concerned with basic research as well as with the immediate solution of minor problems at hand. I would also point out that there needs to be done a better job of presenting this field as a potential career area to the young high school student. In these days of increased national interest in science, it is our experience that many students receive the motivation for a career at the high school (or earlier) level.

SOME RECENT DEVELOPMENTS

As some of you are aware, we have been concerned at Indiana University with the production of a film entitled "A Career in Bacteriology." I am happy to tell you that in this film we have included industrial microbiology as one of the important areas in which a student may expect employment following completion of his training.

I should like to inform you that the Society of American Bacteriologists recently appointed a Committee on Education, of which I am presently serving as the chairman, and has directed this committee to assemble various teaching aids relating to bacteriology for the high school teacher of biology, and in other ways to promote an interest in bacteriology at the high school level. It is abundantly evident that this program is bearing fruit.

I might also tell you of another recent development which will bring greater emphasis to microbiology at the high school level. I refer now to the series of 120 films which are being developed by the American Institute of Biological Sciences for use as a possible core of a tenth-grade biology course. I am happy to tell you that one complete unit--12 films--will be devoted to microbiology. The students who have the opportunity to take such a course in the future will be, in my opinion, far more interested in microbiology and better trained when they come to the college program than those of us who had the traditional high school course of yesteryear.

TRAINING FOR A CAREER IN INDUSTRIAL MICROBIOLOGY: INDUSTRIAL VIEWPOINT

H. B. Woodruff

Merck Sharp & Dohme Research Laboratories, Rahway, New Jersey

In presenting the case for a career in industrial microbiology, one is immediately placed on the defensive. Wherever we turn, we hear the disadvantages of industrial microbiology, the grimy and weed-strewn grounds which appear like a state prison, the routine manipulations required, the conversion of a promising scientist into a plodding drudge. One of our speakers today has stated privately that he would not encourage a brilliant and ambitious student to attach himself to most commercial organizations. When it comes to the advantages of employment in industrial microbiology, financial gain is presented grudgingly, but even this is counterbalanced by the argument that academic life provides certain built-in luxuries which no amount of money will purchase.

How shall I answer these charges? I know they need not be true. I have had as great challenge and stimulation from conducting research in an industrial laboratory as I would have had under any other circumstances. I am certain that several members of my department feel the same. I have friends in our factory and in our sales organization who have had a fully satisfying life applying their knowledge of microbiology in these areas. Yet I agree that there is much basis for the charges against industrial microbiology. The routine, the drudgery, the dissatisfaction are apparent throughout any organization. We should examine why this is so and then take a positive approach to the problem of industrial microbiology.

I believe the dissatisfaction with industrial microbiology is due to misfits, people with improper training who have been forced into an improper position for their personality and interests. In any industrial research laboratory there are people who are fully qualified as salesmen or factory supervisors and who could be extremely happy in these areas. But these are people who were trained as research scientists, who throughout their college training heard degrading remarks about salesmen, and who have been forced into research to do an inadequate job, to be unhappy, and to be the drudges. There are others who have all of the qualities to be excellent research men, but who fail to reach satisfaction because they have had an inadequate trade-school type of applied training. A career in microbiology, whether industrial or academic, is dependent on education.

Our universities have three great responsibilities: First, to lead students into the correct fields for their qualifications, while at the same time--this is very important--recognizing all fields of microbiology for which they are training men as of equal social status. Secondly, the universities have a responsibility to train the students adequately and correctly for the fields they select. Finally, universities have the responsibility of guiding students into appropriate jobs, and of giving accurate and honest references, to be certain that the men are started out in an effort where they can be successful and will be happy.

8

I should like to digress here and write further on this third responsibility of job counseling and honest references, for I feel strongly about it. For many years industry has been the dumping-ground on which professors can drop their inadequate and ill-prepared students. The frankly dishonest reference or the tongue-in-cheek variety seem to be the accepted thing when a student is recommended for a position in industry. I am certain that the same reference letter is seldom written for an unqualified man who applies for a research position in a university as for one applying for a research position in an industrial laboratory. Frankly, this dishonesty has forced many large industries to select new employees from a very small number of universities, those where the major professors are personally known and trusted; or, in the rare case where a man is selected from an unknown department, the major professor's recommendation is discounted and the selection is based on a personal interview and on the grade record of the individual in his undergraduate education.

Following this statement of the responsibilities of our academic institutions, let me proceed with a positive approach and consider these problems: first, the opportunities for a microbiologist in industry, and, second, training required for microbiologists to be successful in their industrial careers.

I am sure that Dr. Porter and Dr. McClung are far more qualified than I to list industrial opportunities, because they know where their men have gone over many years. I must speak from narrow experience in one area, the pharmaceutical industry, an industry which employs large numbers of microbiologists. In order of importance, the opportunities which I see in industrial microbiology are: (1) microbiological research; (2) microbiological control (this offers many opportunities for research in analytical methods and techniques); (3) either the supervision of microbiological production processes, or the application of principles of sanitation in the subdividing of pharmaceutical products, the preparation of food products, or in waste disposal; (4) sales of pharmaceutical products, of products involved in microbial deterioration or of the many other products resulting from microbial growth or having some influence on microbial growth; (5) the special responsibility of sales service in which aid is given to an individual who purchases a microbial product; (6) patent law, the writing of patent drafts on microbiological processes; and (7) the position of administrative assistant who aids management in companies by contributing information and background on microbial processes.

These are only seven areas of industrial possibilities. They can be subdivided further and I am sure there are others which I have not mentioned but, even among these seven, it is a mistake to think of them all as areas which are opportunities for the "microbiologist." I believe the training for certain of these areas should be radically different from the training for others of the areas. In fact, it would be preferable to use two different names for people who have had the two types of training, but it is difficult to devise proper names. For the purposes of the present discussions, let us consider the opportunities in terms of Application Microbiologists and of Research Microbiologists. Those whose intention it is to enter research laboratories of industry or conduct research in analytical methods and supervise the control over microbial products will fall in the category of Research

Microbiologists. All the other opportunities for industrial employment can be grouped together under the term Application Microbiologists.

Because of the variety of opportunities available in industry, let us consider the Application Microbiologist first. What are his qualifications and what should be his training? First, I have said that I believe that the university has a responsibility for directing students into the appropriate courses to train them for industrial opportunities. How does one search out and find men who will make good Application Microbiologists? Preferably, these people are extroverts, for in sales, in production, in patent law, or in administrative work they will be dealing with other people, primarily people who are not scientists. These Application Microbiologists must be people with a very broad interest in science, people who wait for and each month read avidly such journals as Scientific Monthly, who enjoy the romance of science, possibly who find time to read science fiction. I know of one of the highest-paid lawyers in New York, a man who has handled many patent cases on fermentation processes, who is one of the best examples of an Application Microbiologist. It is true that he is a lawyer but his undergraduate training was in science; he has a wide knowledge of microbiological processes (incorrect, to be sure, in the minor details). He has a plush, expensive office on Wall Street, overlooking New York harbor, with a large bookcase containing, not law books (he can go to his own law library for those), but every issue of paper-backed science fiction that has appeared in the last ten years. It is men with such a widespread and general interest in science who make the best Application Microbiologists.

To work in these applied areas—production, sales, or administration—what training should men have? At the undergraduate level, I believe a single year of biology, two years of chemistry, and possibly physics, would be desirable; but the specific training should be in the applied aspects of science, i.e., in applied microbiology, for it is the useful aspects of science in which Application Microbiologists show an interest and this interest should be broadened by telling them of every application of microbes which it is possible to fit into the time available in an undergraduate program. These Application Microbiologists are the people who will be expected to transmit bits of scientific information to others, to help make a sale of a pharmaceutical product to a doctor, to convince a paint manufacturer that a fungicide will improve his product, to inspire operators to efficient work in conducting a multi-thousand gallon fermentation tank, to solve a customer's problem, or even to ferret out the scientific angles which will convince a patent examiner that a major scientific advance has been made and that a patent should be granted to cover a new product.

The primary responsibility in the careers of these men in industry is not as scientists, it is not as microbiologists—it is to deal with people. If Application Microbiologists need any one solid groundwork in their education, it is in psychology, and courses in psychology should be included in their program. Unfortunately, today most people who are being trained for the field of Application Microbiology are taught by college research professors who have little understanding of psychology, in fact, little respect for it. Frequently, a good high school biology teacher with a bit broader experience could do a better job in training Application Microbiologists than many college professors, because the high school biology teacher is an educator by profession, having an understanding of the probable interests and potentialities of all students.

The Application Microbiologist should in the course of his training have occasional contact with the solid university research man. But this contact should be just sufficient to whet his scientific interest. The individuals being trained as Application Microbiologists should be encouraged and be stimulated, by being fed fascinating bits of information concerning science, and given a broad cultural background. They should not be looked down upon because they are not promising research scientists, but they should be taught microbiology and general biology as cultural subjects to broaden their background, and to make their lives in industry fuller and more interesting.

In contrast to this positive approach, where does industry today get its Application Microbiologists? Usually, they are selected from men thoroughly trained in basic science, as Dr. McClung has recommended, essentially trained as research scientists because this is the only university course open to them. In such a specialized course, these men having general interests either get such low grades that they are unable to obtain a research job, and thus are forced into the application field with a feeling that it is a second-best choice; or these individuals somehow get into research, possibly with the help of an inaccurate reference letter such as I mentioned previously. Once in industrial research, they prove to be misfits and are thrust out into the application area after having proved to be failures. These individuals enter the field with a psychological disadvantage from the start. Improperly trained, looked down upon by their peers who are in research, these people enter the field where they should be most happy and could make a real success, hiding the fact that they are microbiologists because of a false shame.

Next, what are the opportunities for microbiological research in industry? I see no difference between successful scientific research in industry or in the university. The primary objective of the research man is the search to expand knowledge. The area of effort undertaken by this individual is of far less importance than the desire of the individual to learn by trial, by experimentation. The university has been extolled as a center of research because only here, it is said, can a man do what he wishes, can a man become an expert in a narrow field. For a few men who, by personality, are qualified, even forced, to work as individuals, not in cooperation and collaboration with others, the university does offer a unique opportunity. However, the narrow specialist in the university is the complement of the routine drudge in industry—both make a minor contribution to be sure, but the contribution is very slight in comparison to that made by the broad, well-grounded research man who is capable and will make advances in a wide range of areas of microbiology.

The true scientist will find a scientific approach to any problem, to any assignment, whether it be in an institution laboratory, a university laboratory, or in an industrial laboratory. In the pharmaceutical industry the man assigned to a screening program for new antibiotics will design his screen so that each day's operation will tell him something new, furnish him with a new fact of knowledge, which in the end will teach him about new kinds of microorganisms, their properties, new approaches to finding antibiotics, or possibly even the significance of antibiotics in microbial metabolism. On the other hand, if the Research Microbiologist is engaged in a basic study of protein synthesis by microorganisms, or on some other even more fundamental problem which no industry could afford to pursue on its own, the Research Microbiologist will again find new knowledge from each day's experimentation and add this knowledge together piece by piece until a new idea is proved.

In microbiological research, it is the man who makes the job and not the job which controls the man. There is little fundamental difference in the research of the man engaged in research in industry or in the university. It is secondary factors which should control which field a Research Microbiologist enters. Such things as the desire to teach and train will lead one to the university. Interest in seeing fundamental discoveries converted into applied successes leads to the industrial laboratory.

It was stated earlier that the Research Microbiologist must have different qualities and a radically different training from the Application Microbiologist. It is the responsibility of the university to ferret out those who should be trained as Research Microbiologists. These men should show an interest in factual matters, an interest in the basic mechanisms underlying life. They should recognize areas where we are lacking in precise knowledge and have a desire to find the answers, to solve problems by experimental trial and error. What training does such a man need? He must be trained in the basic sciences, with emphasis on chemistry—at least one type of chemical training for every year that he is in college.

His biological training should be in the basic aspects, the controlling principles of genetics, evolution, and classification, with laboratory studies which illustrate these principles, and with possibly one quick course serving the applied aspects of microbiology.

Finally, the Research Microbiologist must receive the stimulation of working with an established research man in the laboratory day in and day out. There can be no education factory here, no mass-scale university approach. Small classes, especially in the later undergraduate years, are essential and, during the graduate period, the student must learn by precept, by working directly with his proctor. So, again, you see I am thrusting the responsibility back upon the university.

In closing, let me say that I believe we have compounded further the error of our approach in considering careers in microbiology as being divided into academic careers or industrial careers. Our careers should be considered correctly in terms of research microbiology as a career, of application microbiology as a career, and possibly, of teaching as a career. The personality qualifications of the individual will determine in which one of the three areas—research, application, or education—that individual can prove successful. The training for the three career areas should be radically different. Finally, each career in itself should be held up as a desirable goal with good status. Once individuals receive the proper training, and then proper recognition is given to the value of microbiology as a cultural subject in preparing people for industrial opportunities outside of research, then in large measure will the stigma now attached to industrial microbiology disappear.

SPACE AGE MICROBIOLOGY

Dr. Alton E. Prince, presiding

Panelists: Dr. P. C. Trexler
Mr. Alan M. Shefner
Dr. John M. Leonard
Dr. Albert R. Krall
Dr. B. Kok
Dr. Walter M. Bejuki

INTRODUCTION

Alton E. Prince[*]

Materials Laboratory, Wright Air Development Center
Wright-Patterson Air Force Base, Ohio

Microorganisms of our time have evolved from types which can be seen as fossils in very ancient Devonian red sandstone deposits (Gwynne-Vaughan and Barnes, 1927). Whether these were fungi or algae makes little difference because, more than anything else, their appearance indicates that the present-day morphology of these organisms was already developed more than six hundred million years ago. This is evidence that living microorganisms have survived all these years without obvious morphological change; therefore it is believed that no new environments will be found on this planet, Mars, or perhaps, even Venus, that will entirely prohibit growth of all microorganisms as we know them. Because we are thinking about the exploration of space and its contents, let us examine briefly the conditions as they are known at present for Venus and Mars.

The surface of Venus has been reported to have conditions all the way from all desert (Jones, 1952) to all ocean (Moore, 1958). I prefer the theory that Venus is in a kind of carboniferous era and, if so, is covered with jungles with all sorts of unknown fungi. Actually, the surface of this planet has never been seen, so that all theories have to be discounted considerably.

Much more is known about the surface of Mars. Strughold (1953) is of the opinion that low forms of both plants and cold-blooded animals could exist on Mars. The atmosphere of Mars (Levitt, 1956) is mostly nitrogen with some carbon dioxide and possibly a trace of oxygen and water. Summer temperatures on Mars in tropical regions are believed to go as high as $+20$ C. The atmospheric pressure at the

[*] Present address: Aerospace Medical Division, Wright Air Development Division, Wright-Patterson Air Force Base, Ohio.

surface of Mars is equal to an altitude of about 55,000 ft on earth. While higher plants and animals as we know them could not exist on Mars, one can agree with Strughold that at least the lower forms of plants, namely the fungi and algae as we know them, could do reasonably well on this planet.

With this brief soar into space and the future, we can now return to our discussion of present-day facts, and for this perhaps we should keep at least one foot on *terra firma*. It has been known for many years that low temperature by itself will not kill fungi. As early as 1913, Buller and Cameron exposed *Schizophyllum commune* to −190 C for 21 days without obvious damage. In 1930, Faull exposed dry spores of *Neurospora crassa* to temperatures from −170 to −190 C for one hour without evident damage to germination. In fact, quick-freeze drying under vacuum is now a common way to preserve fungi. As for the other extreme, the resistance of thermophilic bacteria to +100 C is well known. *Aspergillus niger* apparently can survive temperatures of about +65 C (Wolf and Wolf, 1947). It is common knowledge that some bacteria can adapt even to the water in a swimming-pool-type nuclear reactor. These are given as examples of known conditions that microorganisms can withstand.

I have reported preliminary experiments indicating that lyophilized spores of *A. niger*, with other selected fungi and bacteria, withstood vacuums from 1×10^{-5} to 5×10^{-7} mm Hg for 32 days without apparent damage to viability. Additional unreported experiments with bare spores of *A. niger* indicate that this fungus can withstand low temperatures of about −60 C in a vacuum of 1×10^{-4} mm Hg alternating with ambient conditions for four days without apparent damage to viability. In any experiments conducted we should be more than casually alert to note data which do not exactly fit present-day facts, because such data may serve as a basis for new concepts. We read in our daily papers about rockets being sterilized before being shot at the moon and beyond. This is very important, but I am more concerned about what new forms may be brought back to earth from these other planets. With the bacteria and fungi, I include in thought, at least, the virus. In this introduction, I have taken liberties that could not be taken by the authors who follow because each has a specific area to discuss. A series of papers should be presented on every subject, but each author has been selected because of his knowledge in his particular phase. It is believed that together, these papers form a sound basis for further discussion and experimentation in preparation for the time when someone will return from foreign planets with information and specimens to tell us what is really there.

REFERENCES

Buller, A. H. R., and Cameron, A. T. 1913 On the temporary suspension of vitality in the fruit bodies of certain *Hymenomycetes* . Proc. Trans. Roy. Soc. Canada, 6, 73-78.

Faull, A. F. 1930 On the resistance of *Neurospora crassa*. Mycologia, 22, 288-303.

Gwynne-Vaughan, H. C. I., and Barnes, B. 1927 The structure and development of fungi. The University Press, Cambridge, England.

Jones, H. S. 1952 Life on other worlds. English University Press, London, England.

Levitt, I. M. 1956 A space traveler's guide to Mars. Henry Holt Co., N.Y.C.

Moore, P. 1958 The planet Venus. The Macmillan Co., N.Y.C.

Strughold, H. 1953 The green and red planet. The University of New Mexico Press, Albuquerque, New Mexico.

Wolf, F. A., and Wolf, F. T. 1947 The fungi, Vol. II, John Wiley and Sons, N.Y.C.

GNOTOBIOTICS IN RELATION TO SPACE BIOLOGY

P. C. Trexler

Lobund, University of Notre Dame, Notre Dame, Indiana

The prospect of examining the moon and neighboring planets for the presence of living forms certainly provides one of the most interesting phases in the development of man's knowledge. There is some evidence of life on Mars and it seems likely that the first examination of specimens will be made by soft-landed robots (Technical Panel on the Earth Satellite Program, 1958). However, someone with a training in biology will have to actually explore the planet to obtain a thorough knowledge of its biota. This adventure imposes staggering demands upon a great many sciences and technologies and forces development of the in-between areas. The possible returns from such an adventure are also imposing, not only in terms of the knowledge gained about extraterrestrial nature but through the effort required to get there. The fields of electronics and metallurgy already have been enriched as a result of the basic research required by their participation. A prolonged space voyage will require a much greater insight into bioecology (Tobias, 1959) than we now possess and the design of a probe will require a reexamination of the fundamental properties of living matter, to mention two areas in which biology will profit from the preparation required.

Gnotobiotics is but one of the many disciplines of biology that can contribute to this adventure and will also be stimulated by the effort. Since this discipline is not widely known at present, it may be best to first define the term and then discuss its relation to biology before considering the space aspects. The word *gnotobiotics* is derived from the term "Gnotobiota" meaning as its roots indicate, *known-biota* (Ward and Trexler, 1958), i.e., organisms grown by themselves or in association with known kinds of organisms. Pure cultures and germfree animals are gnotobiota as well as associations of them. So far, the use of the term has been confined to mammals and birds. It is especially useful in designating associations of pure cultures of bacteria with "germfree" animals to avoid an illogical terminology. The term gnotobiotics designates the entire discipline involved in the production and use of gnotobiotes.

Germfree mammals (guinea pigs) were first reported in 1895. They were more of a laboratory curiosity than a useful research tool. Reproduction in a germfree mammal, the rat, was reported by Reynier *et al.*, in 1946. At the present time gnotobiotic rats, mice, and a few other species are usable research tools, capable of large scale production and use in any laboratory equipped for sterile operations (Trexler, 1959). They are now produced at the rate of several thousand per year at the University of Notre Dame alone. Three other colonies are in existence and several commercial breeders are interested in their production and sale as soon as there is a market for them.

How do these animals fit into the scheme of biological research? Often the first reaction of biologists is that the gnotobiote is a highly specialized animal, useful in a few areas, such as the investigation of the role of the intestinal flora, the study of the etiology of certain diseases such as dental caries and the *in vivo* pure culture of viruses and other parasites. On the other hand, the viewpoint that the gnotobiote, potentially at least, is the ideal experimental animal seems quite attractive. Microbiota most likely play a role, which is usually ignored, in the majority of investigations that terminated in the death of the animal. There are many examples of latent infections in laboratory animals evoked by stress or stimuli, such as *Pseudomonas* activity following the radiation of mice, bartonellosis in rats following splenectomy, and PVM virus in mice following intranasal instillation of serum (Dubos, 1958). Obviously a variation from normal in an experimental animal caused by an agent other than the one under test can but detract from or invalidate the results of the experiment. In this respect laboratory investigations differ from field or clinical trials which utilize animals and environments as they are found. A gnotobiote, i.e., a pure animal or one associated with a selected series of pure microbic cultures, is a simpler biological system than the conventional animal, and therefore a better laboratory animal for about the same reasons a pure grade of a chemical is preferred as a reagent over the crude source material.

The development of gnotobiotics has not depended so much upon new concepts as upon the availability of new materials which permit the devising of convenient methods and procedures. The traditional sterile or aseptic technique depends upon an open system for the control of contamination. Direct contact contamination is controlled by a rigid discipline as far as the handling of materials is concerned. Airborne contamination is controlled by careful manipulation and either the elimination of drafts or the controlled movement of sterile air. While these methods with an assist from antibiotics have been satisfactory for pure culture of microbes, tissues, plants, and the smaller animals, they have not been useful with mammals and birds.

The closed-system apparatus used for the rearing of germfree mammals and birds consists essentially of three elements: a container for enclosing the controlled or sterile area; a means for manipulation within the area; and a means for the introduction and removal of materials. The construction of the container for enclosing the sterile environment, which we term "the chamber," may vary from stainless steel, sturdy enough to withstand 20 lb steam pressure, to light-weight plastic film which can be either sterilized within an autoclave or with germicide. Manipulation is ordinarily accomplished through shoulder-length rubber gloves attached to the wall of the chamber or complete sterile garments which will permit an investigator to walk around in the sterile chamber. Air is sterilized either by incineration or filtration, introduced into the chamber and removed through filters or a simple trap. There are three methods of introducing sterilized objects: the sterile lock, germicidal dip bath, and split-seam transfer. The sterile lock consists essentially of a double-doored chamber serving as a vestibule. Objects introduced into this sterile lock can be sterilized by means of steam under pressure or the surface of presterilized materials can be sterilized by means of a gaseous germicide such as peracetic acid. The germicidal dip bath or trap is used to pass materials into the chamber after submersion for a sufficient time to obtain sterilization. The bath can also serve as a simple liquid door provided presterilized materials are handled with sterile technique, the liquid serving merely as a seal. The third method, usable only with plastic film, is termed the split-seam transfer. A sterilized package is cemented or heat-sealed to the wall of the plastic film isolator; the cement or fused seam serves to immobilize all contaminating organisms. A passageway is then cut with a hot-wire cautery. After passage of the materials the opening can be sealed with cement or a heat-sensitive tape.

There appear to be two aspects of space biology to which gnotobiotics may make a significant contribution: the prevention of biological contamination attending space travel, and the use of gnotobiotes for satellite research or space voyages. A committee established by the International Council of Scientific Unions (Hughes, 1959) has defined contamination as the spoilage of materials for future investigations and recommended procedures to reduce the hazard. Experience with sterile operations certainly has demonstrated that closed systems are required to maintain sterility for long periods. Since sterility itself cannot be demonstrated experimentally, but only the absence of various organisms as shown by accepted tests, the procedures must incorporate a safety factor. Apparently, the necessary desiderata are approached more closely by the maintenance of continuous colonies of gnotobiotes than in any other operation. The sterile isolators or rooms are essentially islands which are kept free of biological contamination and can serve as test areas for space biologists. The apparatus and operations resemble those used for the control of dust and gaseous contamination and should present few difficulties to trained technicians. It may be well to use the term "gnotobiotics" to designate these procedures in order to distinguish them from the usual inadequate "sterile technique" associated with the laboratory and the operating room.

There are two phases to the sterilization of a space probe: the sterilization of the contents and the sterilization of the surface after it leaves our biosphere. In the simplest case the probe could be assembled and sterilized by heat or possibly ethylene oxide much as flasks of media or canned foods, allowing an appropriate safety factor. A prolonged residence at 150 to 165 C is undoubtedly the simplest sterilization procedure, though it is unlikely many can be so treated. The alternative is to sterilize the components individually by suitable means and assemble them within a closed sterile or gnotobiotic system. The size and complexity of assembly determines the type of isolator required. A large probe could require a rather complex assemblage of apparatus not usually associated with sterile practice. However, the engineering problems seem to be rather simple in comparison with the other aspects of space exploration.

Since the problems involved in the design, construction, and launching of space probes are complex enough as it is, the advisability of adding to these problems can be questioned. The danger of irretrievably losing the opportunity to investigate forms of life that may exist on the moon or the planets has been pointed out by Lederberg and Cowie (1958). The cost of taking the necessary precautions may be viewed eventually in terms of other considerations, including national security and prestige and the needs of other sciences. Perhaps some experiences with gnotobiotes may be of value in this consideration. Aerobic spore formers such as *Bacillus subtilis* occur in large numbers in the gut tract of conventional guinea pigs without causing harm. However, when pure cultures of aerobic spore formers are

introduced into the environment of a germfree guinea pig they usually kill the
animal within 48 hr. This is similar to the introduction of foreign organisms that
produce so many of our pests. We are all familiar with the problems that arise
from upsetting the delicate ecological balances in our biosphere. Conceivably a
single contaminant from earth could overrun and destroy many forms of life upon
another planet. Such an occurrence could not be interpreted as implying a higher
development of life upon our planet nor does it indicate how the extraterrestrial
organisms would fare in the presence of a balanced earthian biota.

While it is not possible to estimate the value of scientific information acquired
by a trip to Mars, it hardly seems likely that it will be economically feasible to
retrieve minerals or other materials from such a distance. However, it is not at
all unlikely that organisms evolved in the harsh environment of Mars may provide
a useful crop in the cold and arid regions of earth. Such an organism, if found,
might be of considerable value to man, but could be exterminated through irre-
sponsible contamination.

The second phase of the sterilization of space probes involves surface sterilization after leaving the earth's biosphere. Undoubtedly, the surface of the probe
will become contaminated as it is placed upon the launching pad and as it makes
its way through the atmosphere. There are at least three approaches to the problem. The probe could be housed in a pod or capsule in order to preserve the
sterility of its surface and contents. The contaminated protective pod could be
discarded after leaving the biosphere. Secondly, it may be possible to coat the
surface of the probe with a protective layer which could be destroyed at a predetermined distance. Perhaps some derivative of a solid propellant might serve
as the basis for such a coating. As an alternative, perhaps some of the exhaust
gases from the rocket engine could be used to raise the skin surface to sterilizing
temperature without harm to the contents. Finally perhaps a technique can be devised for using the radiant energy in space to obtain surface sterilization. The
sterilization of space probes seems entirely feasible with our present knowledge;
however, considerable research and development will be required to determine the
most practical methods.

The contamination of the earth by extraterrestrial life carried by a returning
space probe must also be considered. While some bacteriologists minimize the
danger of extraterrestrial organisms establishing themselves in the face of competition from earthian forms (Bisset, 1959), experience with intercontinental contamination by organisms of many levels of development indicates that the danger
is real. Three centuries of sickness and death on both sides of the Atlantic followed the discovery of the New World (Editorial, 1959). The necessity for strict
quarantine is recognized by all modern nations. It is impossible to anticipate what
suitable niche in our biosphere may be found by an organism with an entirely
different origin and evolutionary history. For example, much of the organic material necessary for man's existence is preserved by drying. Martian organisms
may need far less moisture for growth. Their metabolism may differ so as to
make the poisons and preservatives now used ineffective. Composition, metabolism, and methods of control must have the highest priority in any investigation
of extraterrestrial forms of life. If interplanetary exploration becomes a reality,

it probably will be necessary to establish satellite quarantine stations. Conceivably, the contamination of the earth by forms of life from another planet could be as destructive to life as we know it as an all-out nuclear war.

Removal of contamination for a spacecraft's return trip involves essentially the same steps as in leaving the earth. However, there are several complicating factors not present for the original flight. Initially, and probably for a considerable time thereafter, not enough will be known about the resident forms to establish sterilization procedures with confidence. The entire operation will have to be carried out without the help of assistants that remain. These difficulties do not seem to impose any insurmountable obstacles. Conceivably a robot could place samples in a sealed container, load the return vehicle and, if necessary, remain behind. The loading port could be seared by the exhaust gases of the rocket and the entire surface of the craft decontaminated by one of the methods suggested for leaving the earth's biosphere. The craft could be received at a quarantine satellite (Buchheim, 1959) and the contents studied with appropriate precautions.

The exploration of a planet requires that the crew of a spacecraft leave through a decontamination lock such as is now used with our sterile rooms. The decontamination procedure in leaving the craft could be simplified if no spore formers were taken along. The explorers would return to their craft through a lock, the inner barrier of which would be transversed by a split-seam transfer so as not to depend upon the action of germicides. The design and engineering problems involved seem rather simple in comparison with those of getting there and back.

At the present time it hardly seems sound to flatly suggest that gnotobiotes be used in satellite experimentation. This primarily is due not to technical difficulties with their use but because there is not enough information accumulated to interpret the results of such experiments. The cost of satellite experimentation in terms of scientific effort as well as finances demands the use of the best animal for the experiment. Eventually gnotobiotes seem destined to be the animal of choice for many purposes because they are a simpler biological system than the conventional animal. The absence of an indeterminate microbiota should simplify the interpretation of long-term experiments and probably aid in the design of apparatus for the maintenance of the proper environment by eliminating microbic spoilage.

The use of gnotobiotes in satellite experiments could be challenged upon the basis that the results of such experiments are primarily intended to help establish the conditions necessary for a trip by man, and he certainly will not be a gnotobiote. There are at least two considerations to this proposition. What evidence do we have that the influence of the microbiota of one group of animals is comparable to that of another group and a different species? Perhaps a laboratory animal with the microbic associates of man rather than those typical of the species may serve as a superior experimental model. Secondly, in preparing man for an extended space voyage it seems appropriate to ask what biological associates should be taken along. Certainly all pathogens, parasites, and pests should be excluded. However, a pathogen is determined not only by the species but also by stress, resistance, and the past experience of the host. Both *Escherichia coli* and *Staphylococcus* species have caused many deaths in man but are part of the normal complement of microbes. At present, we do not know whether or not microbes are of benefit to man in the maintenance of an individual's health. Super-infections, following antibiotic therapy, indicate the need for research in this area (Kersten, 1958). Do some species, present in all individuals, contribute to man's susceptibility to radiation sickness, aging, or other diseases? This information would be most useful to space travelers as well as to the rest of the population. It seems unlikely that such information can be obtained without the general use of gnotobiotes as laboratory animals. Undoubtedly, once man acquires this knowledge, he can do something about it. Sieburth (1959) has observed antarctic birds that lack an aerobic intestinal flora, probably caused by a naturally occurring antibiotic in their diet. The interrelationships among the microbes that inhabit the surface and cavities of our bodies, the development of a variety of sterilizing agents, and new materials of construction present an unsurpassed challenge for improving man's lot. Certainly, at present we live in an unexplored microbic jungle and have succeeded merely in controlling the more obvious "predators."

Undoubtedly space exploration will provide another stimulus to explore the foundations of biology and medicine. Judging by the contributions biology and medicine have made in the past to mankind's welfare, the results of another phase in the development of these subjects could very well justify the expenditures made, no matter what is found upon the moon or the planets. The cross-fertilization between biology and the different sciences, engineering, and the technologies concerned seems likely to yield the first benefits.

REFERENCES

Bisset, K. 1959 Microbes to the moon. The Listener, 61, 410-411.

Buchheim, R.W.   1959   Space handbook: Astronautics and its applications. House Doc. No. 86, U.S. Gov't Printing Office, Wash., D.C.

Dubos, R.J. 1958 Infection into disease. Perspectives in Biol. & Med., 1, 425-435.

Editorial. 1959 The gnotobiotic future. World Med. J., 6, 119-120.

Hughes, D.J. 1959 Contamination of the moon. Science, 130, 340-342.

Kersten, H.H.   1958 Intentional replacement of bacteria following antibiotic therapy. J. Iowa State Med. Soc., 48, 240-243.

Lederberg, J., and Cowie, D.B. 1958 Moondust. Science, 127, 1473-1475.

Reyniers, J.A., Trexler, P.C., and Ervin, R.F. 1946 Rearing germfree albino rats. Lobund Rep. 1, 1-44, Univ. Press, Notre Dame, Ind.

Sieburth, J.M.   1959   Gastrointestinal microflora of antarctic birds. J. Bacteriol., 77, 521-531.

Technical Panel on the Earth Satellite Program. 1958 Research in outer space. Science, 127, 793-802.

Tobias, C.A. 1959 Space Science. Science, 129, 906-910.

Trexler, P.C. 1959 Progress report on the use of plastics in germfree equipment. Proc. Animal Care Panel, 9, 119-125.

Ward, T.G., and Trexler, P.C. 1958 Gnotobiotics: a new discipline in biological and medical research. Perspectives in Biol. & Med., 1, 447-456.

ADAPTATION OF MICROORGANISMS TO RADIATION

Alan M. Shefner

Biological Research Section, Armour Research Foundation, Chicago, Illinois

Adaptations of microorganisms to particular environmental stresses have been frequently observed and may be initiated by a variety of causative factors. If the stress imparts a selective advantage to variants in the culture, or if it induces an increased frequency of mutations to forms with a selective advantage, the original strain may vanish with time and be replaced by a strain better adapted to the changed environmental conditions. It is an example of such a stress and response that I intend to consider.

It has been known for some time that various species of bacteria have a wide range of sensitivities to radiations. In addition to the more obvious differences in radiosensitivity between spores and vegetative cells it has been shown that radiation may affect the various strains of a particular bacterial species to varying degrees.

A number of strains of *Escherichia coli* have been subjected to radiation, and it has been demonstrated that *E. coli* strain B is relatively radiosensitive. A radiation resistant mutant of *E. coli* B was isolated by Witkin in 1946. The mutant strain, designated as B/r, could be distinguished from the *E. coli* B by the difference in their response to ultraviolet radiation.

Witkin (1946) demonstrated that the ultraviolet radiation was acting as a selective agent for the more resistant B/r forms which occur as spontaneous mutations in the *E. coli* B culture. However, at higher doses a considerable increase was found in the frequency of the radioresistant mutants, and it appeared evident that the selecting agent was also the inducing agent and that the number of induced mutations far outweighed the number of spontaneous mutations (Witkin, 1947).

Thus it was logical to expect that *E. coli* B placed in an environment that included exposure to ionizing radiation would be able to adapt to this environmental condition, or more properly, to expect the resistant cells, which have a high selective advantage, to steadily increase in frequency in the culture.

Gaden and Henley (1953) tested this proposition by irradiating cultures of *E. coli* B with Co^{60} at a dose rate of 80,000 r/hr for periods of up to 3 hr daily. Samples of the irradiated cultures were transferred to fresh broth, incubated for 17 hr and again irradiated. They found that the increase in radiation resistance as the number of irradiations increased is quite marked.

The same workers irradiated strain B/r in a similar fashion and found, that after only 17 irradiations a strain was derived from strain B which was even more radioresistant than that designated as B/r.

Witkin and others have investigated the differences in response of *E. coli* strains B and B/r to various agents. Bryson (1949) found B/r to be more resistant to the lethal actions of substrates which had been previously irradiated with UV or which had been treated with hydrogen peroxide. Clark (1952) has investigated the cata-

21

lase activity of *E. coli* B and B/r by measuring the rate of oxygen output from a dilute hydrogen peroxide solution in a Warburg apparatus. He found that the B/r strain has a much higher catalase activity than strain B. This was also reflected in the growth of the two strains in nutrient broth to which varying concentrations of hydrogen peroxide had been added. The B/r strain is much more resistant than the B strain to higher doses of hydrogen peroxide.

Thus it had been shown that the radiation resistance of *E. coli* B cultures would rise in response to repeated exposures to radiation and also that the resistant strain of *E. coli* designated as *E. coli* B/r also had an increased resistance to the presence of hydrogen peroxide in the culture medium.

Work has been conducted at our laboratories in an effort to combine and extend these observations.

The cultural and exposure conditions used varied somewhat from those of Gaden and Henley (1953).

E. coli B grown in tubes of nutrient broth at 37 C was washed and resuspended in M/15 phosphate buffer, pH 7.0, and samples were removed for later use in plate counts. The suspended cells were then irradiated with Co⁶⁰ at a dose rate of approximately 26,000 r/hr for 4 hr. Dilutions of both the irradiated cel ⁻ ⁻nd the nonirradiated controls were than plated out on nutrient agar and incuba... ...e remaining irradiated cells, constituting the bulk of the irradiated sample, were then centrifuged down, resuspended in fresh nutrient broth and incubated at 37 C for approximately 48 hr.

This procedure was repeated through eleven irradiations. Figure 1 shows a plot of the surviving fraction of cells following each irradiation. The surviving

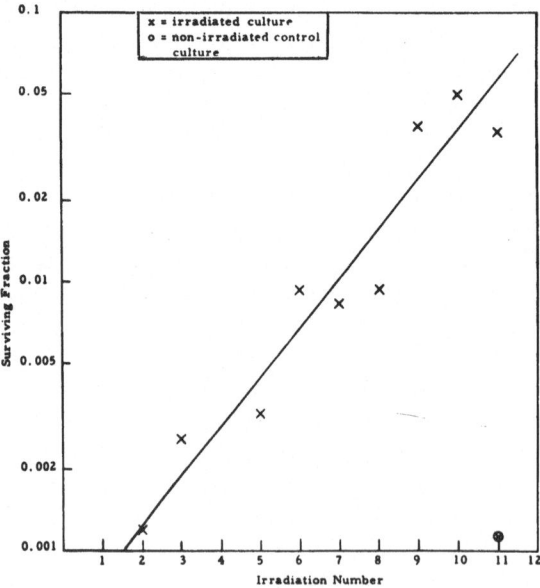

Fig. 1. Change in surviving fraction of E. coli following repeated exposures to gamma radiation.

fraction is the number of viable cells per milliliter following irradiation divided by the viable cells per milliliter in the respective control. An experimental control culture of *E. coli* B was carried through the same procedures during the course of the experiment without, of course, being irradiated. At the eleventh irradiation this control was also irradiated and the surviving fraction agrees with the result obtained in the original irradiation.

The percent of organisms surviving the test irradiation dose has increased approximately thirty-fold over the course of the eleven irradiations.

From the cells plated out after the eleventh irradiation, ten single colony isolates were picked and checked for species purity by streaking on eosin methylene blue agar, by microscopic examination, and by their ability to ferment particular carbohydrate substrates. The radiation resistance of each of the ten isolates was then determined and the surviving fractions ranged between 2 and 5%.

Eight of the ten isolated strains together with *E. coli* B and authentic *E. coli* B/r were also grown in nutrient broth to which was added amounts of hydrogen peroxide ranging from 0 to 75 ppm. Light transmittance of 24-hr-old cultures was measured at 540 mμ in an Evelyn colorimeter with an uninoculated broth tube set at 100% relative light transmittance. As is readily seen in Table 1 all of the R-11 isolates are at least as resistant to hydrogen peroxide as is strain B/r and all differ markedly from the response shown by *E. coli* B.

Table 1. Growth of *E. coli* Strains at Varying Hydrogen Peroxide Concentrations*

Strain	Hydrogen peroxide, ppm									
	0	4	8	12	16	20	40	50	60	75
B	64	77	78	87	84	100				
B/r	55	62	75	61	71	66	90	92	91	100
R-11-1	65	63	62	67	67	67	88	90	88	90
R-11-2	60	63	62	67	66	62	87	89	86	88
R-11-3	61	65	63	68	68	73	92	89	90	89
R-11-5	55	59	60	62	64	69	89	88	88	87
R-11-6	64	64	66	67	65	68	89	88	87	97
R-11-7	60	63	64	68	74	75	90	90	86	87
R-11-9	64	66	68	67	70	71	84	85	87	87
R-11-10	60	63	69	73	65	71	86	90	95	96

*Uninoculated blank tubes set at galvanometer deflection of 100 (= 100% relative light transmission.)

The degree to which the ability to resist radiation and to withstand the presence of hydrogen peroxide in the media are interrelated has yet to be conclusively determined.

During our studies we also conducted a parallel program of radiation exposures using a different species of bacterium, *Serratia marcescens* strain 9986. Somewhat more ambiguous results were obtained which are shown in Fig. 2. If the first point is considered to be aberrantly high, a least square regression line is

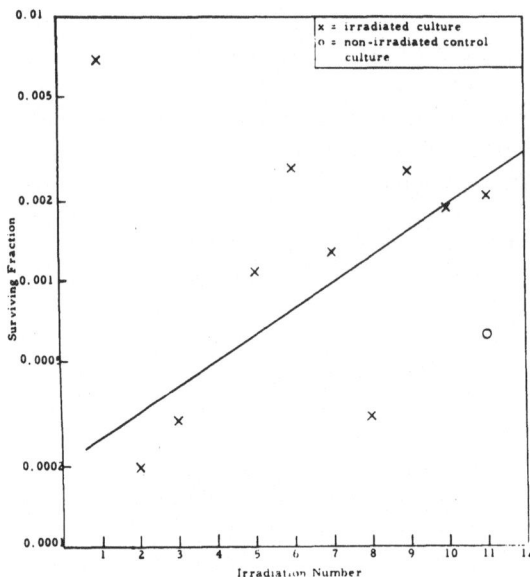

Fig. 2. Change in surviving fraction of S. marcescens
following repeated exposures to gamma radiation.

obtained which shows a gradual increase in the radioresistance of the culture.

Before general conclusions are drawn concerning the readiness with which radiation-resistant strains may be obtained it should be noted that Gaden and Henley (1953) found a decrease in radiation resistance in *E. coli* strain 15 after five successive irradiations.

These results may be applicable in pointing out certain of the problems to be faced in space microbiology. If photosynthetic gas exchangers are used, the algae may be subjected to a variety of environmental stresses including radiation, toxic vapors, and temperature fluctuations, among others.

The responses to these stresses may take many forms. To be most pessimistic, the stress environment could lead to the death of the culture. If some degree of protection can be provided for the organisms, or if the stresses are less toxic in nature, they may still be sufficient to act as a selective force on the culture. The selection of variants more resistant to the stress or the initiation of adaptive responses may lead to a decrease in the photosynthetic rate of the culture. On the other hand the adaptive response, which could include a detoxification of atmospheric contaminants, might not affect the gas-exchange capacity of the culture, but might render the algae unfit for consumption. Other responses might render the algae more susceptible to invasion by microbial contaminants with subsequent deleterious effects. Another possibility which must be considered is that the environmental factors would lead to an increased frequency of mutation in the culture with the mutant cells having less desirable characteristics.

The conclusion one can reach is that even if these fears are overdrawn, it still would be advisable to determine the effects of possible environmental stresses on the functioning of the gas-exchange system. If it is determined that a particular environmental factor which may be difficult to engineer out of the system is likely to be a hazard, it may be possible to select for strains more resistant to this factor. In addition it may be important to use algal strains with a low spontaneous mutation rate.

REFERENCES

Bryson, V. 1949 Influence of irradiated media upon *E. coli* resistant to radiation and nitrogen mustard. Abstracts of papers, 49th General Meeting of the Society of American Bacteriologists.

Clark, J. B. 1952 Catalase activity in *Escherichia coli*. J. Bacteriol., 64, 527-530.

Gaden, E. L., and Henley, E. J. 1953 Induced resistance to gamma radiation in *E. coli*. J. Bacteriol., 65, 727-732.

Witkin, E. 1946 Inherited differences in sensitivity to radiation in *E. coli*. Proc. Natl. Acad. Sci. U.S., 32, 56-58.

Witkin, E. 1947 Genetics of resistance to radiation in *E. coli*. Genetics, 32, 221-248.

ALGAE AND SUBMARINE HABITABILITY

John M. Leonard

U.S. Naval Research Laboratory, Washington, D.C.

The Chemistry Division of the Naval Research Laboratory has been studying submarine atmospheres for almost thirty years. The studies have included chemical means of carbon dioxide removal, alkaline metal oxides as sources of oxygen, and physical techniques for submarine air analysis. Such work becomes more and more important with the ever-increasing significance of the submarine in national defense. Studies of several new aspects of submarine habitability are being made; as a small part of the program, the writer has explored the literature to evaluate the possibilities of algae as means of long-term oxygen supply and carbon dioxide removal.

When one remembers that the earth with its envelope of air is indeed a "spaceship" of plants and animals, it is natural to regard their association as a key to the major atmospheric problem of the nuclear submarine. Carbon dioxide is one of the principal end products of plant and animal catabolism; its production, through the consumption of atmospheric oxygen, is a most important source of energy for all living things. A reverse process, by which oxygen can be restored to the air and the oxidized carbon can be reduced, is essential because our planet is a closed system. The principal agent in the reverse process is of course, the green plant. The reaction is endothermic and the energy for it comes from the absorption of sunlight. With some gross simplifications, the two reciprocal processes may be represented as follows:

$$C_6H_{12}O_6 + 6O_2 \longrightarrow 6CO_2 + 6H_2O \tag{1}$$

$$6CO_2 + 6H_2O \longrightarrow C_6H_{12}O_6 + 6O_2 \; . \tag{2}$$

In each reaction, the steps are many and complex. They are imperfectly understood, even at best. The photosynthesis reaction (2) shows the carbon being incorporated into a simple sugar, a possibly misleading simplification. All plant tissue must be formed by some variation or extension of this scheme. The synthesis of protein might be illustrated by the formation of the dipeptide, alanylalanine, with urea as the source of nitrogen:

$$CO(NH_2)_2 + 5CO_2 + 4H_2O \longrightarrow C_6H_{12}O_3N_2 + 6O_2 \; . \tag{3}$$

Nitrate may also serve as the nitrogen source:

$$12CO_2 + 4NO_3 + 12H_2O \longrightarrow 2C_6H_{12}O_3N_2 + 21O_2 \; . \tag{4}$$

In summary, the material products of photosynthesis are oxygen and more plant. The plant meets its energy needs through a reversal (1) of the photosynthesis process. The utility of the operation to human and animal economy depends upon (2), (3), and (4) exceeding (1).

Application of photosynthesis to the submarine demands, in principle, that a portion of the vessel be a sort of greenhouse in which the plants are illuminated by artificial light and through which the air of the submarine is circulated. Space limitations of course call for a high rate of photosynthetic activity per unit volume of equipment. The higher plants are eliminated at once because of the functional specializations of their tissues. Only the chlorophyll-containing parts make a photosynthetic contribution; supporting and vascular tissue, for example, would be parasitic. Only the simplest plants, those with the highest fraction of photosynthetic tissue, deserve serious consideration. These, most notably, are the green algae.

Sunlight experiments in the mass cultivation of algae are not new. The nutritional excellence of the algae, especially their high protein content, has long commended them to serious study. The topic is treated at length by Burlew, 1953.

It is evident that a photosynthesis air-purification unit would be essentially an algal "farm" or "factory" with a major product--algae--which would have to be disposed of. (The interplanetary traveler of the future will probably eat it; the modern submariner is unlikely to be in such straits.) Algae may finally prove to be unfeasible as a means of submarine air purification, but they should not be dismissed without careful experimental inquiry.

In June, 1956, the Office of Naval Research sponsored a conference of some of the nation's leading authorities in the fields of photosynthesis and submarine air purification. Though the estimates of space and power requirements differed considerably, the conference was unanimous in regarding the process as seemingly feasible and worthy of thorough exploration. Even the highest guess of the power requirement--1000 horsepower--did not seem to alarm those acquainted with the capabilities of the nuclear submarine's power plant (Office of Naval Research, 1956).

Some of the important features of a photosynthesis gas exchanger are compactness, tolerable power consumption, long-term dependability, and adjustability to the varying respiratory needs of the crew. The unit should be as nearly automatic as possible so that the vessel is not burdened with additional personnel. Human breathing requirements--and therefore the over-all capacity of the exchanger--can be estimated rather well. In turn, the nitrogen and mineral needs of the unit and its output of algae can be computed with reasonable accuracy. Figuring the actual space and power needs of the device is quite another matter. Even the estimate of the theoretical energy requirement is crude enough. The writer offers his estimates, based largely upon "handbook" type data.

The daily energy need of the average man ranges from about 2200 kcal for a sedentary worker to 4500 or more for a man performing hard physical labor. A value of 3000 will suffice for a round-figure estimate. In normal metabolism a liter of oxygen is equivalent to about 4.8 kcal and the respiratory quotient R.Q. (CO_2 produced/O_2 consumed) is about 0.83. Therefore,

$$\frac{3000}{4.8} = 630 \text{ liters } O_2/\text{man}/24 \text{ hr}$$

$$630 \times 0.83 = 523 \text{ liters } CO_2/\text{man}/24 \text{ hr}$$

This 523 L of CO_2 contains 285 g carbon which is incorporated into algal tissue which will average 50% carbon, dry weight basis, and about 3% ash (Burlew, 1953). Because these are rough approximations, such niceties as figuring the carbon content on the dry, ash-free basis, and ash on simple dry weight are here ignored. The exchanger for a 100-man submarine would, therefore, be producing about 60 kg or 130 lb of dry algae per day. On the assumption that none of the ash is recoverable the system would require about 4 lb of minerals daily.

Obviously the system must operate more efficiently than the algal "farms" where space is of no great importance and where the caprice of sunlight is tolerable. Under laboratory conditions one strain of *Chlorella* shows (Myers, 1946) a "daily growth constant" of about 0.84 which means that the algae double themselves about every 8 hr. Under steady-state conditions, a unit producing 60 kg algae daily (20 kg/8 hr) must, of necessity, contain 20 kg cells, dry weight. A concentration of 12 g/L is fairly strong, but by no means unthinkable. This works out to 1600 liters (57 ft^3) of algal suspension alone.

This writer visualizes the unit as a thin vessel, conforming to the contour of the submarine hull and penetrated lengthwise with a large number of tubes, very closely spaced, somewhat like the fire tubes of a boiler. The tubes, made of transparent plastic, would contain tubular lighting elements. The air would be bubbled in through the bottom; the algal suspension would circulate around and among the tubes. It is likely that the overriding consideration of volume will be not the volume of the suspension, but rather how closely the requisite number of light elements can be packed. As will be seen later, a large amount of energy must be put into the system via a multiplicity of relatively low-wattage light sources.

Most of the power need of the unit is, of course, for illumination. It is an unfortunate coincidence that this most important requirement is the most difficult to calculate. Part of the trouble is due to the fact that the quantum efficiency of the photosynthetic reaction has not been rigorously determined. It must suffice here to say that the question may have no simple numerical answer. Recent conservative estimates (Emerson and Chalmers, 1955) are in the range of 8 to 12 quanta per molecule of oxygen produced. Other reports (Burk *et al*., 1951) argue for much lower quantum requirements, 4.0, 4.8, even as low as 0.95.

The entire visible spectrum is effective in photosynthesis, but quanta in the red region are sufficiently energetic. With the higher frequencies, most of the excess energy appears as heat, a small amount as fluorescence. For red light ($\sim$7000 A), the energy is about 40 kcal/einstein, which amounts to 480 kcal/mol of oxygen on the conservative estimate of 12 as the quantum requirements. This 480 cal represents the luminous, photosynthetically useful energy that must be put into the system to produce one mole (22.4 L) of oxygen. As noted above, the "heat equivalent" of oxygen in normal metabolism is about 4.8 cal/L--about 107 cal/mol. This means that the useful, radiant input must be 480/107 or 4.5 times the energy requirement of the crew. Obviously the conversion of electrical energy to useful radiation must be achieved as efficiently as possible. The nuclear submarine may have power in abundance, but space for a large cooling plant is another matter entirely.

Most photosynthesis studies in the laboratory have been conducted with incandescent lamps whose output of visible energy is 10 to 12% of the power input

(Knowlton, 1949). From spectral energy distribution data and the known quantum energy requirement of photosynthesis, it is apparent that about 85% of the visible energy (or 8.5 to 10% of the input) is available for the reaction.

Incomplete data (luminosity values only) suggest, to this writer at least, that the efficiencies of mercury vapor lamps would be rather low. A luminous output of 40 lumens/watt of input is a fair estimate; but the mercury spectrum has almost none of the red light which is so efficient in photosynthesis, and it is rich in the shorter wavelengths with their quantum-energy surfeits.

The sodium vapor lamp, with its energy right in the middle of the visible spectrum, may have some possibilities. An output of 45 lumens/watt of input is attainable. One lumen corresponds to $1/650$ watt at 0.555μ, the wavelength of maximum sensitivity of the eye. The luminosity curve for the normal eye indicates a relative luminosity of about 0.70 in the vicinity of the principal sodium lines (0.59μ). This gives an estimate of 0.0022 watt/lumen or 0.0022×45 or 0.10 watt of luminous energy per watt of input power. The quantum energy of yellow light is about 48 cal, of which $40/48$ or 83% is directly usable. Multiplying 83% by 0.10 equals 8.3%, the fraction of the power input available for photosynthesis.

Red neon tubes seem to have attractive possibilities because the orange-red portion of the spectrum is utilized so efficiently in photosynthesis. Outputs of 12 to 16 lumens/watt look poor because of the low sensitivity of the eye in the long wavelengths; the useful output may be quite high. However, the writer has been advised that the presently available neon sources are not very effective. The matter may deserve close scrutiny.

One type of hot-cathode fluorescent lamp, presumably typical, puts out 8.2 watts of light from a 40-watt input, for an efficiency of 20.5%. The output is about 43 lumens/watt. Of the 8.2 watts of visible energy, about 40%, or 3 watts, is in the frequencies above the quantum energy corresponding to 40 cal. So only 5.2 watts (13%) of the 40-watt input is useful in photosynthesis. Sylvania "VHO, cool white" fluorescent lamps have an actual output of about 50 lumens/watt, though there is one 8 ft tube for which a higher efficiency is claimed. This indicates a photosynthetically useful output of about 15%. The total allocation of energy would be approximately as follows:

1.	Convective and conductive heat	50 %
2.	Nonvisible radiant energy	25 %
3.	Visible radiation	25 %
	a. Photosynthetically useful	15 %
	b. Nonuseful	10 %

Most of item 1 can be eliminated by blowers before it reaches the suspension; item 2 plus item 3b will enter the suspension and must be dissipated from it. As already noted, the useful radiant input must be about 4.5 times the metabolic needs of the crew, viz., $4.5 \times 3000 \times 100/0.15 = 9 \times 10^6$ cal per day, which is 440 kw or 550 horsepower, the power requirement of fluorescent lights. This power estimate is based upon a low, conservative figure of $1/12$ as the quantum efficiency and upon the use of presently available light sources. As noted already, a quantum efficiency of $1/8$ is not unreasonable and this revises the power estimate down to slightly less than 300 kw. And all the while, the estimates of Warburg and his associates can be a fillip to the engineer's hopes.

As stated above, about 35% of the 9 million calories will enter the suspension as photosynthetically useless radiation. The writer estimates that only about 2% of this waste energy will be dissipated in vaporization of water from the algal suspension. Even this 2% will reappear wherever the moisture is condensed. There will be an input of heat that could warm the suspension about 4 C in a minute. The writer regards refrigeration and heat-exchanger problems as beyond his competence. But dissipation of this energy requires a temperature gradient--the higher the better. Obviously the common algae whose growth optima are in the range 25 to 30 C are unsuited. Kratz and Myers (1955) have described a new blue-green alga, *Anacystis nidulans*, with a temperature optimum of 41 C; Sorokin and Myers (1953), a *Chlorella* with an optimum of 39 C. Whether they have the other requisite characteristics remains to be seen. It seems to the writer that a search for algae of even higher temperature optima should be undertaken, even though the problem of air cooling and dehumidification would be aggravated by their use.

It has been noted many times that the alga chloroplast, under continuous illumination, become saturated at relatively low light intensities--even as low as 100 ft-c. A thin layer of the green algal suspension absorbs most of the incident radiation, the highly efficient red especially. This appears to be the dilemma: If the cells closest to the surface have optimum illumination the rest are in comparative darkness--respiring parasites that are doing no photosynthesis. On the other hand, "dazzling" the surface layer, even to the extent of damaging the cells, is the price of admitting a photo-useful amount of light to the cells beneath. The happy solution is that the chloroplast can tolerate high-intensity light when it is delivered in short bursts (Burlew, 1953). In fact, radiation is used most efficiently when there are alternate periods of light and dark. There are apparent benefits, whether the alternations are of the order of milliseconds or of hours. An optimum of 1 millisecond of light alternating with 10 to 20 milliseconds of darkness has been noted (Phillips and Myers, 1954). It appears that a proper pattern for stirring a suspension of suitable opacity could provide these alternations. And the system can tolerate no dead spots; unlighted cells are not photosynthesizing but they are consuming oxygen and producing carbon dioxide just the same. Good stirring is only one of the besetting problems.

In addition to high-temperature tolerance, there are several other requisites. It is unfortunate that *Chlorella* is deficient in all of them. These properties are nonsticking, nonfoaming, and long-term stability in continuous culture. The tendency of *Chlorella* to coat the walls of containers and other submerged surfaces would obstruct light transmission, among other things. Myers (Office of Naval Research, 1956), regards surface treatments of the containers as mere temporary palliatives. Foam control probably can be worked out. The last factor, long-term stability, deserves particular mention. Most of the algal characteristics cited so far are based upon experiments of relatively short duration. Might not Myers' *Anacystis*, for example, shift its growth optimum downward from 41 C in an experiment lasting several months? Might not the growth rate, even the apparent quantum efficiency, change with many weeks of continuous culture? It is highly likely, for any biological response is the resultant of many reactions, each with its own co-efficient.

The growth rate of the alga is the most important single variable; gas exchanges are inextricably bound to the rate at which the organism grows. In addition to securing algae which are naturally fast growers, nutritional means for stimulating growth should be thoroughly explored. Considerable work of this sort has been done with the higher green plants, not nearly so much with the algae. Brannon and Bartsch (1939), for example, report that 33 ppm phenylacetic acid stimulated *Chlorella* growth over 260%. Yvon (1956) found another of the classical growth promoters, alpha-naphthylacetic acid, to stimulate *Chlorella* at 10^{-5} to 10^{-6} molar doses in moderate light, to retard growth in intense light. On the other hand, Lhotsky (1954) observed no stimulation from either alpha-naphthylacetic or 3-indolylacetic acids. In the present connection, studies should include gibberelin, plant hormones, vitamins, and any other possible growth promoters.

As suggested already, there may be no presently known species of alga which is completely satisfactory for the purpose. At best, there is none which cannot bear improvement. The quest for new species might well go beyond the sampling of ponds and puddles. Chemically- and radiation-induced mutations could produce the optimum combination of physical and physiological characteristics. The idea is scarcely novel, even with the algae. Claes (1954), for example, has used x-ray mutation of *Chlorella* to study the biosynthesis of carotenoids. Siedel (1954) has reviewed recent work on *Chlorella* mutants. Muller (1954) treats the general topic of producing mutations by radiation at some length.

Serious biological and engineering difficulties confront the development of a photosynthesis gas exchanger for the nuclear submarine. But the prospects appear good enough to warrant a serious experimental effort. The work could be channeled as follows:

1. As many organisms as possible should be given a simple, short-time laboratory screening for desirable characteristics, e.g., high growth rate, high-temperature optimum, high-intensity light tolerance. The organisms should be sought through direct collecting and by solicitation from other laboratories. "New" species can be sought by irradiation and chemical treatment of cultures at hand.

2. Algae which qualify under the screening tests should be examined for the stability of their desirable characteristics in protracted, continuous-culture experiments in which the gas exchanges would be monitored.

3. With a few test species of algae, chemical growth-stimulators can be assayed in the same fashion as the organisms above.

4. A pilot-plant exchanger, big enough to reflect the problems of a full size unit, should be undertaken with one of the presently available species. The species would probably be supplanted by more suitable species as they are found, but the sooner the biological engineering problems are faced, the better. This important element of the task should not be undertaken without competent engineering assistance, right from the start. Perhaps it would be more accurate to say that it is an engineer's job at which the biologist and the biochemist can help.

REFERENCES

Brannon, M. A., and Bartsch, A. F. 1939 Influence of growth substances on growth and cell division in green algae. Am. J. Botany, 26, 271-279.

Burlew, J. S. 1953 Algal culture, from laboratory to pilot plant. Carnegie Inst. of Washington Publication 600.

Burk, D., Hendricks, S., Korzenovsky, M., Schocken, V., and Warburg, O. 1949 The maximum efficiency of photosynthesis: a rediscovery. Science, 110, 225-229.

Burk, D., and Warburg, O. 1951 Reaction involving one quantum, and energy cycle in photosynthesis. Z. Naturforschung, 6b, 12-22.

Claes, H. 1954 Biochemical synthesis of carotenoids in *Chlorella* mutants. Z. Naturforschung, 9b, 461-470.

Emerson, R., and Chalmers, R. 1955 Transient changes in cellular gas exchange and the problem of maximum efficiency of photosynthesis. Plant Physiol. 30, 504-529.

Knowlton, A. E. 1949 Standard handbook for electrical engineers. McGraw-Hill Book Co., N.Y.C.

Kratz, W. A., and Myers, J. 1955 Nutrition and growth of several blue-green algae. Am. J. Botany, 42, 282-287.

Lhotsky, S. 1954 Algae as testing material. Ceskoslov. Biol., 3, 45-48.

Muller, H. J. 1954 Radiation biology. Vol. 1. McGraw-Hill Book Co., N.Y.C.

Myers, J. 1946 Culture conditions and the development of the photosynthetic mechanism. III. Influence of light intensity in cellular characteristics of *Chlorella*. J. Gen. Physiol., 29, 419-427.

Office of Naval Research. 1956 Conference on photosynthetic gas exchangers. Symposium Report ACR-13.

Phillips, J. N., and Myers, J. 1954 Growth rate of *Chlorella* in flashing light. Plant Physiol., 29, 152-161.

Siedel, W. 1954 The biosynthesis of chlorophyll. Angew. Chem., 66, 735-738.

Sorokin, C., and Myers, J. 1953 A high-temperature strain of *Chlorella*. Science, 117, 330-331.

Warbug, O., Geleick, H., and Briese, K. 1951 The separation of photosynthesis and back reactions. Z. Naturforschung, 6b, 417-424.

Yvon, A. 1956 Action of α-naphthaleneacetic acid on growth of *Chlorella pyrenoidosa*. Compt. rend., 242, 1205-1207.

STUDIES ON ALGAL GAS EXCHANGERS WITH REFERENCE TO SPACE FLIGHT

A. R. Krall and B. Kok

Research Institute for Advanced Studies, a Division of the Martin Company
Baltimore 12, Maryland

In this paper, we shall consider problems that may arise in placing man in a small, completely enclosed space for a long period, and our belief that an algal gas exchanger may assist in solving these problems. We shall give data on growth of a large algal culture, demonstrating that such a culture will grow reliably for long periods of time and will generate enough oxygen and food to satisfy a man's requirements. It will do this within a reasonable volume using a relatively small amount of electrical energy.

Starting from the viewpoint of the contained man, we have outlined a closed ecological system as shown in Fig. 1. The first thing the man will run out of is oxygen. At the same time he will pile up a toxic concentration of carbon dioxide. Later, water and food supply will become critical and at some time toxins will accumulate to greater than critical levels. It will be necessary to recycle most of

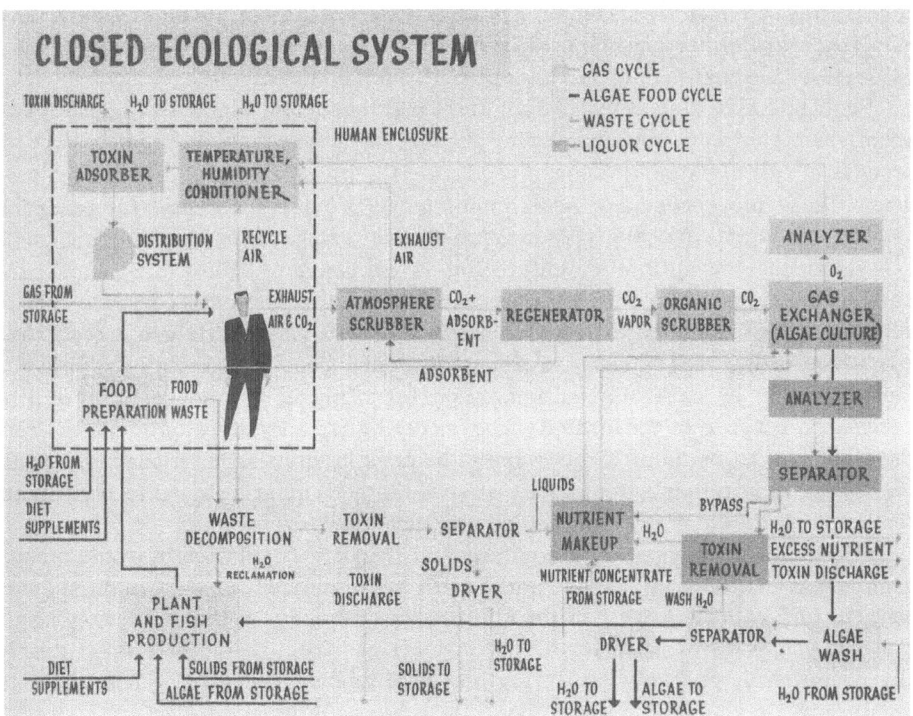

Fig. 1. Schematic diagram for a closed ecological system.

the organic material if the system is small and man is to remain in the system for prolonged times, say of the order of months or years. Think of the difficulties of reutilizing finger-nail clippings and hair in such a system. Ingram (1958), at Columbia, has done a thorough survey of problems such as these that are likely to be encountered in a closed ecology.

Our work on these systems originated with our interest in algal culturing as a means of removing carbon dioxide and regenerating oxygen. Connecting the algal culture and the man is not as simple as hooking two pipes between the two components. In the schematic diagram shown in Fig. 1, we have attempted to provide for every difficulty we could foresee. It will probably be necessary to remove noxious organic and inorganic materials from the atmosphere that are put there by the man or by his machines. Short-chain hydrocarbons such as ethylene, produced by both man and machine, are known to exert profound effects on the growth of higher plants. For example, at concentrations of 0.01 ppm ethylene can promote adventitious root growth on tomatoes. There are other things that are not shown specifically. For instance, some investigators (Gafford, 1959) have found that an algal culture operating in a closed gas system produces traces of carbon monoxide. Our own work has not shown this, possibly because we so far have used only open, rapidly growing cultures. We have shown in other work (Krall and Tolbert, 1957; Chappelle and Krall, 1959) that actively growing such plants and cell-free preparations from those tissues consume carbon monoxide while older tissue does not. Absorbers either of a combustion type or some regenerable type may be necessary for long tours, in order to cope with such contaminants.

Problems may also arise in the culture nutrient cycle, since all algae secrete some soluble organic material during growth, shed some insoluble material in the process of cell division, and leave products of autolysis of dead cells in the medium. These may cause excessive foaming and provide substrates for bacterial growth. Some algae may also poison themselves by excretion products. Pratt (1943) has shown that an antibiotic, chlorellin, is excreted by *Chlorella* Spoehr *et al.*, (1949) believe that chlorellin is an oxidized fatty acid. The extent and the type of such excretions would be expected to vary widely with different algae and with the conditions under which they are grown. It looks possible to minimize many of these troubles by careful selection of species of algae. In a spaceship it will be necessary, because of the limited water-carrying capacity, to reclaim water from the algal growth medium. If this were to be done by distillation alone, for example, power requirements would be excessive. Methods must be found to regenerate the liquid phase which are simple and use little power.

Production of oxygen by the culture will require regulation by some means. This means that both inlet and outlet gas will be monitored for carbon dioxide and oxygen, and rate of growth of the culture regulated, for instance by varying the gas pumping rate or light level. The effluent gas from the culture must then be air-conditioned, probably by drying and cooling before returning it to the man.

If the system is to be completely closed, the man must use the solid output of the gas exchanger as food, either directly or indirectly. Fink in Germany and Tamiya's group in Japan have done extensive work on the use of green algae as food. Professor Tamiya has found that blanched green algae are very palatable. It has been proposed to convert some of the plant material to meat in order to supplement the man's diet. This, however, would be a very wasteful process; only 30-50 percent of algal material can be converted to meat usable by man. The process also imposes an extra respiration load on the algal system, demanding a great increase in volume and weight.

The waste materials from the whole system will have to be fed back through some sort of sewage disposal system and detoxifier to the gas exchanger. This may be either a chemical or a biological system.

Bacterial decomposition might be feasible but would probably create many new problems. A bacterial flora capable of decomposing all wastes in such a system would be very complex. It may be possible to use partial bacterial decomposition of organic wastes followed by chemical oxidation of residues. Either method will impose an extra oxygen demand on the algae because waste products are highly reduced. All these things make the problem look rather complex. In practice it may turn out to be more complex or very much simpler.

After considering a diagram such as Fig. 1, it is possible to list a few factors to consider in the selection of an algal species for use in a gas exchanger. Table 1 is such a list. We will discuss growth rates together with illumination response later. It is necessary that the assimilatory quotient, the ratio of carbon dioxide consumed to oxygen produced, be the same as the respiratory quotient of the system to prevent accumulation of either gas. There are means of adjusting this quotient by varying the reduction level of mineral nutrients presented to the culture. If one presents nitrogen as ammonia to a culture, the algae fix more CO_2 per oxygen molecule evolved than if nitrate is used as a nitrogen source. Kok (1952) gives figures for the ratio of CO_2 fixed/O_2 evolved: 0.8 using ammonia; 0.82, using urea; and 0.62, using nitrate. The respiratory quotient of a man is about 0.82, very close to the assimilatory quotient of algae using urea. It would be advantageous, therefore, if the algae could utilize urea, as excreted by man. In a closed system, the quotient will tend to take care of itself if the nitrogen waste is highly oxidized the quotient will tend to take care of itself. If the nitrogen waste is highly oxidized before putting it back into the algal culture, thus removing a good bit of oxygen from the atmosphere, it will then cause the algae to produce an excess of oxygen,

Normal, rapidly grown algae have a very high nitrogen content—about 10% of their dry weight; this is about double that of man. In a fully closed system we will

Table 1. Algae Selection Factors

High Growth Rate.

Suitable Assimilatory Quotient.

Compatible Nutritional and = pH Requirements.

Illumination Response.

Desirable Engineering Characteristics.

Tolerance for System Poisons.

Minimum Algal Noxious Byproducts.

have to match the nitrogen cycle as well as the carbon cycle. It may be necessary
either to process the algae before consumption or to grow them under partial
nitrogen starvation. Algae can be grown containing as little as 5% nitrogen without
appreciable loss in their photosynthetic efficiency (Bongers, 1956). The algae
should also contain an amino acid content which is qualitatively and quantitatively
sufficient for adequate nutrition of a man for fairly long periods.

Our next consideration for selection is that nutritional and pH requirements
of the culture be compatible with the system and that concentration ranges over
which nutrients can vary be rather wide. If the algae require a very high pH for
growth, much of the CO_2 in the system might be tied up in the culture. On the other
hand, an organism requiring a low pH would require a culture medium with very
poor CO_2 buffering and might require extensive gas-liquid mixing to give a high
enough rate of CO_2 absorption to allow maximum growth rates. Mineral nutrients
required should not be toxic to man since the algae are to be eaten.

Another consideration on the selection list is that the species have desirable
engineering characteristics. This means that it must not be damaged by the ma-
chinery used, that it not settle out quickly, and that it not foul the system by
excessive foam excretions, or by adhering to the walls of the system. These properties should also aid in making it possible to grow the maximum amount of algae
in the minimum volume, a necessary prerequisite for keeping weight at a minimum.

We will now dwell a little further on the light-conversion factor—partly since
this aspect of algal growth and its implications are not commonly met in industrial
microbiology. Response of the cells to illumination varies with both intensity and
wavelength. Algal photosynthesis responds to light as shown in Fig. 2. At zero
light intensity photosynthesis is negative—the algae are respiring. On increasing
the light intensity, we reach a point at which photosynthetic oxygen evolution equals
the respiratory uptake rate. Photosynthesis increases linearly [to about 500 ft-c
(foot-candles) in ordinary algae] then gradually levels off. Above this saturation
level, which is about one-tenth to one-twentieth of sunlight intensity, the same
fraction of incident light is absorbed as is absorbed at lower levels, but it is not

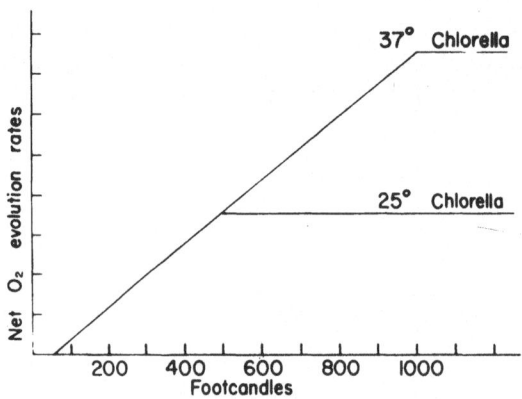

Fig. 2. Response of photosynthesis to varying light intensity
(idealized plot).

utilized in photosynthesis; the excess absorbed energy is dissipated as heat. The maximum efficiency at which algae can convert absorbed light into cellular material is about 20%. Using this figure, it is possible to construct a curve of efficiency of utilization of light versus light intensity. Figure 3 is such a plot, describing the response of a very thin suspension of algae. Efficiency would be zero at the compensation level, 20% at 500 ft-c, 10% at 1000 ft-c, and down to 1% at 10,000 ft-c, the intensity of sunlight. (We will not discuss the photoinhibitory effect at present, solarisation, which may occur in light of this and higher intensities.) In a dense suspension of algae, where the light is weakened as it penetrates the culture and is used with increasing efficiency, conversion of bright light is more efficient. The data of Fig. 2 are for a *Chlorella* strain which grows optimally at 25 C and which was cultivated in relatively weak light; saturation sets in at 500 ft-c. Thin suspensions of high temperature strains of *Chlorella, Scenedesmus,* or of *Anacystis nidulans* (a blue-green alga), growing optimally at from 35 to 42 C, if cultivated in bright light, do not saturate until over 1000 ft-c (Sorokin and Kraus, 1958). Apparently, light limitation can be relieved; mainly by adaptation to higher intensity and to a lesser extent by higher temperature. Unfortunately, in thick suspensions—required to ensure complete absorption of the light—individual algae tend to adjust their chlorophyll content and saturation level to the average rather than the maximum light intensity. The main advantage of presently known high temperature strains, therefore, lies not so much in their ability to utilize strong light as in their wider temperature-tolerance range.

The best conversion figures obtained so far in full sunlight are about 8 to 9% conversion of absorbed energy to cell material (about 4% if considering total incident solar irradiation including infrared). A twofold gain in efficiency of utilization of sunlight is possible if the light-saturation problem could be circumvented in some way. Most of the efficient artificial sources of light pose the same prob-

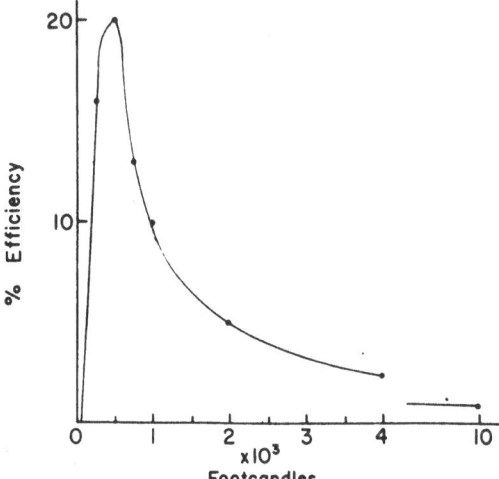

Fig. 3. Photosynthetic efficiency of thin algal suspensions under various light intensities.

lem, some to a higher degree than sunlight. One relatively simple solution with sunlight would be to use a "diffuser," a corrugated illuminated light entrance—a somewhat self-defeating measure when used with artificial light.

Let us consider a diagram of the photosynthetic system:

$$\text{Light} \rightarrow \boxed{\text{chlorophyll}} \quad \xrightarrow[\text{I}]{\text{Chemistry}} \quad \text{A} \quad \xrightarrow[\text{II}]{\text{Chemistry}} \quad \text{B}$$

A quantum of light is absorbed by the chlorophyll. Its energy is then residual in the chlorophyll in some excited state for a short while. The energy moves out into another state probably by driving a chemical reaction by "Chemistry I." It may then move to another state, possibly causing carbon dioxide fixation or oxygen evolution by "Chemistry II." These latter reactions are enzymatic in nature and temperature-dependent. We can demonstrate that the limitation of photosynthesis is in these enzymatic steps. If light is fed to the system in short, very bright flashes, with relatively long dark periods between flashes, such that the integrated light level is below saturation, the rate of oxygen evolution is the same as if continuous light at the integrated level had been supplied (Emerson and Arnold, 1932). This means that somewhere in the system big packets of light energy can be taken up, stored in a reservoir, and utilized slowly. The size of this reservoir has been determined; one light quantum is stored per 200-500 chlorophyll molecules. According to present concepts, these groups of pigment molecules cooperate as a "photosynthetic unit," funneling absorbed quanta into their common (enzyme) outlet.

This mechanism presents intriguing possibilities to the engineer. By using intermittent illumination, i.e., flashing lamps, or intermittency achieved by vigorous stirring of the culture, the cells can take up high-intensity light and utilize it efficiently. This mechanism suggests that use of a thick and very turbulent suspension might yield efficient utilization of bright light. The required intermittency patterns, however, are too rigorous to make this approach attractive from an engineering point of view. Too much power is required to achieve the necessary stirring rate.

It appears much more attractive a possibility to search for an alga with low chlorophyll content per cell, i.e., one in which the photosynthetic unit has perhaps 50 chlorophyll molecules. Such a photosynthetic unit would be only one-tenth as likely to intercept a light quantum in a given time at a given light intensity as a unit containing 500 pigment molecules and would therefore saturate with light at perhaps 5000 ft-c. We intend to follow this line of work in the near future.

We now come to the discussion of the few more or less preliminary experiments carried out by us to look into the feasibility of algal cultures as gas exchangers. In our work we have used a few common algae which have been studied extensively in other laboratories. These may not conform to all the requirements laid out above. These requirements can be specifically outlined only when the final system to be used is being designed. For our experiments we have constructed a tank of lucite $42 \times 36 \times 16$ in. Figure 4 is a picture of this tank. Through the ends of the tank we ran 276 42-watt, ¾-in.-diameter fluorescent lamps which were sealed in place with rubber O rings. The system was operated at 8 kw elec-

Fig. 4. Experimental gas exchanger.

trical input. The tank holds 48 gal (18 L, or 6 ft^3) of culture solution. Eight cubic feet are occupied by the bulbs. Cooling is necessary. Twenty, 50-ft loops of ¼-in. Tygon are used, of which 15 are on all the time and five are regulated by a thermostat. Temperature is controlled within one degree.

Air enriched with 1 to 2% CO_2 is bubbled through holes in tubes at the bottom at a rate of 2-3 ft^3/minute. Culture fluid was pumped out the bottom and back through a spray system at the top at a rate of about 10 gal/minute. The whole culture was thus recirculated every five minutes. This air-liquid countercurrent kept down foaming at the top of the culture and allowed no opportunity for settling-out of the cells. Fouling of the tank by cells growing in corners or on flat surfaces presented only a slight difficulty. The unit was operated on a batch procedure, about 25 to 30 gal out of the 48 being discarded each morning and replaced by fresh nutrient solution. A densitometer system was developed and tested by correlating its readings with the packed cell volume measured after centrifugation. The densitometer was placed in the culture circulation system and cell density followed continuously during the later stages of this work. Densities varied from 0.3 to 0.4% packed cell volume just after adding nutrient, to 1.5 to 1.7% just before changing nutrient the next day. Figure 5 shows a few plots of 24 hr runs. The big jump in density at from 5 to 6 hr in the uppermost plot was observed frequently. It may indicate a sudden acceleration of growth on the addition of nutrient, or could partly be the result of synchronized cell division in this period. Since the densitometer operates partly on a scattered light and partly on an absorbency basis, simultaneous breakage of capsules would give an apparent increase in cell density without an actual increase in cell volume.

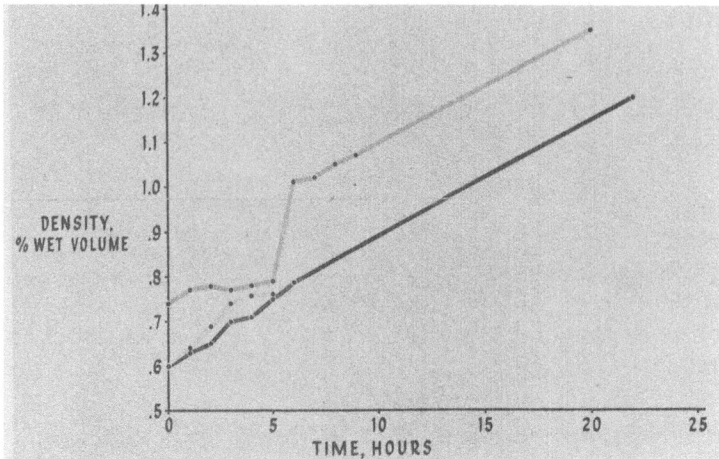

Fig. 5. Characteristic growth curves for Scenedesmus growing in
the experimental gas exchanger.

In the diluted culture, light absorption is about complete within 1/2 cm of the bulbs; at high densities all the light is absorbed within much shorter distance. The bulbs are 1 cm apart. This means that at high cell densities average illumination of the individual cells is quite low. Under such conditions, algal respiration is a considerable fraction of the energy and gas balance of the system. Subtracting respiratory rate from the photosynthetic rate considerably lowers the over-all efficiency. This limitation may be illustrated another way. Since increase in optical density is approximately linear with time over a 24 hr period and total cell mass doubles in this time, the increase in cell concentration per cell at the end of 24 hr is only half that which it was in the beginning, proof that cells are working only half as hard at high densities as at low. Myers and Graham (1959) found a distinct optimum in the cell density required for maximum efficiency in the utilization of sunlight intensities. This optimum was at a point slightly below complete absorption of light. It therefore is not unreasonable to expect that our growth rates can be considerably increased and stabilized by holding the optical density constant at its optimal value; a value which will be different for each different light intensity used. A continuous dilution procedure or continuous removal of cells accompanied by reutilization of nutrient solution is planned in future runs.

Table 2 lists the mineral elements in the nutrient solution used in these experiments. This is a quite conventional formula. The nutrient is brought to pH 7.5 before adding it to the culture. We have measured depletion of nitrogen, phosphate, magnesium, and iron. Iron dropped to very low levels at times; growth was very much slower when this occurred. Use of EDTA will solve this difficulty. Nitrate is essentially completely utilized in one day's operation. Measurements such as these will become very much more critical when we begin operating the system with a closed nutrient cycle. Copper, for example, could quickly rise to toxic levels if added at twice the rate at which the cells remove it from the nutrient.

Table 2. Mineral Nutrient Composition (ppm in the final medium)

NO_3	1430	Bo	2.7
K	750	Cl	0.65
PO_4	480	Mn	0.5
SO_4	202	Zn	0.5
Na	267	Cu	0.4
Mg	50	Co	0.04
Ca	5	Mo	0.015
Fe	5		

No precautions are taken to keep the system sterile except to keep the tank and the nutrient bottles covered. No detailed studies on condition of the culture in this respect have been done. Only a few bacteria were visible under the microscope. The culture has never had a bad odor nor foamed excessively, nor has the tank been fouled by mold growth. We have grown two algal strains; *Anacystis nidulans*, a high-temperature blue-green alga (Kratz and Myers, 1955), and *Scenedesmus* strain 3 (Kok and Van Oorschot, 1954), a high-temperature green alga. *Anacystis* grew somewhat cleaner, with almost no foaming and negligible deposits on the tank wall, cooling tubes, or lamps. *Scenedesmus*, we believe, gave higher yields on the average than did *Anacystis*. One possible explanation for this is that the fluorescent lamps had a phosphor that emits the most light at 630 mμ, exactly the wavelength where phycocyanin, the accessory pigment of *Anacystis*, absorbs most strongly. Such a coincidence in emission spectrum of the source and absorption spectrum of the cell would give highly efficient transfer of energy from the lamp to the algae and would cause light saturation effects to appear at lower net energy input levels. The first layers of cells absorb most of the light giving poor penetration into the suspension. The surface intensity of the bulbs used is about 2000 ft-c. Thus the first layer of algae is operating in an area of the efficiency curve where we have rather poor utilization of light. *Scenedesmus* does not contain phycocyanin and therefor such effects would not be expected. Temperature of the culture did not appear to be critical between 32 and 38 C (in any given period it was regulated to within one degree).

The longest run with *Anacystis* was 25 days, with *Scenedesmus*, 80 days. Daily performance of the culture was the same at the end of these runs as at the beginning after the culture had achieved full operating density. Cultures were started by serial transfers from an agar slant through liquid cultures of increasing volume up to 10 L and then to the tank. The operating level of the tank was from about 0.4 to 1.5% cells. Two or three days were usually all that was required for the tank to reach this density after inoculation.

Although not of high precision, the data in Fig. 5 can be used to compute yields and efficiencies. The tank produced on an average, about 300 g dry weight algae per day. Maximum yields were about 500 g dry weight per day, about 2.5 to 5 lb net weight or enough to satisfy a man's dietary requirements.

Gas exchange data were collected using the Haldane apparatus for spot analyses. Oxygen evolution data show that from 1.0 to 1.5 ft^3 of oxygen was produced per hr. Using the previously noted value of 0.62 for the ratio of CO_2 to O_2 exchange using nitrate, these values are equivalent to the production of from 340 to 510 g dry algae per day a good check on the observed growth figures. This represents from 1700 to 2850 kcal energy stored as cellular material per day. The total energy input to the tank was 8 kw-hr electricity, equivalent to 164,000 kcal/day. A reasonable guess for the efficiency of the electricity to light conversion in the cold-cathode lamps used is 15%. This gives a daily usable light input of 24,600 kcal. Daily fixation of 1700 to 2850 kcal, amounts to a light-to-cell material conversion efficiency of 7 to 11.5% or an electricity conversion efficiency of from 1.05 to 1.7%.

Another basis for comparison is furnished by cell production per square meter of illuminated surface. Algal production of 25-30 g of algae per m^2/12-16 hr day has been reported for sunlight (Kok, 1952). Our tank has 14.6 m^2 of lamp surface. At 500 g/24-hr day we produced 34.2 g/m^2 per 24-hr day. These data were obtained with no attempt at optimizing the system. The bubbles in the system were large and, because of consequent lower diffusion rates, probably gave a suboptimal carbon dioxide level. The surface intensity of the lights was 2000 ft-c, higher than optimum; nutrient levels, especially iron and perhaps other micronutrients, were probably not optimal. We did not search for optimal cell density for maximum efficiency of light conversion.

With more careful controls, it is reasonable to expect at least a constant maintenance of the highest observed yields. Most probably we can do much better even with the present tank. We chose fluorescent tubes because their brightness is about right and they are convenient and reliable. Tank design certainly was not optimal, but the system has proven a convenient and reliable work-horse for further nutritional and atmospheric recycling studies. Other light sources, such as luminescent flat plates, in combination with design improvements, may doubtless improve efficiency and decrease weight and size. These would demand a thorough study in themselves. The 8 horsepower and 20 ft^3 that our system uses to provide oxygen for one man may be the maximum requirements. Using well substantiated laboratory data about these same figures were predicted by participants in a conference held to determine feasibility of use of photosynthetic gas exchangers in submarines (ONR Symposium, 1956).

The most important conclusions that can be drawn from this rather preliminary work is that we have an algal system capable of producing enough oxygen to sustain a man (1 ft^3/hr) and that it will do this reliably, hour after hour, day after day for long periods of time. The longest experiment was terminated after 80 days only as a matter of convenience. The cultures appeared to be in as good a condition at the end of these periods as at the beginning. This performance was achieved using personnel untrained in biological work, who used no sterile technique. At times the culture was inadvertently subjected to thermal or chemical shock and recovered quickly. Large cultures appear to be more stable than small ones and more reliable.

It may be asked how else an ecology could be closed. Chemical means for gas regeneration and absorption have been devised and are utilized in submarines. It seems extremely unlikely that a material as complex chemically as a man's diet can ever be synthesized from the waste products of a man by other than biological means. I would like to quote Professor Marvin Johnson who taught me about biochemical aspects of fermentation. "If it has more than six carbons linked together, bugs do it cheaper." Chemical methods of atmospheric regeneration have been said to be more reliable. Our culture has proven remarkably rugged. Breakdown of controls, stoppage of pumping, and exposure to contaminants, did not kill it. We believe an algal system is far more compatible with a man than, say, an electrolysis system which would produce hydrogen. The safeguards on that system would probably be much more complex than on ours.

It is admitted that our system is bulky. It is hoped that refinements in engineering can greatly reduce this bulk. It is possible, however, to show that storage of food and oxygen becomes even more bulky after a short time. Cruises of the atomic submarine are limited by their oxygen-carrying capacity, for example. We have estimated the time period of use at which our system becomes more economical than a storage system using CO_2 absorbers. At our present low rates, our system becomes more economical at from 30 to 60 days. With the advances in culture techniques that we confidantly expect to make, and with engineering development of smaller, lighter equipment, we can reasonably expect to cut this time from two- to fourfold. Thus, if a man wants to visit Mars, he probably will carry one of our tanks with him.

ACKNOWLEDGMENTS

We wish to acknowledge the efforts of Charles Friese of the Martin Company, who made the detailed engineering design and operated the culture tank.

REFERENCES

Bongers, L. H. J.   1956   Aspects of nitrogen assimilation by cultures of green algae. Ph.D. Thesis, Wageningen.

Burlew, J. S. Ed.   1953   Algal culture from laboratory to pilot plant. Carnegie Institute of Washington Publication 600, Chapter 17.

Chappelle, E. W. and Krall, A. R.   1959   Enzymatic uptake of carbon monoxide by cell-free extracts of spinach leaves. Abstracts 136th Meeting American Chemical Society. 71c.

Emerson, R., and Arnold, W. A.   1932   A separation of the reactions in photosynthesis by means of intermittent light. J. Gen. Physiol, 15, 391-420.

Gafford, R. D., cited in Wilks, S. W.   1959   Carbon monoxide in green plants. Science, 129, 964-966.

Ingram, W. T.   1958   Orientation of research needs associated with environment of closed spaces. Preprint 57-17. Proceedings of the American Astronautical Society, 4th Annual Meeting N. Y. (Jan. 30, 1958).

Kok, B.   1952   On the efficiency of Chlorella growth. Acta. Bot. Neerl., 1, 445-67.

Kok B. and van Oorschot, J. L. P.   1954   Improved yields in algal mass cultures. Acta. Bot. Neerl., 3, 533-546.

Krall, A. R. and Tolbert, N. E.   1957   A comparison of the light-dependent metabolism of carbon monoxide by barley leaves with that of formaldehyde, formate, and carbon dioxide. Plant Physiol., 32, 321-26.

Kratz, W. A. and Myers, J. 1955 Nutrition and growth of several blue green algae. Am. J. Botany 282-287.

Myers, J. and Graham, J. R. 1959 On the mass culture of algae. II, Yield as a function of cell concentration under continuous sunlight irradiance. Plant Physiol., 34, 345-52.

ONR Symposium 1956 Conference on Photosynthetic Gas Exchangers, Report ACR 13. Office of Naval Research, Washington.

Pratt, R. 1943 *Chlorella vulgaris*. VI, Retardation of photosynthesis by a growth inhibiting substance from *Chlorella vulgaris*. Am. J. Botany, 30, 32-33.

Sorokin, C. and Kraus, R. 1958 The effects of light intensity on the growth rates of green algae. Plant Physiol., 33, 109-113.

Spoehr, H. A., Smith, J. H. C., Strain, H. H., Milner, H. W., and Hardin, G. J. 1949 Fatty acid antibacterials from plants. Carnegie Institution of Washington Publication 586.

Tamiya, H., Iwamura, T., Shibata, K., Hase, E., and Nihei, T. 1953 Correlation between photosynthesis and light-independent metabolism in the growth of *Chlorella*. Biochim. et Biophys. Acta, 12, 23-40.

THE MICROBIOLOGICAL CHALLENGE IN SPACE

Walter M. Bejuki

Prevention of Deterioration Center, National Research Council, Washington, D.C.

It is the purpose of this paper to crystallize, within the allotted time, the possible potential position of the microbiologist on that team of scientists and engineers who are currently applying their several disciplines in meeting the challenges represented by today's astronautical frontier. As nautical became prefixed with astro-, so biology, zoology, botony, and microbiology can be expected to grow into the new experimental and applied areas which are connoted in the prefix astro-. Astromicrobiology, like any of the other new areas so created, will seem a little uncomfortable at first. This concept in microbiology may be subject to criticism, but evaluated it must be. The evaluation best comes from the microbiologist himself. It may be recalled that at the symposium of May 14-17, 1958, on "Possible Uses of Earth Satellites for Life Sciences Experiments," the biologist considering satellite uses was cautioned. My own notes on this read, "Biologists not yet ready to go into orbit. Keep your feet on the ground." From this memo we can abstract and build on the idea "not yet ready." What the state of readiness is, for the microbiologist in particular, represents an early challenge. What he does with his own state of readiness once he has determined it represents a portion of the opportunities associated with the expanding effort and must be his own decision. Opportunities here, however, do father a continuing host of additional challenges.

The nonbiological use of satellites and rockets, reflecting the engineering capabilities of contemporary missiles, has been reviewed and published in the proceedings of the symposium "Scientific Uses of Earth Satellites" (Van Allen, 1958). The 34 papers contained therein all concern themselves with the geophysical aspects of space probing and illustrate well the fundamental instrumentation, telemetry, and other factors involved in sensing and transmitting back to earth the data collected.

Approximately two years later, Dr. Pickering, presenting Dr. Froelich's paper, "The Army Rocket, Satellite, and Space Flight Program," at the "Biological Uses .... Symposium," indicated the following five factors as governing our ability to use successfully earth satellites in scientific exploration. These concern the adequate development of:

1. A rocket cluster of adequate size
2. Guidance and control systems
3. Environmental control
4. Communication
5. Production and operational capacity

The 15 intervening months have seen some marked advances in many of these areas. A summary of this progress has been reported by Medaris (1959). This list of space probe achievements, conducted for the National Aeronautics and Space Administration, can well be supplemented by achievements of the other services. In 1959 it is expected that, under the auspices of the Advanced Research Projects Agency, a cluster of booster rockets will be developed for Project Mercury which will produce a thrust of 1,500,000 lb. This can indicate for us what payload potentials exist. A good rule of thumb indicates that the launch to weight ratio may be estimated as 1000:1 and proved valid for example, in the Vanguard. This represents an advance in capabilities particularly since pre-Vanguard ratios were more nearly 2000:1. Assuming however that a 1,500,000-lb thrust is achieved, a payload of only 1500 lb would represent our capability in Project Mercury, where we expect to carry a man and all his supporting gear. The significance of biological probing, therefore, where microorganisms are proposed to act as environmental sensors looms high, in relationship to available payload. Properly miniaturized experiments can utilize with maximum efficiency any space on the payload dedicated to biological research. There is no contention here implied that data obtained with microorganisms as probes have a high transfer value in terms of studying spatial environmental effects on man. On the other hand, it is generally contended that man's sensitivity to various types of radiation is higher than that of any other organisms, including other primates and therefore he would actually be his own best test animal. Microbiological space probes must be evaluated with the same reservations that exist for all bio-assays. Beyond this, the academic "need to know," what extraterrestrial environmental effects will be on biological systems is basic to the growth of biology.

The effect of environmental factors on bacteria and fungi is a study of fairly long standing and suggests parameters may well be used as a basis for comparative studies of limiting environments. The limiting environments would involve such factors as temperature, including high and low values, moisture, pressure, as in vacuums producing sundry outgassing phenomena. Outer space explorations would take us through various types of atmospheres. These may be enumerated as:

1. Chemical
2. Radiations
   a. nuclear, alpha, beta, gamma, x-rays
   b. solar
   c. cosmic
   d. magnetic
   e. auroral
   f. electrostatic
3. Particulate
   a. space dust, meteoritic grit
4. Gravity-associated
   a. zero and near zero
   b. accelerated, shock
5. Planetary (various combinations)

Table 1, for which I am indebted to Erik Linden of the Signal Corps, is an abridgement of an excellent compilation made by him in his studies on the effects of high altitude environments on materials. The biological implications in the column "Remarks" reflect areas of biological significance which we can profitably consider. In passing, it may be well for us as microbiologists to recognize that here in the geophysical arena a nomenclature problem exists too, in relation to the naming of the various zones enveloping the earth. The system advanced by the International Union of Geodesy and Geophysics has been invoked here.

Table 1. Altitude Characteristics (after Linden)

Miles (Approx.)	Remarks	Sphere
10	O_2 respiration limit	Strato-
10	Temperature is -80F	Strato-
12	Water in biosystem boils	Strato-
15	Ambient air insufficient for pressurization	Strato-
13-26	Ozone filters dangerous ultraviolet light	Strato- & Chemo-
20-50	Ultraviolet light intense, energy and chemical activities high O_2, N_2 — monatomic, corrosive	Chemo-
23-27	Biologically effective ultraviolet light	Chemo-
30	Temperature is +30F	Meso-
25-72*	Meteor-safe zone ends	Meso-
50-240	Gases ionized by ultraviolet of sun, reflects radio waves (Layers) D — 50 miles E — 100 miles F — 180 miles F_2 — Variable; seasonal, sun spots, flares	Iono-
55	Temperature is – 95F	Thermo-
80-100*	Vision changes from atmospheric to space optics	Hetero-*
100*	Zone of complete silence	Hetero-*
120-140*	Aerodynamic support ends. Centrifugal force required.	Hetero-*
120*	Aerodynamic heating ends. Heat transfer by radiation only.	Hetero-
600*	Outer space: molecules can escape earth's gravity	Exo-
1,400-40,000	Van Allen Layer (Two belts of electrons or protons, possible solar origin, producing x - rays on striking satellites. Layer 1 starts at 1400 mi, 200 mi thick. Layer 2 starts at 8000 mi, 4000 mi thick. Between them lies a region of no radiation. Beyond 40,000 mi manifestations are absent. There appear to be voids over the poles.)	Exo-

* — and beyond

A closer look at the altitude characteristics should encourage the microbiologist. The hazards indicated here in terms of temperature, anaerobic conditions, chemical effects, desiccation, solar, and other radiations are not ones with which he is unfamiliar. Under laboratory conditions these influences have been thoroughly surveyed in several areas. Data of this kind represent an excellent information resource. In other areas, for example, the new environmental reference frame represented by zero or near-zero gravity, will have to be evaluated by the microbiologist *even as it must be evaluated by every other discipline* represented on the astronautical team. In the compound appraisal of gravity, particularly in the applied areas no discipline has a long history of cognizance, much less a history of achievement. Theoretical support according to Surosky, Hill, and di Rende (1959) is lacking, and the environmentalist seeking help with his applied problems must continue to be empirical until significant breakthroughs occur. Biological history is rife with empiricism. Biology with its innumerable coexisting and interacting systems has used empiricism well and will continue to do so.

Experiments in gravitational effects involving greater and lesser values than g can well be considered by the microbiologist in conjunction with his allied engineer. The technology and energy resources required to study plus and minus gravity values as affecting biological systems may be considered more feasible in the microcosm of the microbiologist than in a representative of the macrocosm. Time factors such as generation times favor the microbiologist and as our ability to extend experimental periods of both high and, particularly, low g values increases, the effect of gravity on microbial life cycles should be contributory to biological advances. The free fall experiments, simulating weightlessness, again may be more easily achieved with microorganisms. Although experiments of this kind are difficult to imagine, they have been done, in high g values with other animals (Surosky *et al*., 1959), and would seem to be less difficult with the smaller living systems.

In anticipation of that time when planetary visits will be practical--and the astronauts are optimistic--the environment and atmospheres of the explored planets will need biological evaluation which must be made by biologists. Here it is expected that physical traits again will be a mixture of environmental hazards differing specifically from the terrestrial one but involving chemical atmospheres, temperature variations, desiccation, radiation, and other characteristics. The introduction of terrestrial life forms into these atmospheres, and their success or failure, is not only a challenge in adaptiveness to the organism, but an evaluation of the capability of microorganisms to meet this challenge is, in turn, a challenge to the microbiologist, and well within his proper cognizance.

With the above introduction let us leave the consideration of physical properties and continue further into the microbiological challenges. The latter can be divided arbitrarily into the open space ecological system as we have considered above, and the closed system which we will consider later. Apart from these two, but supplementary to both, I would like to mention an area of microclimatology and microecology.

The behavior of organisms, and this can well include the microbial forms, in relation to their microclimate, has been considered by Geiger (1957) as micrometeorology. For certain reasons which we cannot discuss here in full, this may be an unacceptable designation, particularly since, in some of the Dubos (1959) studies, considerations allied to those of Geiger have been discussed under the concept of bioclimatology. Both Geiger and Dubos summarize well the responsiveness of organisms to meteorological and geophysical influences. From his own experiences the microbiologist knows too that the organisms with which he is familiar have wide sensitivity ranges and specific tolerances that are often governed by easily recognized influences. Not infrequently, however, the causes for specific responses remain elusive. This challenge, namely, the need to mobilize and expand the knowledge of the influence of environmental factors on microorganisms would yield parameters of value in both terrestrial and extraterrestrial probing.

Microbiological challenges in closed ecological systems lie in the following areas:

1. Gas exchange microbiology
2. Waste disposal or modification
3. Food production, waste utilization
4. Microbiology in psychobiological support

An excellent source document in the energy requirements in waste handling in a closed system has been prepared by Slate (1957). Palevsky (1957), in a recent progress report on handling air contaminants in a biosatellite, sets out very well the optimum atmosphere necessary for maintaining life and sedentary work in confined space. Parameters in terms of temperature, humidity, air motion, foreign matter, odors, bacterial population, and sanitation are indicated. Ingram (1957) deals with skin excretion and sweat biochemistry in terms of specific chemical compounds. The microbiological implications here in terms of odor and possible utilization of excreted compounds in a balanced independent ecology seem obvious. In the over-all problem, Ingram points out that in considering today's needs in terms of sustaining man in a space capsule, investigations progressively reveal that questions and needed answers multiply. No longer is the man in a box only the clean-cut model which stimulated Einstein in his thinking, but it is also a model for one of today's important challenges to all of biology—microbiology not excepted. Apparently, where man goes his microbes go also.

Masson (1957) in his review covers methods of obtaining oxygen from carbon dioxide. Applied biology in terms of the algologist receives its fair share of attention. Whether oxygen recovery must be mediated through chlorophyll-containing organisms is a question the microbiologist should explore. Miller (1959), in reporting on the Space Logistics Conference of 1958, points out that Roadman indicates the weakest link in space travel will be man. Requiring ideally 283 lb to maintain himself in space for 7 days the present equipment for this purpose is actually a 707-lb requirement. Reduction of this weight penalty must surely be considered by the closed-space ecologist. Perforce the microbiologist and others have become ecologists.

Within this framework it seems readily apparent that industrial microbiology in its fermentative, synthesizing, and organic material utilization should be carefully reviewed for applicability. Even the psychobiological support value of microbiology as an entertaining, diversionary device should not be dismissed lightly. One calls to mind the history of Woronin (Large, 1940) with his fungus gardens and laboratory in his bedroom where he lived, almost a recluse, studying and enjoying his rusts, nitrogen fixing organisms, and club root-causing fungi.

The challenges associated with astrobiological exploration are as follows:
1.   Extent of biosphere, limiting factors, duric forms;
2.   Origin of life, comparative planetary bioevolution, life precursor biochemistry, sampling procedures, contamination prevention.

The extent of the biosphere has long intrigued microbiologists and aerobiologists. Distribution patterns, dissemination studies, epidemiology, and etiology have long speculated upon and in part explored, higher and higher regions of our atmosphere. Lindbergh, among others, you will recall, exposed slides for microbiologists over northern areas of the globe. Have these studies indicated just where the biosphere ends? Are some of the factors listed in our earlier environmental considerations limiting in nature? Probably, but how, where, and why, have not all yet been answered completely. With the coming of the space vehicle, satellite, and ultimately fixed space stations, operational bases will exist for pursuing duric forms into areas never before seriously considered. If the terrestrial biosphere is traversed and new planets are approached, where will evidences of the Martian or Venusian biosphere be detected? When they are detected how will they compare with each other and with the terrestrial biosphere? We tend to think in anthropomorphic terms, but microbes came before man, and it seems that precursor biochemistry, in the framework or our knowledge of organic evolution on the earth, went from molecules to microorganisms, probably plant forms. It would be a reasonable preliminary assumption to anticipate a similar pathway on other habitable planets. Doesn't the microbiologist belong on the front line of these explorations? Tikhov (1955) answers this question in the affirmative.

Before we dwell any longer in outer space, however, I would like to return, for a moment, to some rather mundane but important aspects of space microbiology as it is manifest here on earth and now. This has to deal with the application and use of existing microbiological knowledge. Those who participated in the symposium sponsored by our society last Christmas at the AAAS meeting in Washington will recall that Dr. Ezekiel reviewed the equipment hazards of microbiological origin which he had witnessed in Navy installations. He indicated at that time that this hazard should be evaluated in terms of our space-exploring vehicles. Fortunately, this is being done. Unfortunately, some of this evaluation work being done at the Engineers Research and Development Laboratories at Belvoir, and Army Ordnance Corps at Detroit Arsenal, indicates that materials susceptible to microbial degradation are still being incorporated into missiles and components, and failures from this cause can be anticipated. Before the microbiologist shoves off into outer space he should stop long enough to address the design engineers. I would suggest the title "Applied Microbiology in Design Engineering."

In discussing microbiological potentials and contemporary experimental design requirements here we can illustrate, by drawing from microbiological studies, the existent environmental parameters that have already been explored by the microbiologist. Long before Sputnik, studies on the biology of existing earth forms were producing data that can well serve us for comparative purposes now. Through the years some interesting data have evolved. Some instances of extreme tolerances have been listed for us by James (1955).

Lichens: (algae and fungi) survive as adults down to −80 C; at −30 C still photo-synthesize slowly.

Rotifers: in antarctic lakes tolerate alternate freezing and thawing without damage, endure −78 C for hours. Desiccated, tolerate electric ovens at 170-200 C for 5 minutes, 151 C for 35 minutes, liquid helium (−272 C) nearly 8 hr.

Tardigrades: found all over world up to 10,000 ft; under dry conditions, ball up and float around for long periods.

Protozoans: tolerate temperatures down to −30 C; others are found in hot springs at 64 C.

Tumor cells: −253 C.

Normal Skin: −150 C.

Mice: live and breed at −21 C.

Arctic mammals: −50 C for long periods.

Bacteria, marine: pressures equal to 1000 atm.

Ascaris eggs: pressures equal to 800 atm, centrifugation for 1 hr at 400,000 atm.

Snails: indefinitely at an oxygen partial pressure of 2 mm.

Bacteria: sulfur, learned to live without oxygen 800 million years ago and have not yet found it necessary to convert.

To these instances we can add many additional parameters particularly from the field of industrial microbiology, using the organisms with which we are most familiar. Contemporary data on the effect of high vacuum on spores of *Aspergillus niger* and several organisms have been provided by our sumposium chairman, Dr. Prince. He, too, is able to report high tolerance values in *A. niger*. By using the excellent bibliography by Sparrow *et al.*, (1958) we can isolate some representative radiation studies on *Aspergillus* and *Chaetomium* spp. The parameters revealed here can be of great use in space probing. Joan Munro Ford (1948) alone, and with Kirwan (1949), in her ultraviolet and x-ray irradiation of fungal spores has characterized mutation and lethal effects based on thousands of spore irradiations. These data from controlled laboratory conditions can provide a comparative base for irradiations made in outer space.

The work by Berk (1952, 1953) on irradiation of deteriorative fungus spores represents good quantitative data, and his review (1952) of the work in this field allied to his own strengthens the position of the microbiologist. The biochemical fundamentals of ionizing radiations on biological systems is adequately introduced by Kuzin (1956) and affords an excellent base for comparative radiobiochemistry. Buchwald and Weldon (1939) have established stimulative dosages for *A. niger*, and Weldon in 1938 and 1940 has catalogued effects of cathode radiation in a vacuum. Dickson (1932) reports the effects of x-radiations on ten fungi initially and later (1933), concentrating on *Chaetomium*, has quantitatively evaluated saltant production in relation to accessory factors such as UV light, heat, and substrate biochemistry. Kresling and Shtern (1936) also establish radiobiochemical criteria which may well be utilized in studying comparative radiation effects: terrestrial versus outer space exposures. Stapleton and Martin (1949) compare lethal and mutagenic effects of alpha particles, fast neutrons, gamma rays, and x-rays, thus contributing good physical biological values for comparative purposes. Additional studies (Swanson, 1952; Swanson *et al.*, 1948) concern themselves with mediation of radiation effects and interaction of radiation with other influences.

From this brief review just on the two genera *Aspergillus* and *Chaetomium* it can reasonably be expected that by mobilization of the data on the similar test organisms, for example those used in specifications testing, an excellent lattice of parameters, traits, response criteria can be made available and will make the microbiologist and his organisms a welcome asset on the astronautical team. The geophysicist and others will continue to map out the radiation zones and other environmental characteristics of space, but they await the biologist, for an evaluation of these regions, in terms of effects on living systems.

Contemporary experimental design requirements, then, should consider simulation of known environments. Dr. Prince, for example, uses a high-altitude environment chamber in studying low-pressure effects. Similar single and combined environments must be created or mobilized and where no data exist, experimental work conducted. Various atmospheres in simulation chambers can be produced and evaluated in terms of tolerances and adaptability of microorganisms. Many of these systems need not be very elaborate. Tikhov (1955) makes reference to the presence of microbial life on the barren and waterless soils of the Sahara. Whatever water was present there to support this activity was undetectable even with special instruments. How xerophilic or xerofacultative are our desert organisms? Doesn't this warrant some further laboratory studies? Some of our microbiologists at Randolph Field are doing this but apparently under conditions not quite as rigorous as the Sahara. The dry desert and other type environments may very well lend themselves to simulation.

Much of the needed knowledge can be mobilized from past studies; additional and new parameters that can be simulated (and let us not be deceived into thinking this is an easy matter) will govern new experimental designs on earth-bound laboratories. In designing microbiological experiments in space vehicles, miniaturization and small cubage and weight provisions should be kept constantly in mind. As more cubage becomes available, new phases can be introduced, providing the general rule of miniaturization and microminiaturization is kept constantly in mind. This requirement should not embarrass the microbiologist—his field lies in the microcosm. In satellite experiments telemetering is of prime importance. Biological experiments will have to express data in terms of what may be transmitted by radio, video, and voice signals.

In reviewing the material for this paper it was interesting to me, and I believe it will be interesting to you, to note in a historical sense and for other purposes, those documents that have come to hand indicating in an approximate way, how America is dealing with the biology-in-space problem. This, of course, is supplementary to space medicine, flight physiology, and similar areas characteristic of man in space.

　　1.　March 26. 1958.　"Introduction to Outer Space," a statement by the President and introduction to outer space. An explanatory statement prepared by the President's Science Advisory Committee.

2. May 14-17, 1958. Symposium "Possible Uses of Earth Satellites for Life Sciences Experiments," National Institute of Biological Sciences, National Academy of Sciences and National Science Foundation.

3. August 3, 1958. "National Academy of Science Establishes Space Science Board," "including both the physical and the life sciences."

4. December, 1958. "Microbiology in Outer Space Research," A symposium sponsored by the Society for Industrial Microbiology and its Washington Section, and the American Astronautical Society under the auspices of the AAAS and AIBS.

5. February 9, 1959. "Academy Research Council, Armed Forces Join in Study of Biological Effects of Space Flight, Announcement of Organization of Armed Forces--National Research Council Committee of Bio-Astronautics.

6. April 29, 1959. "Symposium on Problems in Space Exploration," 96th Annual Meeting of the National Academy of Sciences.

7. August 22, 1959. "NASA Names Advisers on Health Issues," The National Aeronautics and Space Administration [announces a] new bio-science advisory committee, on problems associated with manned space flight.

In conclusion, I offer "Enter the Space Doctor, Dr. James Absolum Whalebottom."

Dr. Whalebottom in reading this manuscript has graciously summarized for us in highly characteristic style—using free verse—his conclusions. This piece is entitled "Microbiologia Mutabilis," and follows.

Microbiologica Mutabilis

Herr Microbiologicus, I said to him one day
The end is here; no more we cope alone with these
Thy fungicides and vats and carbon links,
Thy enzymes one and all, thy rets and brews
Thy copper eights and quinolates, thy phenols
Yellow, white, and gray, and blue.
Thy cheeses, yes, and yeasts we need, and probably
Thy algae too, but all in all, the greatest
Need is this, that thou through thy menagerie,
Go far aloft, where apogee and perigee
Have much to tell, if we but can
Convince thee that thou must, at last,
Disjoint thy scope, and lo, in spite of many decades past
Which have conditioned thine objectives toward the bowels of earth,
Assembly redesign, so now she skyward pokes—
To tell us this through your comrades dear
Penicillum, Aspergillus, Rhizopus, Chaetomium,
Lord how these names of thine, strain my simple runic rhyme!
Yet these I say must go aloft for thee;
Harnessed then to our telemetry, in such a way
Through thy shrewd design that soon we little information lack

Of dangers, dreams, hypotheses, deep spread throughout the perigees.
Now fear not friend, they go not but to die
But rather go as the canary goes—down into the mine
To probe ahead the miner's foes, of dank and gas and goodness knows.
But tools these are thy microbes fine,
Which for the future we must design
To serve us well in space confined.
No whale, no cow, no man or dog
Are we yet prepared to send aloft
But well it seems, the vanguard might be
Your unpronounceably labeled coterie.
I've heard thee speak of duric forms
Of xerophiles and thermophiles and, if thou search a bit
Could not thou turn up too some cryophiles
Which, sent aloft could tell us things
By tape and otherwise, what cares we need to exercise
Before, we ourselves, cope with destiny and impending zero gravity?

REFERENCES

Berk, S. 1952 Biological effects of ionizing radiations from radium and polonium on certain fungi. Mycologia, 44, 587-598.

Berk, S. 1953 The effects of ionizing radiations from polonium on the spores of *Aspergillus niger*. Mycologia, 45, 488-506.

Berk, S. 1952 Radiation mutants of *Aspergillus niger*. Mycologia, 44, 723-735.

Buchwald, C. W., and Wheldon, R. M. 1939 Stimulation of growth in *Aspergillus niger* under exposure to low-velocity cathode rays. Am. J. Botany, 26, 778-784.

Dickson, H. 1932 The effects of X-rays, ultra-violet light, and heat in producing saltants in *Chaetomium cochliodes* and other fungi. Ann. Botany (London), 46, 389-405.

Dickson, H. 1933 Saltation induced by X-rays in seven species of *Chaetomium*. Ann Botany (London), 47, 735-754.

Dubos. Rene J. 1959 Problems in bioclimatology. NAS-NRC news bull.

Ford. J. M. 1948 Lethal mutations produced by ultraviolet and X-ray irradiation of fungal spores (*Chaetomium globosum*). Australian J. Expt. Biol. Med. Sci., 26, 244-251.

Ford. J. M., and Kirwan, D. P. 1949 Mutants produced by x-irradiation of spores of *Chaetomium globosum* and a comparison with those produced by ultraviolet irradiation. J. Gen. Physiol., 32, 647-653.

Geiger, R. 1957 The climate near the ground. Harvard University Press, Cambridge, Mass. p. 494.

Ingram, W. T. (no date) Orientation of research needs associated with environment of closed spaces. AFOSR Rept. No. TW 58-106.

Ingram, W. T. 1957 Report on skin excretion. AFOSR Rept. No. TW 58-260. October 1957.

James, P. F. 1955 The limits of life. J. Brit. Interplanet. Soc., 14, 265-266.

Kresling, E. K., and Shtern, E. A. 1936 The effect of radium and ultraviolet rays on the growth, biochemical properties, and variation of *Aspergillus niger*. Zentr. Bakteriol. Parasitenk., Abt. II, 95, 327-340.

Kuzin, A. M. 1956 Reviews on radiobiology. Institute of Biological Physics, Academy of Sciences, U.S.S.R., Moscow. Translated from Russian by David Franklin. AEC-tr-3353.

Large, E. C. 1940 The advance of the fungi. Henry Holt & Co., N.Y.C.

Masson, H. J. 1957 Report on study of methods for obtaining oxygen from carbon dioxide. AFOSR Rept. No. 57-379.

Medaris, J. B. 1959 Satellite to the sun. Army Information Digest, June, 10-18.

Miller, W. O. 1959 A hard look at space logistics. Missiles & Rockets, April 6, 17.

Palevsky, G. 1957 Report on handling air contaminants resulting from a closed ecological system. AFOSR Rept. No. TN 58-269.

Slate, L. 1957 Report on thermal energy exchange with specific application to waste handling in a closed ecological system. AFOSR Report No. TN 58-268.

Sparrow, A. H., Binnington, J. P., and Pond, V. 1958 Bibliography on the effects of ionizing radiations on plants. Brookhaven National Laboratory, Rept. No. 504 (L-103).

Stapleton, C. E., Martin, F. L. 1949 Comparative lethal and mutagenic effects of ionizing radiations in *Aspergillus terreus* Am. J. Botany, 36, 816.

Surosky, A. E., Hill, D. A., and di Rende, J. B. 1959 Gravity-zero gravity-environmental continuum. Presented for Institute of Environmental Sciences, Chicago.

Swanson, C. P. 1952 The effect of supplementary factors on the radiation-induced frequency of mutations in *Aspergillus terreus* J. Cellular Comp. Physiol., 39, 27-38.

Swanson, C. P., Hollaender, A., and Kaufmann, B. W. 1948 Modification of the X-ray and ultraviolet induced mutation rate in *Aspergillus terreus* by pretreatment with near infrared radiation. Genetics, 33, 429-437.

Tikhov, G. A. 1955 Is life possible on other planets? J. Brit. Astr. Assoc., 65, 193-204.

Van Allen, J. A. 1958 Scientific uses of earth satellites. U. of Michigan Press, Ann Arbor, Michigan.

Wheldon, R. N. 1938 Changes observed in cultures of *Aspergillus niger* bombarded as spores with low voltage cathode rays. Mycologia, 30, 265-268.

Wheldon, R. N. 1940 "Mutations" in *Aspergillus niger* bombarded by low voltage cathode rays. Mycologia, 32, 630-644.

SELF-SANITIZING AGENTS FOR FABRICS

Dr. Charles C. Yeager, presiding

Panelists: Dr. L. S. Stuart
Eugene N. Kramer
Jay C. Chapin

INTRODUCTION

Charles C. Yeager

Technical Director, Scientific Oil Compounding Company, Inc., Chicago, Illinois

The textile industry in its desire to create new markets and new sales appeal for its products, has developed great interest in purifying or sanitizing chemicals. One of the most successful germicidal programs ever developed involved the use of biocidal agents in bath soaps. Capitalizing on the consumer demand thus created, manufacturers were quick to see the textile field as a fertile area for purifying or sanitizing finishes.

The rapid development of interest in the field has created problems never before faced by either chemical manufacturers or by the textile industry itself. Essentially, the development of sanitizing agents for textiles should proceed along the same general lines as the highly successful fungicidal programs of the postwar era. Characteristics accepted as essential for a good fungicide are even more necessary for a sanitizing agent, because of the probability of intimate skin contact.

Unfortunately this area has attracted charlatans and "gimmick" salesmen in large numbers. The gullibility of many manufacturers has engendered falsification of test data to the point where few legitimate suppliers of biocidal agents know where they stand. It is discouraging to see general acceptance of materials which, because of their lack of efficacy, should not even be considered. Compounds that have long since been discarded as completely worthless by governmental scientists are not only accepted but eagerly sought for by consumer goods suppliers.

This symposium can be properly divided into three main areas of consideration. First we must consider the merchandizing viewpoint: the reasons behind this consumer demand and the problems to be considered in the promulgation of a sanitation program. It is only apropos that this viewpoint be presented by a representative of the leading consumer supplier.

57

Once the need for such a finish is defined the problems involved in development and testing of sanitizing finishes must be considered. In this regard the author of the following paper, an authority on formulation and testing techniques, will discuss in detail the desirable physical, chemical, and biological properties of antimicrobials, as well as the testing procedures currently used to screen candidate compounds.

Finally we will consider the governmental viewpoint. Since the use of sanitizing agents is considered economic poison by governmental decree, we will ask a leading authority representing the U.S. Department of Agriculture to discuss the historical significance of self-sanitizing agents and present his viewpoints on current developments.

PROBLEMS IN THE SELECTION, TESTING, AND FORMULATION OF ANTIBACTERIALS TO BE USED AS TEXTILE ADDITIVES

Jay C. Chapin

Scientific Oil Compounding Company, Inc., Chicago, Illinois

A great deal of confusion seems to exist regarding the use of antibacterials as textile additives for sanitary finishes. A variety of compounds have been proposed for such finishes, among which are representatives of some of the major groups of antibacterial chemical agents (1957).

Among the phenolics, for example, are found hexachlorophene and orthophenyl-phenol; representing the organometallic complexes we have tributyltin oxide and phenylmercuric lactate. Various quaternaries also are currently being used, and the antibiotics are represented by neomycin.

All of these materials differ considerably from each other, both chemically and in their antibacterial activity, and yet for various reasons they are considered by their manufacturers, when used in special formulations, suitable for the purpose the merchant has in mind when he asks for an antimicrobial finish.

Such chemical types as the sulfonamides and dyes and others have not been mentioned because the criteria which govern the selection of the antibacterial agent for commercial textiles involve the solution of a number of problems not necessarily related to the biological activity of the material.

To discuss some of these problems in detail, it is well to begin with the considerations which lead a manufacturer to choose one particular antibacterial in preference to another. This will be followed by some of the problems encountered in the treatment of textiles with an antibacterial, then with the test methods used for determining the effectiveness of an antibacterial, and finally with the advantages and disadvantages of various types of emulsion and solvent systems used for preparing the antibacterial in a form suitable for application.

The desirable chemical and physical properties of the textile antibacterial are relative to the purpose intended, and undesirable characteristics may often be overcome by suitable manipulation.

The well-known masking of an objectionable odor with an odorant might be cited as an example. A more direct method would be to substitute the reactants used in making the bactericide. The group of the molecule which carries an odor can sometimes be replaced by a less objectionable homologous group.

Thus, for example, a quaternary ammonium compound which has been obtained by the reaction of naphthenic acid as the anionic portion of the molecule might be substituted with another fatty acid without losing any of the effectiveness of the compound, but with a loss of the objectionable odor.

A more difficult problem, which is of importance to the user of the treated product, is that of toxicity. Some of the most potent bacteriocides possessing a broad spectrum of activity are very toxic to warmblooded animals.

Two widely used antibacterials, phenylmercuric lactate and tributyltin oxide, are quite toxic; nevertheless, they enjoy wide application and are used in garments in intimate contact with the skin.

These two materials can be used because they have high activity, and consequently may be used in very small quantities. Frequently, they are applied together with a binder, such as a thermoplastic or thermosetting resin, which immobilizes the greater portion of the active principle, releasing it slowly as the continuity of the film is broken through laundering and use.

In order to use the more toxic materials with safety, very precise and exhaustive testing must be carried out to establish proper use levels. To be on the safe side, any candidate antibacterial should be finally checked out, at a minimum of five times the intended use concentration, by patch tests on human subjects. The results should be negative for both primary irritation and sensitization.

A very troublesome problem--in my experience the most troublesome—is the color change in fabrics which may be caused by the additive under certain conditions. It is imperative that the chemical selected produces no change in either the color or the hand of the fabric.

None of the antibacterials already mentioned will produce appreciable discoloration of the treated fabric, unless there is exposure to sunlight or, more specifically, the ultraviolet portion of the spectrum. With phenolics and other organic chemicals, a darkening effect will be observed, and with certain quaternaries a bleaching of dyed material. The darkening is often intensified by excessive accumulation of the active principle on the surface of the fabric, and at other times may result from improper application or too rapid drying.

In many cases the darkening can be reduced considerably, if the treater follows his application with a mild scouring of the treated fabric. But no way is known to prevent the bleaching action of dyed goods by quaternary compounds.

The ultraviolet-screening agents seem to have little effect in retarding the bleaching. A great advantage of many of the organometallic compounds is that they show little darkening on exposure to an ultraviolet source over extended periods of time.

The antibacterial, after being applied to the fabric, must, of course, be reasonably stable and not volatilize under the heat of the treater's dry cans or curing ovens. Many phenols have an appreciable vapor pressure, and in selecting among the many available phenolic derivatives, care must be exercised when long-term protection is desired. The darkening of phenolic compounds is also accompanied by photodecomposition on ultraviolet exposure. We have studied the loss of certain phenols on ultraviolet exposure and have found that losses of the active principle from the treated fabric can be quite rapid.

Degradation is also noted with tributyltin oxide and phenylmercuric lactate, in the latter case with the release of mercury vapors.

From time to time, candidate materials are tested which possess all of the desired qualities, but which are virtually insoluble in all of the commercially available solvents. Under these conditions, no satisfactory concentrate of the material can be made except in the form of a dispersion. Many molecules which contain thio groups and metal fall into this insoluble category.

Unfortunately, concomitant with solvent insolubility, desirable characteristics for a textile additive may exist, such as light and heat stability, water solubility, and low toxicity. Our experience has been that most textile treaters do not like dispersions, although a well-made dispersion, properly applied, may be equally as effective as a solvent treatment or an emulsion. In cases of this kind, the method used in preparing the dispersion is of paramount importance.

The screening of candidate antibacterials can be long and tedious work, and any procedure along this line should be designed to give as much information as possible with a minimum of manipulations.

Since ultimately the antibacterial is intended for textile use, very few *in vitro* tests are performed in our laboratory, and the compound is applied directly to the fabric after solubilization in a suitable solvent. Some chemicals prove to be extremely insoluble, but usually some solvent is found that can give the 0.25, 0.50, and 1.0% concentrations used in the initial screening. Cotton sheeting, because of its close weave and ability to lie flat on the agar test plates, is most frequently the fabric used. Diaper fabric also is a very satisfactory material, since when purchased it is free from additives, such as sizing and softeners. In any event, the fabric used should be as free as possible from extraneous material, since the substantive and the bactericidal properties of the tested material may be seriously interfered with by the presence of additives.

After application of the chemical, the fabric is air-dried and then conditioned in various ways. A portion of the fabric is oven-dried for three minutes at 320 F to determine volatility and discoloration under heat. Another portion of the fabric is exposed in a fadometer to ultraviolet light, and another portion is put through a number of washing cycles to determine substantivity.

Activity tests against both gram-negative and -positive organisms are then made of the fabric conditioned in the various ways and compared in activity with the unconditioned fabric.

A tentative test method for detecting the antibacterial properties of fabrics was recently published by the AATCC (Schwartz, *et al.*, 1959). In evaluating this method, a number of laboratories ran tests, using a series of fabrics which had been treated with various antimicrobials at different concentrations. Close agreement was obtained, especially with those fabrics containing a high concentration of an effective bactericide, such as phenylmercuric lactate. Results, however, were not so uniform with fabrics containing only a very small amount of the active principle. The method gives the result in terms of a zone of inhibition. If no zone is present, the sample is removed, and the contact area is examined for the degree of inhibition present under the sample.

A serious criticism of the method is that it depends on the ability of the active principle to leach out of the fabric into the agar medium, while actually the main interest is the ability of the treated fabric to inhibit growth in the interstices of the fabric itself.

The streak plate method has been suggested to eliminate leaching as a consideration. In contrast to the seeded agar plates, where the bacteria are distributed throughout the medium, with the streak plate method bacteria are present only on the agar surface.

When the fabric is placed on the surface, it blots up the bacterial inoculum, and leachability of the active principle becomes of minor importance. A series of interlaboratory tests recently conducted by the AATCC Committee for the Detection of the Antibacterial Properties of Fabric, however, have shown that in comparing fabrics where the concentration of the active principle is very low, the streak plate test can also give confusing and difficult-to-interpret results.

A point that might be worth mentioning here is the distinction between sterilization and disinfecting. Terms such as "self-sterilizing" have helped to confuse this distinction, especially in the mind of the consumer.

Self-sterilization, in the final sense of the word, should never be implied as a property of treated fabric. Rather, the impression should be given that a treated fabric will reduce significantly the number of microorganisms present in the fabric under specified conditions.

The AATCC Committee is at present considering the investigation of a more refined method of testing fabrics. The aim will be to determine a standard of performance of treated fabric as measured by the ability of the fabric to reduce an introduced bacterial population to a reasonable base level.

The proposed test will follow essentially the quantitative method for evaluating fabrics developed by Majors (1959). Originally, this method was developed by Latlief et al. (1959) for determining the production of ammonia from nutrient broth and urine by *Proteus mirabilis*, from treated diapers.

In the test, diaper fabric was saturated with the inoculated broth-urine solution, and incubated in wide-mouth jars. After the incubation period, the formation of ammonia is determined by an acid-base indicator present in the medium. With Majors' method the test is made objective by measuring the titratable alkalinity.

This method has been extended to include organisms other than *Proteus mirabilis*. The more sensitive methods of detecting antimicrobial effectiveness are necessary when determining the presence of the active principle after the fabric has been washed a number of times.

Many fabrics sold today claim to be durably antibacterial. Various figures are given by different manufacturers as to the durability of their product. We believe that a minimum of 20 washing cycles is a realistic goal.

There are two ways to treat cotton fabric with an antibacterial to give it some permanency through washings. One method is the physical binding of the chemical to the fabric with a resin binder, either thermosetting or thermoplastic. The other way is to treat the fabric with a chemical substantive to cotton; for example, a quaternary ammonium salt. Both methods can be combined because substantive cationic substances are exceptionally amenable to fixation to the fabric by resin binders.

Both of these methods have obvious disadvantages. With many goods, where it may be desirable to have an antibacterial finish, a resin additive may be undesirable. This is true of merchandise such as sheets, pillowcases, and diapers.

Certain quaternary ammonium compounds are capable of remaining in the fabric through many launderings without the addition of a binder, but they rapidly lose their effectiveness when washed. The loss of the active principle is attributed to the neutralization of the active cationic portion of the molecule by the presence of free alkali in the soap. The speed at which the neutralization occurs can be decreased by making the quaternaries less reactive. The lowered reactivity can be accomplished by substituting the anionic portion of the molecule, which is usually chlorine, with a long-chain fatty acid (Erskine 1953).

In conclusion, a discussion of proper formulations of antibacterials may be in order. After finding a material which meets the stipulated requirements, it must be put into a form suitable for application by the treater. An essential step, therefore, in the development of a commercially usable bactericide is the proper formulation of the material. Proper formulation can increase the inherent effectiveness of the material, whereas poor formulation can decrease it. A proper formulation can be easily applied and will produce a uniform and effective finish; poor formulation will often result in uneven deposition of the biocide, with a subsequent reduction or spottiness of effectiveness.

Concentrations of the active principle in an organic solvent requiring further dilution with an organic solvent find very little application. The form more acceptable is a water-soluble or water-dispersible concentrate.

As the majority of antibacterial agents are more or less insoluble in water, solubilization in an organic solvent is necessary. The resulting "oil" can then be emulsified or made into an emulsifiable concentrate in a number of ways, all of which have advantages and disadvantages. The foundation of a good emulsion system is a stable "oil" solution. It requires solubilization of the antibacterial in a water-insoluble system, which is not only stable by itself, but remains stable after emulsification. There should also be incorporated into the "oil phase" for bactericides crystalline in nature, nonvolatile solvents or carriers, which will prevent recrystallization or blooming of the fabric after treatment.

Recrystallization is highly undesirable, as it results in discrete particles on the fabric rather than in a continuous film. The nonvolatile solvent selected, of course, must in no way affect the fabric or the activity of the chemical.

The choice of emulsifiers is rather restricted, since it is generally detrimental to the treatment to have any rewetting property put into the fabric by residual emulsifier, especially where a permanent type of treatment is sought.

Subsequent washing of the fabric accelerates the leaching of the antibacterial. The leaching effect can be overcome to some extent if a resin binder is applied simultaneously with the antibacterial.

With some types of fabrics, notably cottons and woolens, retention of the antibacterial through laundering can be improved by using cationic emulsifiers. Certain phenolics can be made more durable to fabric by the use of cationic emulsifiers, especially when more catonic emulsifier is used than that required to make the emulsion. In some way, the emulsifier appears to give a substantivity characteristic to the phenolic and at the same time to retard neutralization by the soap used in the laundering cycles.

From what has been said so far, it is apparent that the proper selection of an antibacterial for textiles involves the solution of a number of problems, among which (although this begins to seem incidental) is that the compound be a good antibacterial; that it does not adversely affect the appearance, or hand, of the fabric; that it be durable through laundering, nontoxic and capable of formulation.

Because of these problems, no single antibacterial in use today is best or ideal for all textile uses. The problems, however, do limit the types of compounds that may be tried and indicate the considerations in the preparation of new compounds.

REFERENCES

Erskine. A. B. 1953 Quaternary ammonium naphthenate and method of making the same. U. S. Patent 2,645,593, Berkeley, Calif.

Latlief, M. A., Goldsmith, M. T., Friedl, J. L., and Stuart, L. S. 1959 Bacteriostatic, germicidal and sanitizing action of quaternary ammonium compounds on textiles. J. Pediatrics, 6, 39, December.

Majors, P. A. 1959 February 9. Evaluation of the effectiveness of antibacterial finishes for cloth. Am. Dyestuff Reporter, 48, 91-93.

Schwartz, I. *et al.*, 1959, January 12. Detection of antibacterial property of fabrics. Agar plate method. Am. Dyestuff Reporter, 48, 8-9.

1957, November 30. Expanded market for residual bactericides on textiles. Chem. Week, 91, 97-100.

FACTORS INVOLVED IN TESTING PRODUCTS CARRYING RESIDUAL GERMICIDAL CHEMICALS

L. S. Stuart
Bacteriology Section, Pesticides Regulation Branch, Plant Pest Control Division
U.S. Department of Agriculture, Washington, D.C.

Germicidal and fungicidal chemicals are currently being added to paper, textiles, upholstering substances, paint, plastics, rubber, cement, waxes, and other materials with certain professed objectives. These might be listed as follows:

1. The preservation of the material itself and articles made therefrom from direct and indirect attack by bacteria and fungi.

2. Odor proofing materials and articles made therefrom by preventing the growth of bacteria and fungi on or in the material and/or commonly associated soils such as perspiration in the case of wearing apparel or moisture condensates on surfaces in humid areas.

3. The provision of a measure of protection to perishable commodities from microbial contaminants carried by environmental surfaces which might subsequently cause deterioration and spoilage.

4. The provision of a measure of health protection to man or other living animals through control of infectious bacteria and fungi carried by inanimate materials or articles fabricated therefrom.

5. The prevention of the production of body irritants and/or nuisances other than objectionable odors by suppressing the growth of bacteria and fungi on or in the material or its associated soils.

These professed objectives should be examined critically before directing attention to the selection or development of laboratory testing procedures. The first objective, that of protecting inanimate materials with germicidal and fungicidal chemicals as residuals which remain in place to attack the invading microorganisms, is universally recognized as possessing great economic significance and is an acceptable commercial practice. The second objective, that of odorproofing materials, is based on fairly substantial but fragmentary evidence that most malodors in clothing, furnishings, and premises arise as a result of microbial activity in associated soils. The third objective, that of protecting perishable commodities from microbial contaminants arising from the environment, has rather common acceptance in the food storage, packaging, and processing fields as a valuable precautionary measure in specific situations. The fourth objective, that of providing a measure of health protection to man and other living animals through the control of infectious bacteria and fungi carried by inanimate materials, is based largely upon assumptions for which scientific validity has not been clearly established.

The first assumption which must be accepted in order to give serious consideration to this last objective is that infectious bacteria and fungi commonly carried by inanimate articles and surfaces do, in fact, initiate infections in man and animals. While it does not seem unreasonable to suppose that this may be true in some instances, there is little or no epidemiological evidence available to show that it is true, or even to define the limited situations within which it might be true. Quite recently (U.S. Pub. Health Serv. 1958) epidemiologists have begun to ask some searching questions relative to the role of the environment in the spread of disease. We can anticipate the development of some reasonable specific limitations with respect to specific diseases, types of surfaces, and environmental situations in this generally accepted hypothesis in the not too distant future.

The second assumption which must be accepted in considering the fourth objective is that in some manner, provision can be made for contact between the infectious bacterial cells or fungus spores and the residual germicidal or fungicidal chemical. In the protection of living plants and animals by residual chemicals, the physiological processes of the host have, in many situations, been found capable of providing the necessary contact between the microorganism and the toxic chemical. This is commonly referred to by the term "mobilization." In the case of the protection of inanimate materials, the residual chemicals that have been provided, if effective, can be mobilized by that amount of moisture necessary for the initiation of the microbial metabolism detrimental to the material. In the absence of this amount of moisture, no microbial deterioration can occur and mobilization of the residual is not necessary. In the same manner, it seems obvious that no mobilization of a residual chemical intended for odor-proofing materials would be necessary in the absence of sufficient moisture for supporting bacterial and fungal metabolism to which the odors have been attributed. It is not unreasonable to suppose, therefore, that the moisture situations inherent in the objectives identified above under 3 and 5 might be such as to provide for effective mobilization of residual chemicals. However, in the case of residues intended for reducing the danger of inanimate materials as carriers of infectious agents, there is no conveniently built-in safety valve with respect to mobilization. Thus, it becomes the critical factor in all practical applications directed toward this objective.

In the theoretical situation of a completely dry living bacterial cell or fungus spore carried by a completely dry surface in the absence of atmospheric humidity, no mobilization of any residual chemical other than a gas could be expected. This, in all probability, constituted the basis for the selection of a series of the high polymers of formaldehyde as the primary residual chemicals of choice for study in connection with investigations by Hoffman, Feazal, and Kaye (1957) of the Chemical Corps, U. S. Army, on the production of self-sterilizing surfaces. A careful study of the published results on this report clearly indicates that no self-sterilizing action was obtained although a self-sanitizing activity was indicated under conditions of relatively high humidity for limited periods of time. In connection with this research project it was found that the mean relative humidity of heated occupied buildings in the city of Pittsburgh, Pennsylvania, was during the winter months less than 20.0% and that no measurable effects on dry bacterial populations could be demonstrated by various residual germicidal chemicals in controlled laboratory studies at relative humidities of 33% or less, the magnitude of the effects increasing as the relative humidity increased above the 33% figure (Hoffman *et al*., 1955). The results also clearly identified moisture as the necessary mobilizing agent for effectiveness with any residual chemical intended for the control of infectious agents carried by inanimate materials.

On the negative side, they indicate that it is not likely that sufficient moisture will be provided by the atmosphere of occupied dwellings to provide for mobilization of residual germicidal chemicals insofar as biocidal activity for infectious bacteria and fungi is concerned. On the positive side, they indicate that under conditions of controlled humidities in the laboratory significant reductions in bacterial populations carried by inanimate surfaces can be obtained by the use of residual chemicals. Lester and Dunklin (1955), and Dunklin and Lester (1959), in rather extensive studies on "Residual Surface Disinfection" using orthophenylphenol preparations, also reported that the ambient relative humidity was the controlling factor in the rate of germicidal action in the laboratory, where under ideal conditions, self-disinfecting surfaces were reported to have been obtained. Of possibly greater significance was their finding of significant reductions in the bacterial contamination of floors, other untreated surfaces, and the air in hospital rooms as a direct result of the use of a germicidal preparation on the floor which provided a residual action. While the data clearly indicate that the practical result was considerably short of self-sterilization or self-disinfection, it must be conceded that reductions of bacterial counts of 95% for floors, 60% for untreated surfaces above the floor, and 50% for the air of rooms cannot be considered as insignificant.

Likewise, the report by Propst (1953), showing that heat sterilized surgical gowns and drapes treated with a residual quaternary ammonium germicide maintained significantly lower bacterial counts during long operations under conditions of high temperatures and high relative humidities, indicates that there may be some significant advantages in the use of residual germicidal agents in this application.

In line with these findings, the very recent reports by Hudson, Sanger, and Sproul (1959a and 1959b) indicate that substantial reductions in the bacterial populations of the hospital environment can be obtained in a maintenance program designed to include a routine providing for the periodic application of contact disinfectants with residual activities in what has been described as a total saturation technique, since floors, walls, table tops, laundry including bedding materials, and attendant apparel, as well as air-conditioning filters received some type of treatment with residual antimicrobial chemicals. The actual data included in these reports were too meager to allow for any final conclusions as to statistical significance, insofar as the residual activity of the chemicals was concerned, but these investigators have clearly gone on record in the belief that they are statistically significant from both a bacteriological and epidemiological standpoint.

In rather sharp contrast to these favorable findings, Colbeck (1958) reported that he could find no residual disinfecting properties with hospital surfaces treated with a disinfectant alleged to be effective in this connection. Also, Phillips (1958) of the Physical Defense Division, U.S. Army, Fort Detrick, Maryland, recently reported that all claims for the production of bactericidal surfaces examined in their laboratories have failed to stand up. Actual use-exposure studies by the Department laboratories with wall paints containing residual germicidal and fungicidal chemicals which provided surfaces that were bacteriostatic, fungistatic, and self-sanitizing in controlled laboratory tests were highly disappointing in that no difference in wall contamination in home bedrooms, small animal rooms in laboratories, public school rooms, or hospital wards was found between areas carrying such paints, over areas treated with control paints carrying no residual germicidal chemicals. In all instances, the counts were so low as to indicate that only in rare instances or in special situations could vertical painted walls be considered significant carriers of infectious bacteria or fungi. In the same negative or seminegative category, the results of actual use tests on sheeting with what appeared in laboratory studies to be an effective residual bacteriostatic and self-sanitizing chemical, showed no differences in the number of viable bacteria carried by untreated sheets and sheets treated with a residual agent after actual use. This was surprising in that it was the belief of the investigators that the relative humidity between the sheets of an occupied bed should be high enough to provide for some mobilization of the residual chemical employed. The foregoing experiences have been cited to illustrate the wide diversity of experimental results bearing on the practicality of objective number 4. It is clearly apparent that careful research will be needed to clarify the current situation.

In the case of objective 5, that of preventing the development of body irritants as the result of the breakdown of materials or associated soils by microbial action, there appears to be, as with objectives 1 and 2, a built-in safety factor with respect to the mobilization of the residual agent. There also seem to be adequate clinical and bacteriological data supporting the practical value of the use of certain residual chemicals in diapers for controlling ammonia dermatitis in infants and similar dermatitis arising from irritants developed in bedding from the secretions of incontinent patients (Benson *et al.*, 1949).

The selection of a suitable testing method for the evaluation of any residual germicidal or fungicidal agent must of necessity take into account intent of use. The selection of the test organism, inoculation procedures, exposure times, temperatures and relative humidities, culture media, and necessary laboratory equipment, will, for the most part, be determined by these objectives.

A very wide variety of testing methods have been employed by investigators to determine value of residual chemicals in the preservation of materials. Those most commonly employed have been characterized by Horsfall (1956) as "poisoned food techniques." Procedure details have been varied according to existing knowledge with respect to the susceptibility of the material under study to microbial damage, the type of damage to be prevented, the availability of pure cultures of organisms implicated in deterioration of the material under study, time, and equipment available to the investigator, and similar considerations.

By way of contrast, two methods, both of which have received rather wide acceptance, will be discussed: Thom, Humpfeld, and Holman (1934) developed a procedure to be employed where the objective is the preservation of cellulosic material from direct attack by bacteria and fungi. This test employs the most commonly encountered fungus capable of attacking outdoor fabrics, *Chaetomium globosum* as the test fungus of choice, and a *Cytophaga sp.*, the most active bacterium that digests cellulose as the test bacterium of choice. An inorganic salts medium is specified and inoculation and incubation procedures are such as to provide pure culture exposures. The results are recorded quantitatively according to losses in the tensile strength of the test materials. This test affords a high degree of precision, but the accuracy of the result interpreted in terms of actual use values is conditioned by such factors as the specific sensitivity of the test organism to individual chemicals, freedom from influence by organisms contributing indirectly to deterioration, such as usually occur in a normal environment, and differences of temperature, and moisture specifications of the procedure from those encountered in actual use. In all fairness, it must be stated that the authors of this procedure clearly recognized these limitations, pointing out that they were choosing between an arbitrary, sharply standardized laboratory technique and costly field trials, which too often yielded indefinite and indeterminate results. It was the hope of these investigators (Personal Commun. 1935) that the basic principles of pure culture testing outlined might be adapted to cover materials other than those of a cellulosic nature. Progress along this line during the past 25 years has been slow and this can be attributed to a variety of factors, including failure to isolate and identify pure cultures of bacteria and fungi having specific significance in the deterioration of other types of materials, difficulties in sterilizing heat-labile materials, the rapid introduction of synthetic products of unknown resistance to microbial attack, and failure to apply newly accumulated knowledge on the biochemistry of microbial metabolism in developing suitable supporting test culture media.

The soil burial test reported anonymously in 1942 (ASTM 1942) has been widely used since that time both as a confirmatory procedure for results obtained in the Thom, Humpfeld, and Holman procedure as well as for primary evaluations on both cellulosic and other types of materials. It is currently a standard method of the American Society of Testing Materials. The essential feature of the procedure is that the material being tested is buried in a biologically active soil with a moisture content permitting active microbial growth. Buried materials are examined at intervals to determine the amount of rotting. The method is commonly considered as providing the most severe conditions for biological degradation. Obviously, the result provides a measure of the direct and indirect effects of the naturally occurring soil flora as well as those of the naturally occurring chemicals. It lacks the precision of the Thom, Humpfeld, and Holman technique, but it does provide a result which may be more accurately interpreted in terms of practical use values, especially in severe situations.

Because of subsequent discussions, it should be mentioned here that in 1939 Burlingame and Reddish (1934) proposed the use of the agar plate diffusion method for evaluating residual fungicidal agents, but that the experience of most investigators indicates that it has very limited value in applications where preservation of materials is concerned.

In evaluating residual chemical treatments as odor-proofing materials, most investigators have employed the agar plate diffusion method with selected strains of micrococci or staphylococci as the test organisms of choice. In some instances, results by this method have been demonstrated to give a close correlation with results obtained in actual use effectiveness studies. It seems reasonable to suppose that for a chemical carried as a residual in a material such as fabric to be effective in preventing microbial activity in absorbed liquid such as body secretions, some diffusion out of the material into the secretions would be necessary. Thus, it is not unreasonable to expect that a method requiring some diffusion from the test material of the residual agent for the establishment of an end point would show some positive correlation with actual use-test results. However, a very active controversy is now being waged over the minimum diffusion necessary to give a practical result in this objective as related to the minimum amount necessary to give a positive result in the agar plate diffusion method. Thus, in many instances, laboratory workers are attempting to set up arbitrary end points for this technique based upon such observations as the extent of growth on or under the material itself in preference to the commonly accepted clear-zone end point. While such arbitrary end points may have some utility in development testing, their use introduces a source of variation greater than that which can be tolerated in a testing procedure employed for referee work. It now seems obvious that if practical values in odor control can be obtained from the use of chemical residuals which will not give a zone of inhibition in the agar plate diffusion method some other type of laboratory test will have to be developed for an accurate evaluation of practical values in this connection.

It should be pointed out that the agar plate diffusion method was initially developed by the Food and Drug Administration to provide a presumptive index to the efficiencies of wet dressings as antiseptics (Ruehle and Brewer 1931). It has been modified and applied in many situations foreign to this initial use. In all applications which have been found to have practical utility, investigators have used clear zones or gaps of clear agar wherein the test organism has failed to grow, in arriving at useful data.

Insofar as odor proofing is concerned, inhibition of bacterial growth in itself can only be a subjective index of possible utility. Thus, test organisms of known odor-producing potential should be given preference in the development of any standard procedure directed toward this objective.

The success of any test method for determining the practical value of residual germicidal and fungicidal chemicals in protecting perishable commodities against contamination by the environment will depend in most instances on the selection of test organisms representative of the type of spoilage or contamination to be controlled, and those test conditions which most closely simulate those under which the product is expected to perform its basic function. Since these will vary from such specific situations as the control of mold growth on the walls of storage rooms held at high relative humidities, such as are employed in the cold storage of shell eggs and the low temperature tenderizing of meat, to packaging and wrapping materials for bakery, dairy, poultry, fish, meat, and fresh and frozen fruits and vegetables, it seems obvious that a diverse set of test conditions must be given serious consideration. In this objective the problem of food adulteration by the residual chemical frequently becomes the controlling factor in acceptance. Unless the antimicrobial chemical has an exemption from the requirements of a tolerance in the specific raw agricultural commodity applications in which it is recommended, or a residue tolerance has been established for the given commodity, then the product is in violation of Public Law 518 (1954), an amendment to the Federal Food, Drug, and Cosmetic Act.

There appears to be no valid reason for acceptance of results obtained in the agar plate diffusion method as evidence relative to the effectiveness of residual chemical germicides and fungicides in providing self-sanitizing, self-disinfecting, or self-sterilizing results in environmental sanitation programs. Claims for such activities are repeatedly encountered, based on such results and such results only. This sophistry has led to a great deal of confusion within the trade, and contributed to a measure of deception, insofar as the consuming public is concerned, that may well discredit the entire practice of applying residual chemicals as an adjunct to environmental sanitation routines. Numerous testing procedures have been proposed that appear to have merit for measuring such activities, but difficulties in correlating the results obtained by these procedures with bacteriological and clinical results obtained in actual use studies have delayed acceptance of the use of residual chemicals as standard methods.

Those procedures with which the greatest degree of success have been obtained involve the inoculation of the treated material itself with bacteria or fungi representative of the infectious species for which control is desired, the incubation of the inoculated material under controlled conditions of moisture availability and temperature for prolonged periods of time with subculturing of the incubated materials at periodic intervals in a suitable subculture medium. Untreated material is used as a control and results with the treated material are corrected according to results obtained with the control.

In tests of this kind, some investigators have insisted that the inoculant be in the form of dry viable cells, and that all moisture be furnished by the atmosphere, using controlled relative humidities. The procedure described by Hoffman, Yeager, and Kaye (1955) can be cited as an example of this type of approach. Insistance on this procedure is based on the belief that the most commonly encountered source of contamination would be dry dust-borne organisms, and that only those materials which can be mobilized by the atmospheric humidity to become effective against this type of contaminant could be expected to possess practical value. Two real difficulties have been found in methods of this type. Since measurements on the effects of the residual chemical must be corrected to allow for the extent of death due to drying and other factors on the control untreated material, and the extent of death with infectious vegetative bacteria in the control frequently exceeds in magnitude the extent of death which can be attributed to the residual chemical, serious questions as to the statistical significance of results cannot be avoided. Also the number of infectious species of bacteria which may be used successfully with the provision of a control viable enough to give any quantitation on the effect of the residual chemical is so small as to limit application in only a few situations of possible public health significance. Gram-negative bacteria behave badly in such tests.

Where a moist inoculant is employed along with controlled moisture and nutritional conditions during incubation, it is possible to establish test controls wherein the inoculant can be held in a lag or stationary phase at a constant level during the exposure period or induced to show logarithmic growth during this period. With such controls, it is a relatively simple matter to determine whether the effect of the residual chemical is static or biocidal, and to evaluate in a quantitative manner any biocidal result which is detected. The number of infectious microorganisms which may be employed is restricted only by potential danger to the operator inherent in any given set of test conditions. Adequate guides have not been developed for interpreting results obtained by such methods in terms of practical use values. However, the greater degree of precision which can be obtained in such procedures indicates that a sharply standardized method of this type offers the best hope for referee evaluations.

Latlief, Goldsmith, Friedl, and Stuart (1951) described a procedure which has been found acceptable for evaluating static activities of residual chemicals intended for preventing the development of ammonia from urea in diapers. Similar procedures can easily be developed for measuring the value of residual agents in preventing the development of other body irritants produced by microorganisms, once the specific irritant is identified.

Care is being exercised to avoid arbitrary regulatory restrictions against the use of residual germicidal and fungicidal chemicals in any application in which there is reason to suppose that practical benefits may occur. However, in many applications, which are currently being proposed, the reasons for supposing that such benefits may be obtained appear to be inadequate. In such instances, existing regulatory authority does not permit countenance of experimentation at the expense of the consuming public.

Some of the factors involved in testing materials carrying residual germicidal chemicals recommended in various applications have been discussed with the objective of promoting a better understanding of the current problems involved and the activities within both research and regulatory groups being directed toward their solution.

REFERENCES

American Society of Testing Materials Standards 1942 Tentative methods of test for resistance of textile fabrics to microorganisms. 3, 1444-1447.

Benson, R. A., Slobody, L. B., Lillich, L., Maffia, A., and Sullivan, N. 1949 The treatment of ammonia dermatitis with Diaperene. J. Pediatrics, 34, No. 1, 49-51.

Burlingame, E. M. and Reddish, G. F. 1934 Laboratory methods for testing fungicides used in treatment of epidermophytosis. J. Lab. and Clin. Methods, 24, 765-772.

Colbeck, J. C. 1958 The use of the laboratory in the epidemiology of staphylococcal disease. Proc. National Conference on Hospital-Acquired Staphylococcal Disease, Atlanta, Georgia, Sept. 15-17, 123-133.

Dunklin, E. W. and Lester, W. Jr. 1959 Residual surface disinfection II. The effect of orthophenylphenol treatment of the floor on bacterial contamination in a recovery room. J. Infect. Dis., 104, 41-55.

Hoffman, R. K., Feazek, C. E., and Kaye, S. 1957 Sporocidal surface coatings. Proc. 43rd Midyear Meeting, Chem. Spec. Manuf. Ass., Inc., N.Y. May, p. 106-110.

Hoffman, R. K., Yeager, S. B., and Kaye, S. 1955 A method for testing self-disinfecting surfaces. Proc. 41st Midyear Meeting, Chem. Spec. Manuf. Ass., Inc., N.Y., May, 76-81.

Horsfall, J. B. 1956 Fungicidal action and its measurement. Principals of fungicidal action. 23-27. Chronica Botanica Co., Wattham, Mass.

Hudson, P. B., Sanger, G., and Sproul, E. E. 1959a A system for control of pathogenic bacteria in the hospital environment. Med. Ann. District of Columbia, 27, No. 2, 68-70.

Hudson, P. B., Sanger, G., and Sproul, E. E. 1959b Effective system of bactericidal conditioning for hospitals. J. Amer. Med. Assoc., 169, No. 14, 1549-1556.

Latlief, M. A., Goldsmith, M. T., Friedl, J. L., and Stuart, L. S. 1951 Germicidal and sanitizing action of quaternary ammonium compounds on textiles, preventing ammonia formation from urea by *Proteus mirabilis*. J. Pediatrics, 39, 730-737.

Lester, Wm., Jr., and Dunklin, E. W. 1955 Residual surface disinfection. 1. The prolonged germicidal action of dried surfaces treated with orthophenylphenol. J. Infections Diseases, 96, 40-53.

Personal Communications, 1935.

Phillips, C. 1958 Sterilization and decontamination. Proc. Conference on Relation of the Environment to Hospital-Acquired Staphylococcal Disease. December 1-2.

Propst, H. D. 1953 The effect of bactericidal agents on the sterility of surgical linen. Am. J. of Surgery, 87, No. 3, 301-308.

Public Law 518--83rd Congress 1954 An act to amend the federal food, drug, and cosmetic Act with respect to residues of pesticide chemicals in or on raw agricultural commodities. Chapter 559, 2nd session, H. R. 7125.

Ruehle, G. L. A., and Brewer, C. M. 1931 U. S. Food and Drug Administration methods of testing antiseptics and disinfectants. Circular No. 198, U. S. Department of Agriculture.

Thom C., Humfeld, H., and Molman, H. P. 1934 Laboratory tests for mildew resistance of outdoor cotton fabrics. Am. Dyestuff Reporter, Oct. 22, 1-7.

U. S. Public Health Service—Communicable Disease Center 1958 Proc. National Conference on Hospital-Acquired Staphylococcal Disease. Atlanta, Georgia, Sept. 15-17.

THE MERCHANT LOOKS AT MICROBIOLOGY

Eugene N. Kramer

Sears, Roebuck and Co., Chicago, Illinois

Society suffered the ravages of microbial damages before true society even existed. Man was struck by plagues and his crops were destroyed by agents which either remained unseen or became visible too late. Almost until modern times his defenses against this invisible enemy were virtually intuitive and it is surprising how successful our predecessors were in learning how to exist simultaneously with their microscopic neighbors.

It is even more amazing that early man was able to harness the useful microbe, to do his will. We certainly can appreciate that early microbiologist who discovered that fermentation produced highly palatable liquids which helped alleviate the rigors of his life. From the nutritional standpoint, the empirical work that went into the production of cheeses and yogurt and the preservation of his foodstuffs is to be commended. Examples of such microbial harnessing are numerous and are being continuously developed.

However, our primary concern here is not the friendly microbe to which we have just alluded. We are concerned with that ancient enemy whose cost to us in lives, property, and money cannot be estimated. It is said that the world is growing smaller but there is a realization that it is also getting more crowded. As a result, the price of supporting microbial parasites is getting to be not only too high, but also too dangerous.

As microbiologists, you are daily making advances in salvaging those things which were previously lost to bacterial and fungal damage. Industry and agriculture have been long aware of microbiological damage and are constantly devising means of controlling it. Agriculture, for example, has combatted the problem both by chemical means and by selective breeding. In industry, millions of dollars are being saved annually through the wise use of the microbiologist's weapons. Fence posts and telephone poles are being treated to prevent their fungal decomposition. Food processing plants are coating their walls with antimicrobial paints to improve sanitary conditions. Bottling companies and fruit and vegetable packers are treating their boxes and baskets to prolong their use. Tarpaulins and rope are now commonly treated. The leather and paper industries are wide users of microbicides with the result that normal allowances which a short time ago were made for waste are gradually becoming the exception. As a result, we are all profiting not only as microbiologists or representatives of industry, but also as consumers. What you are doing and what you can do for us in serving the ultimate consumer is my concern as a representative of a merchandising organization. We know that you can give the customer products which will last longer. We know you have enabled us to give the customer better products which will do a better job.

The merchant has been long aware of the potential damages which could be caused in consumer merchandise by microbial parasites, either from personal experience or as a recipient of a customer's letter of complaint. Until relatively recent times he was powerless to do anything about it. When industry had solutions for him they were either impractical or too expensive.

With the end of World War II much information pertaining to fungal damage and its prevention was released to industry and many newly discovered chemicals were made available. The bulk of this work was utilized for industrial purposes and the consumer was generally unaware of it.

However, today we are in a new era. The consumer, though he may not be fully aware of it, is taking greater advantage than ever of microbiological knowledge. I do not even refer to or include medical science. What I am talking about is the simple fact that people take more baths. They change their underclothing and socks more often. They use more germicides and disinfectants in their household tasks than ever before as a general way of life. The consumer has been made aware of the advantages of using such products as Lifebuoy, Dial, Zest, Praise, and other soaps containing bactericides or bacteriostats. They use more Lysol and Chlorox. They use antibacterial deodorants to combat perspiration odors. To combat dandruff they use readily available medicated shampoos. Floor waxes, laundry detergents. room sprays, scouring powders, rug shampoos, toothpastes, and cosmetics are being promoted and sold with microbiological devlopments which have been built into them.

The consumer public has recently read in his daily newspaper, in the Saturday Evening Post, in Time Magazine, in the Reader's Digest, and in other magazines, about the dangers of bacterial infection and the necessity of protecting against them. As a result, the public is receptive to and is beginning to demand more safeguards against his hidden enemies. In America, and in more and more places in the world, good sanitation is becoming a way of life.

Industry, the microbiologist, and merchandising organizations such as Sears have recognized this desire on the part of the consumer and are anxious to satisfy it. Within the last several years we have seen the development of a new means of fulfilling the trend to greater sanitation in our lines. This latest development is the inclusion of antimicrobial chemicals in the consumer's wearing apparel and other consumer-use merchandise. Such names as Sears Kenisan, Permachem, and Cyana purifying finish, among others, are becoming familiar on consumer merchandise.

In addition there are the manufacturers of basic materials including the Scientific Oil Co., Nuodex, Dow, Sindar, Monsanto, Geigy, and Vanderbilt to mention only a few of the companies actively involved in providing chemicals for sanitary finishes. Such items as undergarments and lingerie, shoes and socks, infants' apparel and furnishings, shirting and blouses, lining fabric, pillows, mattresses and bedding, bathroom accessories, sporting and camping equipment, and toys are examples of merchandise whose use is being prolonged or enhanced through the use of antimicrobial treatments.

There is no question that the number of companies and items involved in the manufacture and promotion of sanitary finishes will rapidly increase in the coming years. The sanitary or purifying finish in textiles and related items is a complementary extension of the consumer's desire for a more hygienic environment. These finishes primarily act as textile antiodorants by reducing and resisting the action of odor-forming bacteria on fabrics. In addition, they will aid in preventing the formation of acids resulting from bacterial breakdown of perspiration and thus reduce staining and prolong the life of the fabric. Though we feel strongly that no positive medical claims can or should be made at the proprietary level, there are, no doubt, certain benefits which will accrue to the consumer as a result of purifying finishes. These finishes in diapers are a deterrent to diaper rash. Similarly, in shoe linings or on hosiery, they can aid in preventing the recurrence of athlete's foot. More generally, the reduction of odors in clothing will certainly aid an individual's mental well-being. The use of chemicals which will reduce or prevent the growth of fungal organisms on many items will certainly be both an aesthetic and financial gain. In Southern coastal areas and not uncommonly in the North, the reduction of fungal growth on clothing and furnishings alone, would make such finishes desirable.

It has been long recognized that in order for a store to sell merchandise, it must be attractively presented to the customer. Packaging and display aside, the merchandise must be clean. With the advent of purifying finishes, the customer can not only obtain visually clean garments from the store but also, clothes which are in a sense microbiologically clean. In the case of try-on merchandise, this is especially important both from the customer's and merchant's viewpoint.

The customer has come to expect something new and something better not only last year, but this year, and next year also. In today's market the customer is highly fortunate in that these developments are forthcoming, and in many cases because of the competitive situation he can get these premium features at little or no added cost. The merchant, to survive, must cater to public demand. At the same time, it is imperative that, in selling a feature that cannot be seen or felt, the merchant be placed in a position of trust where his integrity cannot be doubted. It is all too easy to attach a label to a piece of merchandise and say that it has been atomized, molecularized, or glamorized, when, in fact, the only thing that has been accomplished is the adherence of that label. The manufacturer of purifying chemicals must meet certain basic requirements to protect the customer, the merchant, and himself. Among these are:

1. His chemicals must be non-toxic. Treated fabrics should not cause skin irritation or sensitivity when tested by accepted techniques. The use of cumulative poisons cannot be condoned. The chemicals must be safe in case the fabric is chewed, a not uncommon occurrence with infants and young children.

2. The chemical should not affect any of the physical or aesthetic properties of the merchandise. A good hand must be retained and the color must not be altered.

3. It should be odorless, or at worst, have no unpleasant odor.

4. These chemicals must be capable of being applied to fabric in combination with other textile additives and without radically altering the method of textile processing.

5. They must be stable on the fabric to assure shelf life of the purifying finish.

6. Finally, though it is recognized that while receipt of merchandise by the customer in a sanitary condition is desirable, it is even more satisfactory to have durability to normal washing or cleaning, to give extended protection to the merchandise. The writer personally feels that where merchandise has not had the durable type of treatment there is a greater tendency to the "gimmick" approach and that the merchant and the customer are not getting their full money's worth.

Though the merchant does not usually have trouble in selling things that the customer can see or touch, there is something mysterious about sanitary finishes. We are selling the unseen and must curb our natural tendency to inflate something which is good and useful into something completely out of proportion to its basic character. As has been stated earlier, we are not medicine-men and must keep this in mind when stating our claims.

As microbiologists our tests show us that sanitary finishes can do certain things and won't do others. By the layman, particularly the merchant, there may be a tendency to inflate our findings to a more glamorous level.

The merchant quite naturally desires to get the maximum promotional value from the sanitary-finish feature which he adds to his line. It is our job to guide the merchant as to what he can ethically say and warn him of those things which are not in his province to say. Experience has shown us that the average copy-writer does not have the technical background as yet to write accurate copy dealing with antimicrobial additives. At Sears, we of the laboratory have established certain basic general claims which we feel to be fully descriptive of the properties of these finishes without exceeding good taste and common sense. These may be altered or reworded to fit a particular piece of merchandise, but the final copy is as a rule resubmitted to the laboratory for final verification. In actual sales we have found that the use of factual and conservative claims has not been a deterrent and quite likely has established greater consumer confidence in this new feature.

We feel that the following claims provide a more than adequate selling tool:
1. The sanitary finish provides hygienic freshness to the fabric.
2. It curbs, resists, inhibits, and retards the growth and action of bacteria and germs.
3. It subdues odor development
4. It assures the customer he is getting a sanitary product.

Probably the hardest aspect of this program for the merchant is a truly effective quality control program. Such a program, to be feasible, calls for complete cooperation between the merchant, the chemical manufacturer, and the textile processor. Without such cooperation, a successful sanitary-finish program is virtually doomed at its inception.

There are several ways in which a quality control program can be handled. For example, the chemical manufacturer should be ready to supply the merchant with billing data showing that the textile processor is buying adequate quantities of the chemical to treat his goods. In addition, the merchant himself must be prepared to periodically get sample swatches of fabric from his cutter and/or pieces of treated merchandise for microbiological testing and assay. In most cases the chemical manufacturer should be prepared to bear the brunt of this program, since all too frequently the merchant does not have adequate testing facilities. He can certify the goods after treatment and not release labels and hang tags unless the finish meets specifications. In addition, the independent laboratory is going to be called on more and more to confirm tests and assure the concerned parties that all is well.

It practically goes without saying that we must be wary of the unethical operator regardless of his position. Too frequently, a good concept has been ruined because of extravagant claims or poor performance. This operator loses the least because he has usually invested the least.

To be fair to the consumer, we must not apprach sanitary finishes as a fad. With the consumer's previously mentioned increased knowledge has come a greater sophistication in his or her buying habits. Probably the best example of this was in the not-too-long-past boom and bust in the chlorophyll field. Outrageous claims were made, with the result that anything green assumed amazing therapeutic powers both implied and expressed. Much of the groundwork for an ethical promotion of sanitary finishes has been laid. But to do a better job of selling, the merchant needs chemicals which will give better durability. Nylon, Dacron, and similar fibers have, to date, resisted effective treatment. He needs chemicals which can be applied easily and safely. He needs quick and simple test techniques which will assure him and his customer that they are getting a quality feature. The commonly-used pour plate and streak plate tests are both lacking in giving definitive answers, and we are still looking for better methods.

The advantages of all this to the microbiologist are obvious. There will be expanded outlets for his products and a greater recognition and prestige for himself. The challenge is there. We have primarily referred to clothing here, but ethical applications for sanitary finishes are all about us. It is only for us to continue the job and carry it through.

ANTITUMOR ANTIBIOTICS

Dr. H. Christine Reilly, presiding

Panelists: Dr. John Ehrlich
Dr. Alfred R. Stanley
Dr. Glynn P. Wheeler
Dr. Robert Fuerst

INTRODUCTION

H. Christine Reilly

Division of Experimental Chemotherapy, Sloan-Kettering Institute for Cancer
Research and the Sloan-Kettering Division
Cornell University Medical College, New York

An extensive search is being made among natural products, particulary among
the metabolic products of microorganisms, for agents which have the ability to
retard the growth of experimental tumors and which may aid in the control of
cancer in man. Microorganisms are recognized widely as excellent little factories
for the production of new types of compounds which are beyond the limited imagination of man. Such agents usually exhibit a high degree of specificity in the extent
of their ability to interfere in the growth of various cell systems.

Although the fundamental differentiation between normal and neoplastic growth,
as yet, has not been elucidated, that a difference, or differences, in the metabolic
patterns of these two systems exists is most likely. It would appear that the difference is indeed subtle in nature; only a substance which has a highly specific
type of action may be expected to exert a controlling influence upon neoplastic
development without causing irreparable damage to normal tissue.

The screening for antitumor agents and the control of their development at
present is being carried out for the most part by *in vivo* test procedures. To be
effective, testing must be conducted on a large scale. Because of the expense
involved, economic problems assume gigantic proportions. The appropriation by
Congress of funds for this purpose is aiding immeasurably in the progress of this
undertaking.

The industrial development of antitumor agents of microbial origin, already
burdened with the usual problems of large-scale fermentation and chemical purification, is complicated further by the necessity for following progress with *in*

79

vivo test procedures. Testing is time-consuming, and assays lack precision. In an effort to develop more rapid and more precise assay systems, attempts are being made constantly to find *in vitro* systems, including microbiologic, which may be employed advantageously in this area.

The advent of new compounds which possess interesting biologic effects has opened new frontiers in biochemistry. By the inhibition of certain specific steps in metabolic pathways, hitherto unknown intermediates have been detected and even characterized. Knowledge of metabolism, particularly that of purines, steadily advances.

PROBLEMS IN DETECTION AND ISOLATION OF ANTITUMOR ANTIBIOTICS

John Ehrlich

Parke, Davis & Company, Detroit, Michigan

In any program for new antibiotics, one is faced with problems at every step: in screening, fermentation, fractionation and isolation, evaluation, process improvement and, ultimately, production. Not only in screening, but at every step from first culture filtrate to final product, the crucial problem is assay, the prompt and precise estimation of how much of the desired substance is present. In no sector of antibiotic research is this problem more vexing than in the quest for an antibiotic effective against neoplasms. For here we lack not only efficient test methods but even one wholly effective drug to serve as a standard of comparison for evaluating leads at an early stage.

The most widely used tests measure inhibition of growth of transplantable neoplasms in rodents. Dr. Stanley will discuss the sequential analysis scheme set up by the Cancer Chemotherapy National Service Center of the National Institutes of Health in an attempt to verify the activity of a primary screening lead. It is obvious that if the test results from the screening system are so suspect that replicate testing is necessary to ensure activity even qualitatively, then this test system cannot be expected to serve well for assay purposes where precise quantitative data are essential. Obviously, there are inherent problems associated with any animal test. In addition to frequently imprecise results, a relatively long test period is required. During this period frozen aliquots of the beers on test must be held in storage, subject to deterioration. And at least equally troublesome precautions have to be taken for maintaining large numbers of microbial cultures in order to ensure the viability of the few promisingly active candidates during the long fermentation, testing, and reporting period.

Rodent tests require a relatively large test sample. This requirement poses problems to the chemist who is striving to isolate the active substance. The preparation of 50 to 100 ml of each test sample is a wasteful and arduous, if not impossible, task for the chemist. And the preparation of large numbers of primary screening beers in sufficient volume for three or more tests in rodents presents problems in logistics that assume staggering proportions.

Testing in rodents involves the problem of toxicity. Not only may the antitumor substance be toxic to the host, but frequently crude beers contain one or more additional toxic components. In the isolation process, these materials may accompany the antitumor activity for a time, with the result that a given sample may contain the antibiotic, but its activity be masked by the larger dilution required because of the extraneous toxic materials.

Another problem that confronts us is tumor variability—variation not only in size and rate of growth but also in sensitivity to inhibition. An example of variation in sensitivity can be found in some of our experiences at Parke, Davis & Company in the early days of Chloromycetin.* Early beers and concentrates and even crystalline Chloromycetin showed minimal but reproducible inhibition of sarcoma 180 in mice. Our early elation subsided and finally vanished when subsequent tests proved totally negative. Eventually, it was concluded that the tumor had probably changed sufficiently to become insensitive to Chloromycetin.

Our experience with azaserine provides another example. Early beers not merely inhibited tumor growth but actually caused regression of the tumor implants. Later, when crystalline azaserine became available, such regression could not be demonstrated with the pure compound or with numerous referments, although beers and compound remain potent inhibitors of tumor growth. Did this experience reflect loss of a potentiating component present in early beers, or reduced susceptibility of later tumor implants to regression by azaserine?

It is common knowledge that tumors can vary in unpredictable ways. One can select for large tumors or small tumors, for fast-growing tumors or slow-growing tumors—all from the same parent tumor line. When one is screening hundreds of beers a week in thousands of mice, more frequently than not, one observes considerable variation in the size of tumors in the control group. What does this mean? More important, if the range of tumor weight in control groups is from 400 to 2000 mg, how confident can one be that an average tumor weight of 400 mg in a group of treated mice really indicates suppression of tumor growth?

There are other obvious problems associated with animal testing. I suspect that more than one pharmaceutical company has suffered through periods of enforced idleness waiting for problems of chronic disease in mice or shortages of hybrid mice to be cleared up.

All of these problems compound to make it extremely difficult to screen large numbers of culture beers effectively and to select the leads confidently. Probably the most serious aspect of the whole situation is that it is even more difficult, and frequently impossible, to move the leads along rapidly and efficiently.

For an adequate screening program, thousands of cultures must be maintained in the laboratory while awaiting test results. Hundreds of flasks containing thousands of liters of media must be inoculated, incubated, harvested, clarified, perhaps filtered, bottled, frozen, labeled, and transported to the testing laboratory. Transportation, for companies that do not do their own testing, is, in itself, a major operation involving dry-ice packaging and split-second timing to get the packaged materials to the airport in time to be loaded on the proper plane which may then be grounded for eight or more hours because of a fog!

Obviously, we would welcome a substitute for the mouse tests. The cost, not only in dollars, but in thoroughness of work is unsatisfactory, to say the least. Unfortunately, the question of what to substitute for the rodent tests is almost unanswerable in the light, or should I say darkness, of our present knowledge of cancer chemotherapy. The rodent tests with all their failings have, at any rate, been of sufficient value to permit the detection and isolation of antibiotics such as some of the actinomycins, azaserine, mitomycin, and other substances which do

---

*Registered trade name, Parke, Davis & Company, Detroit, Mich.

exert tantalizing hints of activity in humans. The development of some of these antibiotics, especially in the later stages, was greatly facilitated by the discovery of collateral activity against microorganisms, notably yeasts. But note that antimicrobial activity of a tumor-inhibitory beer can sometimes be a costly red herring, leading one to an antimicrobial by-product of the beer instead of the desired tumor-inhibitory compound. Microbiologic assays can rapidly yield relatively precise results on large numbers of gratifyingly small samples. For this and other reasons, many investigators have searched for some suitable battery of microorganisms growing under varied environmental conditions that could be used as a primary screen or prescreen for detecting antitumor antibiotics. So far, this approach has not met with significant success.

Of course, there have been instances where beers that were totally inactive against the primary tumor spectrum, and that were pursued solely on the basis of antibacterial activity, have yielded crystalline antibiotics markedly effective against neoplasms. Actinobolin is a recent example of this sort. Although actinobolin is a fairly potent antimicrobial agent, relatively large amounts of this antibiotic are required for inhibition of neoplasms. These concentrations are not usually found in crude beers. This brings us to another disadvantage of rodent tumor screening. The rodent itself constitutes a rather high screening hurdle because beers, once injected, are considerably diluted by body fluids. Hence, only those beers that contain either a very potent antibiotic or a high concentration of a less potent substance pass the screen.

The use of tissue culture procedures is, perhaps, the most promising substitute for rodent tumor testing at this time. Procedures are now available for culturing diverse tissues obtained from higher animals in vitro, and such procedures are being constantly improved. Since they measure specific cell sensitivity (or cytotoxicity), they provide probably more precise results than do in vivo tests, and in a shorter period of time. Also they require smaller samples than do rodent tests. However, a tissue of malignant origin growing in vitro is relatively free of host influences and thus differs unavoidably from the same tissue growing in vivo. And inhibition of a given neoplasm growing in tissue culture by no means assures that the same tumor in a mouse would be sensitive, even if it could be exposed to the same concentration of the antibiotic. At our present stage of knowledge, perhaps the most intelligent approach would be a compromise: that is, to perform primary screening of beers against transplantable neoplasms of animal and, where possible, of human origin in rodents or perhaps chick embryos; and to use tissue cultures, where possible, for assay systems to guide fermentation studies, strain improvement efforts, and chemical fractionation and isolation--activities requiring hundreds or even thousands of assays. Occasional critical samples could be tested in vivo to ensure the reliability of the tissue-culture assay.

In view of the lack of data proving direct correlation between tissue-culture inhibition and inhibition of the same tumor implanted in vivo, it would appear premature at this time to perform primary screening in tissue culture, although a tissue-culture prescreen would almost certainly reduce the number of culture beers to be screened in vivo. For the present, it appears wiser to select the best leads from the mouse screening programs and follow them up promptly with tissue-

culture assays wherever possible. Leads could be developed more quickly and
efficiently than with *in vivo* assays, and the *in vivo* test facilities could be more
advantageously employed in primary screening, in checking tissue-culture assay
of key samples, and in secondary testing against a broader spectrum of tumors.
Such secondary testing will often uncover a sensitive tumor more useful for assay
purposes than the original tumor. Equally important, such testing will sometimes
reveal activity against a more challenging tumor, such as a melanoma or a spontaneous tumor of rodents, and hence heighten interest in that lead.

Returning for a moment to the subject of tissue-culture techniques for primary
screening, many workers have observed that cells of normal origin soon lose
their original features when cultured *in vitro* and acquire the more rapidly multiplying, immature characteristics of neoplastic cells. Harry Eagle and others have
sought to compare the effects of numerous tumor-inhibitory substances on certain
cell lines of normal and of malignant origin and in their tests observed no significant differences in response. But some investigators are finding that if they
compare the sensitivity of certain cell lines of malignant origin, maintained in
tissue culture, and of primary cell suspensions from homologous normal tissues,
differential cytotoxicity is frequently observed. If these indications can be confirmed and supported by much more extensive and unequivocal evidence, such
paired systems may yet provide a more rational and attractive basis for using
tissue-culture procedures in primary screening.

In spite of our present problems in the screening and assay areas, we have
been successful in developing several new antibiotics that are encouragingly effective in inhibiting the growth of neoplasms in laboratory animals. In fact, we
have so many such leads that we are confronted with a serious problem. The cost
to take ten new antibiotic beers out of the screening laboratory and follow them
through necessarily elaborate microbiological, chemical, and pilot plant studies,
to produce kilogram quantities, to evaluate them pharmacologically, and then to
mount clinical investigation programs is staggering.

The dollar cost is substantial but the cost in time and in technical man-hours
is almost prohibitive. A substance like azaserine can cost its developers from a
quarter of a million to a half million dollars, and more if extensive clinical
investigation is undertaken. To this must be added the thousands of hours of pre-
cious research time invested. And in the case of azaserine, the Sloan-Kettering
Institute contributed costly mouse assays of over 400 research samples from
Parke, Davis & Company.

It is simply impossible to follow up all the leads intensively and simultaneously.
Obviously, one has to arrange them in some sort of order and assign priorities.
But how arrange them? What are the criteria for selecting the best of the lot? It
is here that we sorely need a material truly effective in the clinic to serve as a
yardstick. I suppose this proves again that nothing succeeds like success. Unfortunately, we have no such yardstick.

Six-mercaptopurine (6-MP) does retard the growth of certain leukemias in
humans but it effects no cure. Furthermore, few beers inhibit mouse leukemia
L1210. And though 6-MP is active against this mouse leukemia and against mouse
adenocarcinoma C755, and gives a good delayed effect against Crocker mouse

sarcoma S180, it is only slightly active in the regular S180 screening test. Hence, if we use 6-MP as our yardstick, will we end up with materials that are as good as but no better than 6-MP? Obviously, we could use several yardsticks, preferably representing different classes of compounds, different modes of action, and different patterns of antitumor activity.

Some workers in the field believe that breadth of spectrum should be a governing factor in selecting materials for preclinical work-up. Others argue that broad-spectrum activity is not a sound requirement and may merely signify inherent cytotoxicity. They believe that good activity against only one tumor is sufficient cause to go all the way through clinical investigation. Who can say which view is right? We know so little about the correlation between the activity of a material against a particular rodent tumor and its activity against cancer in man that we dare not overlook a single possibility. So, once again, we compromise. We admit that we should work on them all; yet there are not enough technicians available for hire at any price to do the job. We have to divide our list of leads into smaller groups and take a few broad-spectrum leads and a few narrow-spectrum leads, work them up, and then wait and see how they perform clinically.

There are decades of work ahead of us before we can replace empiricism with rationalism. But in spite of the problems that confront us, in spite of the frequent feeling of bewilderment and frustration, I am sure that we are more confident than ever before of the ultimate success of our search. Let us hope that it will be soon.

THE CANCER CHEMOTHERAPY NATIONAL SERVICE CENTER ANTITUMOR TESTING PROGRAM FOR ANTIBIOTICS

Alfred Stanley

Cancer Chemotherpy National Service Center, Bethesda, Md.

The submission of antibiotic filtrates on a routine basis to the Cancer Chemotherapy National Service Center screening program was started with one supplier in June, 1956; a second joined in September and by the end of the year six suppliers had submitted approximately 3500 different filtrates.

During 1957 and 1958 five more suppliers joined the program, giving the present total of 11. As of July 1, 1959, approximately 36,500 different cultures or filtrates had been submitted for screening. In all, a total of about 75,000 samples have been received including fermentation research samples, fractions, etc.

Almost all (99%) of the antibiotic filtrates have been received from industrial sources. The cultures used were isolated and the filtrates produced by, and at the expense of the industrial supplier in most cases. There are exceptions to this. The government program provided, in general, only for the testing of the samples submitted. Industrial suppliers include Abbott Laboratories; Bristol Laboratories; Ciba, Inc.; Hynson, Westcott, & Dunning; Parke, Davis & Company; Chas. Pfizer & Company; Schering Corporation; E. R. Squibb & Company; and the Upjohn Company. Nonindustrial suppliers include the Brooklyn Botanic Garden and the Michigan State Health Department Laboratories. Single or small groups of samples have also been received from individuals and other industrial concerns.

The Mouse Screen

In establishing an antitumor screen for large-scale operation, one of the first problems encountered was, of course, that no one knew for certain how to test for effectiveness in patients with cancer except by actual clinical trial. This, of course, would be impossible for the large number of samples to be tested. All major antitumor screens in use were reviewed, as were cooperative, comparative studies conducted by many of the leading cancer research laboratories. These results were reported in what is called the "Gellhorn Report," on the basis of which a three-tumor screen was selected. This screen, using sarcoma 180, carcinoma 755, and leukemia 1210 in mice, had shown positive results on all of the chemicals then known to have clinical effects against certain types of human cancer.

Mouse requirements were the next problem. Sarcoma 180 would grow in the ordinary Swiss mouse, which was in good supply. However, the carcinoma and leukemia required inbred strains of mice or their first generation hybrids. This was selected as the best solution and so the F_1 hybrid of the C57BL/6 Jax and the DBA/2 Jax, known as the BDF_1 mouse, became the second test animal.

These mice, in keeping with all other inbred stocks, were in short supply. Thus, a major problem was the expansion of mouse production facilities, keeping ever mindful the dangers of disease. The breeder of the specific strains received a contract to provide for a large expansion of the production of the two strains. Five satellite producers were then set up under contract. These additional facilities received breeding stock from the prime breeder and produced the F_1 hybrids for the screening contractors. The primary and five satellite suppliers are now producing over 1,000,000 hybrid mice/yr for this program.

Some of the special precautions taken were the establishment of two genetic centers at widely separated universities to provide a source of these strains in the case of catastrophe. Two contractors are provided whose responsibilities are concerned with the health of the mice. One works in the area of ectromelia (mouse plague) and the other in salmonellosis and other diseases. The assistance of the National Research Council was secured in determining requirements and publishing specifications for minimum standards for mouse production.

Even with this mouse production program one of the center's biggest problems has been a shortage of hybrid mice, thus restricting tests using the carcinoma and the leukemia. As an aid in relieving this situation a fourth tumor, the Ehrlich's ascites, has been introduced. This will grow in the Swiss mouse and is used in place of carcinoma 755 with a large number of filtrates. This tumor was selected since it appeared to have approximately the same sensitivity as the carcinoma 755 though it does not necessarily respond to the same materials.

Data Analysis

The next problem was to set the standards for a positive response: To determine what to call active and what to call inactive materials. The optimum procedure would be to pick out 100% of the true actives and discard 100% of the true negatives. Such a precise division is impossible. Thus, a less exact end point was selected, one which would be capable of eliminating most of the negatives with the loss of as few true positives as possible.

The point selected by the advisory committee on screening, as indicating a positive effect, was a difference in tumor weight between the treated and control animals of 58% or more. In other words, if you divide the tumor weight of the test animals by the tumor weight of the control animals it would give what we call a T/C value of 0.42 or less.

A three-stage sequential test was established in which any sample having a T/C value of 0.54 or less on the first test would be retested. If the T/C product of the two tests $[(T/C)_1 \times (T/C)_2]$ was 0.20 or less, it was tested a third time. Then if $(T/C)_1 \times (T/C)_2 \times (T/C)_3 = 0.08$ or less, it was considered to have passed the screen.

Since, if the value 0.42 was used each time, the $(T/C)_1 \times (T/C)_2 = 0.176$ and $(T/C)_1 \times (T/C)_2 \times (T/C)_3 = 0.074$, a near-miss area is "built in" to our screen.

Passing the screen does not mean that a material has definite antitumor activity. This screening of filtrates is an "enrichment process." In the entire group of filtrates as first submitted for screening, indications are that any one filtrate has about one chance in 300 or 400 of having definite antitumor activity. After passing the screen, which has been described, a filtrate appears to have about one chance in five of being definitely active against tumors. The false positives are eliminated by the screening of refermentations.

This three-stage sequence has recently been changed to a two-stage sequence for the antibiotic filtrates. The values used are as follows: $(T/C)_1 = 0.45$ or less and $(T/C)_1 \times (T/C)_2 = 0.20$ or less. Here again the near-miss area is built in.

As an example of the effectiveness of this sequential screen it is predicted statistically that the three-stage sequence will pick up 95% of materials which show 70% or more inhibition (T/C of 0.30 or less) while the two-stage will pick up 90% of them. Each one will pick up only about 1% of those materials showing 40% inhibition or less (T/C of 0.60 or greater).

The big problems are, of course, the false positives and negatives due to variability in tumor weight. This variability can be reduced by using older tumors and matching weights at the start of the test but as older tumors are used, fewer materials show an effect and in the testing of crude beers, materials active after purification may not be present in a large enough concentration to be effective on established tumors. The present procedure will be proved either useful or not as more and more purified antibiotics are subjected to extensive testing.

Additional Screens

Two other screens have been established on a routine basis. They are (1) the use of transplanted human tumors in rats, and (2) tissue culture.

Transplanted Human Tumors

The use of the HS $\#$ 1 and HE$_p$ $\#$ 3 transplanted human tumors as a routine screen with antibiotic filtrates has been established in the program conducted by one of the filtrate suppliers. These tumors are grown in x-ray- and cortisone-treated white rats. It must be recognized that these animals are not completely normal after such pretreatment. This disadvantage is at least partially discounted by the ability to test compounds on a human type of tumor. They are run on as many filtrates as possible in parallel with the mouse screen.

It is too early to draw many conclusions, but in keeping with most other tumor types, they appear to have their specificities; that is, some filtrates may be active on HS $\#$ 1 or HE$_p$ $\#$ 3 but inactive on the mouse screen, and others may be just the reverse, while still others will be active on both.

Tissue Culture

The tissue-culture screening program has been established with four screening contractors and in-plant screens at three of the filtrate laboratories. Ten percent of the original filtrates submitted to the mouse screen are also tested on tissue culture. Follow-up samples on filtrates in which the original beer was positive on tissue culture are also followed on this screen.

This program is just well under way and the comparison of the mouse screen and tissue culture on a large number of samples has not been completed. A preliminary study indicated that if activity in tissue culture at $100\ \mu g/ml$ were used as the cutoff point, 13 out of 99 known active compounds would have been thrown out. Other materials are active on tissue culture and not in the mouse screen; so present indications are that the two types of tests do not correlate completely. The big question is, which one correlates best with activity against human tumors in man? Some materials which show marked tissue-culture activity and no activity against the mouse tumors in our screen are being tested in the clinic. This is the only way in which these questions of correlation can be answered.

Results of the Program to Date

From the start of the antibiotic filtrate program up to the present time approximately 2% of the original beers submitted have passed the screen on one or more tumors. If it is correct that only one out of five of these is truly active against tumors than 0.4% or only four filtrates out of every 1000 submitted have true antitumor activity. However, not all of these will get to the clinic. Technical difficulties will result in some of them being dropped from active interest before they are purified.

There is a long and often a very discouraging period between the time when a filtrate first passes the screen and the time when it is pure enough to be considered for clinical evaluation. Yet this purified product ready for clinical testing is the "pay-off" of the program.

Two filtrates submitted originally in 1956 (by two different laboratories) were purified to such an extent by late 1958 that they were considered for pharmacological studies and clinical investigation. The first one is now in the clinic and the second one is in the stage of preclinical pharmacology. One of the laboratories has started an independent clinical testing of two other antibiotics from the program.

In addition, there are a group of eight to ten antibiotic products approaching a highly purified stage. Some of these are now being studied in tumor screens, in tissue culture, and microbiologically. Some 100 more antibiotics are at some stage in the purification process.

Thus, from the first 10,000 different filtrates submitted to the program (June, 1956 through May, 1957) four are, or we hope soon will be, in clinical study and 17 more are in various stages of purification. This is about 0.2% of the starting materials and is half of the 0.4% anticipated as being active. Approximately 50 additional filtrates are still under consideration but have not been submitted to the chemists for fractionation as yet. It would appear that our experience so far is remarkably close to that anticipated on a theoretical basis.

You can readily see, then, that this program, which has been in operation since 1956, is just now at a stage where it is beginning to supply candidate antibiotics for clinical evaluation. There are long-term projects and the next few years should be most interesting as larger numbers of purified antibiotics become available for clinical trial.

At least two major contributions are hoped for from this phase of the Service Center program. First of all, and most important, the center hopes that out of this program may come a number of new antibiotics having a wide range of antitumor activity. It is realized that one cannot predict the outcome of a screening program looking for chemicals with a specific type of action. The center believes, however, that this is a reasonable approach to the problem because of the seemingly unlimited diversity in the synthetic capabilities of microorganisms.

Secondly, it is hoped that this program will result in a much clearer understanding of the correlation between the many possible screening techniques and clinical effects on the numerous types of human tumors.

This program is completely dependent upon the vision, experience, and technical skill of the scientific investigators throughout the country for its success.

There are a wealth of these abilities contributing to this search for new anti-tumor antibiotics. With their help, many new antibiotics will be made available for clinical evaluation.

MECHANISMS OF ACTION OF ANTICANCER ANTIBIOTICS

Glynn P. Wheeler

Southern Research Institute, Birmingham, Alabama

Perhaps at the beginning of this discussion we should give reasons why we are interested in knowing the mechanisms of action of these antibiotics. On the one hand we are interested for purely academic reasons, that is, to satisfy man's inquisitive nature and to add to his storehouse of knowledge. This inquisitiveness leads us to ask such questions as the following:

Why is a material that is produced by one organism toxic to another organism?

Can these toxic effects be prevented or overcome?

Are the various toxic materials toxic for the same reason?

Why are some organisms not affected by these agents while other organisms are killed by them?

Why are some tissues or cells within an organism affected by the agents while other tissues and cells are not?

How do organisms become resistant to an initially toxic agent?

These and other related questions offer a challenge to the scientist, and seeking answers to these questions is justification enough in itself for the effort that is put into the work. However, in addition to these purely academic reasons for investigating mechanisms of action, we know there are also practical reasons. Knowledge of the modes of action of anticancer antibiotics might make possible more rational use of antibiotics, both individually and in combination, to prolong the lives of victims of cancer. Studies of the mechanisms might also point out biochemical differences between normal and neoplastic cells which might be exploited in various ways to destroy preferentially the cancer cells. Also, if one knows the mode of action of a compound, then it might be possible to design other organic chemicals which would be more effective, and therefore, knowledge of the mechanism aids the organic chemist in designing new candidate anticancer agents.

Some of the methods used in investigating the mechanisms of action of drugs are shown below:

1. Studies of the reversal or prevention of toxicity.
2. Studies of cross-resistance.
3. Morphological examination of treated cells.
4. Studies of the metabolism of radioactive drugs and of the effects of drugs upon the metabolism of radioactive substrates.
5. Studies with enzymes.

No doubt this list is incomplete, but it does point out some of the methods that are used in these investigations. Various biological systems may be used in these studies, and one can often investigate a particular area of metabolism by the proper choice of the biological system. Although bacteria and neoplasms may be considered as quite different in many respects, they nevertheless have a number of metabolic steps in common, and much useful information concerning mechanisms of anticancer agents has been obtained by studying the effects of these agents on the metabolism of bacteria. On the other hand, the fine points of the mechanism might be best learned by studies with isolated enzymes.

The above list of methods will serve somewhat as an outline for the presentation of the results that have been obtained by many workers on the mechanisms of action of several anticancer antibiotics. The information that will be presented will be taken from the literature, and, therefore, this presentation will merely serve the purpose of drawing together information that is scattered in the literature.

The first two materials that we will consider are azaserine and 6-diazo-5-oxo-L-norleucine (DON), structures of which are shown below and which are being considered together in order to point out some of the similarities of their effects and some of the differences.

$$\overset{-}{N}=\overset{+}{N}=CH-\overset{\overset{O}{\|}}{C}-O-CH_2-\underset{\underset{NH_2}{|}}{CH}-COOH$$

Azaserine

$$\overset{-}{N}=\overset{+}{N}=CH-\overset{\overset{O}{\|}}{C}-CH_2-CH_2-\underset{\underset{NH_2}{|}}{CH}-COOH$$

6-Diazo-5-oxo-L-norleucine (DON)

$$NH_2-\overset{\overset{O}{\|}}{C}-CH_2-CH_2-\underset{\underset{NH_2}{|}}{CH}-COOH$$

Glutamine

The structure of glutamine is also shown for the purpose of comparison. The reason for the comparison will become obvious as the experimental results are presented.

Reilly (1958) found that bacteria growing on minimal medium were more susceptible to inhibition of growth by azaserine and DON than were bacteria growing on complex medium. Therefore, nutrients could alter the sensitivity and in some cases prevent the inhibition. Kaplan and Stock (1954) observed that the inhibition of *Escherichia coli* by azaserine could be prevented by phenylalanine, tyrosine, or tryptophan, but Dr. Reilly (1958) found that these amino acids did not prevent the inhibition of *E. coli* by DON. However, she found that a combination of arginine and glutamine did reduce the inhibitory effect of DON. In studies with the yeast *Kloeckera brevis* she (1954) also found that thirteen of the nineteen amino acids tested reversed the inhibition by azaserine. Bennett and co-workers (1956) found that the inhibition of *E. coli* by azaserine was significantly prevented by 5-amino-4-imidazolecarboxamide, adenine, guanine, hypoxanthine, xanthine, methionine, and glutamine. Maxwell and Nickel (1957) noted that inhibition of *E. coli* by DON was prevented by adenine, guanine, hypoxanthine, and the corresponding nucleosides and guanosine (2',3') phosphate. Hedegaard and co-workers (1959) found that the inhibition of *E. coli* by azaserine was prevented by histidine, urocanic acid, and formiminoglutamic acid and they suggested that the latter compound might serve as a donor of amino groups in place of glutamine. Nakamura (1956) found that azaserine inhibited the growth of amoebas, but that DON did not. The inhibition was completely overcome by adenine, adenylic acid, and 2,6-diaminopurine, and partially overcome by serine, methionine, 5-amino-4-imidazolecarboxamide and leucovorin. No reversal, however, was obtained by guanine, guanosine, and guanylic acid. Aaronson (1959) observed that azaserine inhibited the multiplication of *Gaffkya homari* equally as well in the presence of exogenous purines, as when purines were synthesized by the organism. Therefore, he concluded that purine synthesis is not the main site of inhibition of this microorganism by azaserine. He found, however, that glutamine, arginine, and glutamic acid annulled the effects of the inhibitor. Karnofsky and co-workers found that certain purines or their derivatives partially prevented the teratogenic effect of azaserine and of DON upon the developing chick embryo (Dagg, 1956) and prevented the inhibition of development of the sand-dollar embryo by these antibiotics (Karnofsky, 1958). Thiersch (1957a, 1957b) observed that adenine sulfate prevented the destruction of rat litter by DON, but not by azaserine. Clarke, Reilly, and Stock (1957) found that the inhibition of the growth of sarcoma 180 by azaserine or DON could be partially prevented by adenine, hypoxanthine, 5-amino-4-imidazole carboxamide, or guanine, but not by xanthine, adenosine, or adenylic acid. Therefore, studies of the prevention of the inhibitory effects of these compounds indicate that in some organisms these compounds must interfere with purine synthesis, while in other systems this interference is not as critical. It is also notable that in several of these systems arginine and glutamine are effective reversal agents.

It has been observed by Maxwell and Nickel (1957) that *E. coli* that are resistant to azaserine are also resistant to DON, and that those that are resistant to DON are also resistant to azaserine. This cross-resistance is not complete, but it does indicate that these two agents must have a similar mode of action in this biological system. The resistant lines were partially resistant to glutamic γ-hydrazide, an analog of glutamine, and the azaserine-resistant line was resistant to DL-β-2-thienylalanine, an analog of phenylalanine. Potter and Law (1957) have studied the cross-resistance of resistant lines of a plasma cell neoplasm of the mouse and found that an azaserine-resistant subline, a DON-resistant subline, and an N-methylformamide-resistant subline were each resistant to azaserine, DON, and N-methylformamide. These sublines were still sensitive, however, to amethopterin and to 6-mercapto-purine. Therefore, in this neoplasm perhaps azaserine and DON function by a common mechanism at a site that is not affected by amethopterin or 6-mercapto-purine.

Maxwell and Nickel (1954) observed the formation of filamentous forms of *E. coli* when azaserine was present in the medium, but not when DON was present. Examination of these filaments indicated that they were multinucleate and nonseptate. Similar filamentation was observed with nitrogen mustard, triethylene melamine, and 5-diazouracil. A number of other inhibitors, including chloramphenicol, chlortetracycline, oxytetracycline, and streptomycin did not cause formation of these filaments. This would indicate possibly that the mechanisms of action of these antibiotics are different from that of azaserine. Davis and Mudd (1956) observed that azaserine also caused the formation of radially enlarged cells and elongated filaments by *Corynebacterium diphtheriae*. The formation of filaments is perhaps related to inhibition of synthesis of DNA.

A flow sheet of the synthesis of nucleic acids is shown in Fig. 1. This flow sheet shows the pathways of incorporation of glycine and formate into the purine moieties of ribonucleotides and of the incorporation of preformed purines into ribonucleotides.

Bennett *et al.*, (1956) studied the effect of azaserine upon purine synthesis of the tumors, intestines, livers, and spleens of mice bearing sarcoma 180, and found that azaserine inhibited the utilization of formate-C^{14} and glycine-1-C^{14} for purine synthesis, but failed to affect the utilization of adenine-C^{14}, 5-amino-4-imidazole-carboxamide-C^{14}, and hypoxanthine-C^{14}. There was also no effect upon the utilization of formate for thymine synthesis. These results were interpreted to mean that azaserine inhibited purine synthesis in the intact animal at a stage prior to

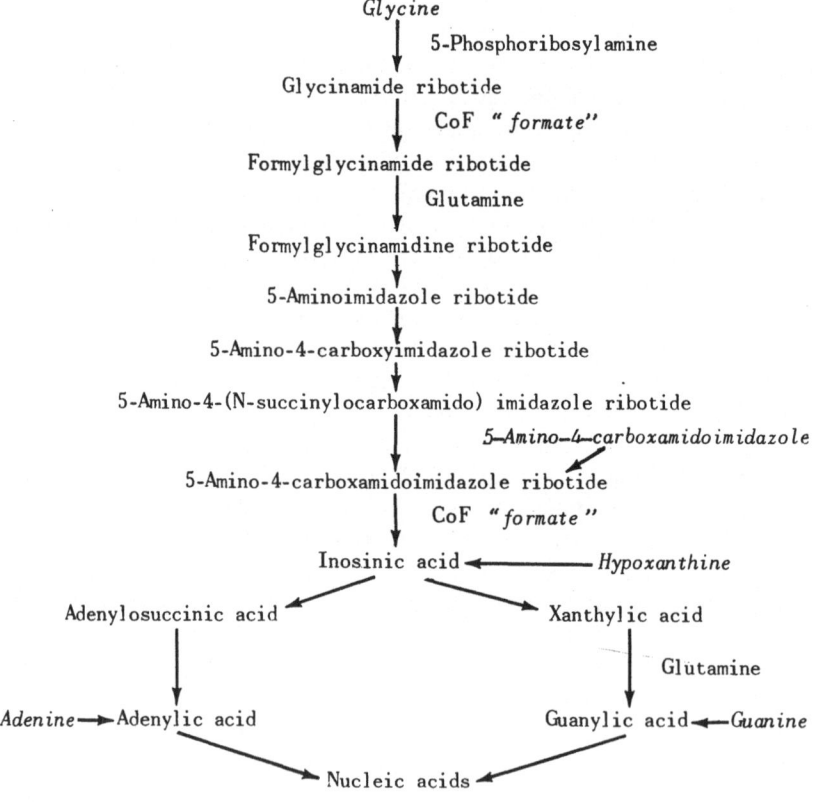

Fig. 1. Biosynthesis of nucleic acids.

the formation of 5-amino-4-imidazole-carboxamide ribotide. Maxwell and Nickel (1957) found that DON interfered with the incorporation of formate and glycine into the nucleic acids of E. coli, but it did not interfere with the incorporation of adenine. Greenlees and LePage (1956) studied the effect of azaserine upon the incorporation of glycine-C^{14} into the acid-soluble adenine and nucleic acid adenine of ascites tumor cells and found that azaserine inhibited purine synthesis and caused an accumulation of a compound that had the properties of glycinamide ribotide. This inhibition could be partially prevented by the administration of glutamine. Henderson, LePage, and McIver (1957) found that there was a correlation between tumor inhibition and inhibition of purine synthesis by azaserine. Barclay, Garfinkel, and Phillipps (1956) observed that DON inhibited the incorporation of formate-C^{14}, glycine-N^{15}, and adenine-C^{14} into RNA and DNA purines, but did not inhibit thymine synthesis.

Tomisek, Kelly, and Skipper (1956), using the chromatographic-radioautographic technique for examining the soluble portions of E. coli cells, observed that azaserine inhibited the incorporation of C^{14} from formate-C^{14}, glycine-1-C^{14}, glycine-2-C^{14}, and serine-3-C^{14} into all of the purine-containing compounds that were observed. This inhibition was accompanied by an accumulation of formylglycinamide riboside and formylglycinamide ribotide. Azaserine did not inhibit the conversion of hypoxanthine-8-C^{14}, adenine-8-C^{14}, and guanine-2-C^{14} into ribonucleotides or other purine derivatives. As this inhibition was observed at concentrations of azaserine that caused only partial reduction in the growth rate of the bacteria, it was concluded that a major site of action of azaserine in E. coli was the inhibition of de novo purine synthesis at some stage subsequent to the formation of formylglycinamide ribotide. Gots and Gollub (1956) found that azaserine inhibited the formation of diazotizable amine by resting cells of a purine-requiring mutant of E. coli. Several amino acids, including tryptophan, phenylalanine, tyrosine, glutamine, glutamic acid, leucine, isoleucine, and norleucine partially prevented this inhibition. These results with E. coli would indicate that azaserine inhibited purine synthesis at some point between the formation of formylglycinamide ribotide and the formation of 5-amino-4-imidazolecarboxamide ribotide.

Using a pigeon liver enzyme system, Buchanan and his group (1958) have shown that azaserine and DON would inhibit the de novo synthesis of inosinic acid, and further studies have shown more specifically that these compounds interfere with conversion of formylglycinamide ribotide to formylglycinamidine ribotide.

Anderson, Levenberg, and Law (1957) observed that azaserine and DON inhibited the conversion of formylglycinamide ribotide to 5-aminoimidazole ribotide in a cell-free preparation derived from the mouse plasma-cell neoplasm 70429. Azaserine also inhibited the incorporation of glycine-1-C^{14} into the acid-soluble purines by this preparation.

In studies with algae, van der Meulen and Bassham (1959) have shown that azaserine and DON cause a buildup of formylglycinamide ribotide, and this accumulation is accompanied by a decrease in the labeling of aspartic acid, glutamic acid, alanine, and serine, and an increase in the labeling of glutamine, α-ketoglutaric acid, citric acid, malic acid, lactic acid, fumaric acid, and succinic acid.

Other steps involved in purine synthesis may also be inhibited by azaserine. Goldthwait (1957) has found that in a pigeon liver, supernatant azaserine inhibited the combination of glutamine with phosphoribosyl pyrophosphate to form 5-phosphoribosylamine. The 5-phosphoribosylamine is the material that condenses with glycine to form glycinamide ribotide, which is one of the first steps in the *de novo* synthesis of purines. Therefore, this inhibition could occur at a step prior to the formation of formylglycinamide ribotide. Bentley and Abrams (1956) have shown that in a rabbit bone marrow extract, azaserine will inhibit the conversion of xanthylic acid to guanylic acid.

Eidinoff and co-workers (1958) have shown that DON will inhibit the conversion of uridylic acid to cytidylic acid, and that this inhibition can be partially prevented by glutamine. Kammen and Hurlbert (1959) found that azaserine would also inhibit this conversion.

Narrod and the group working with him (1959) found that both azaserine and DON caused decreases in the levels of diphosphopyridine nucleotide in the liver, and these workers (Langan *et al.*, 1959) and Preiss and Handler (1958) have shown that azaserine will inhibit the conversion of desamido DPN to DPN. In this latter reaction glutamine served as the donor of the amino group. Therefore, the lower concentration of DPN could be due to the inhibition of synthesis of adenine and also to the inhibition of the conversion of the desamido DPN to the DPN.

Srinivasan (1959) has shown that DON and azaserine can inhibit the conversion of shikimic acid 5-phosphate to anthranilic acid by a cell-free extract of a mutant strain of *E. coli*. In this reaction glutamine is the donor of the amino group.

It is of interest that in each of these reactions glutamine is one of the reactants, and the consistency of the involvement of glutamine in the reactions that are inhibited by these antibiotics is rather indicative that they act as analogs of glutamine. These results indicate multiple sites of action, and it is not yet established which site is the most critical one for inhibiting the growth of tumors.

Much less work has been done on the mechanism of action of some of the other anticancer antibiotics, and little is known concerning their mechanism of action. However, the material that has been published on several of these is reviewed below.

In studies with several microbiological systems, Foley (1956) found that actinomycin D was a competitive inhibitor for pantothenate, and in addition, he found that this inhibitor could be reversed noncompetitively by certain amino acids, dicarboxylic acids, orotic acid, and adenine. As a result, he suggested that actinomycin D interfers with pantothenate-dependent reactions concerned with the biosynthesis and/or utilization of amino acids by these bacteria. In experiments with *Bacillus subtilis*, Slotnick (1957) could not demonstrate reversal of actinomycin D by pantothenate nor by any of the following compounds: β-alanine, potassium pantoate, nucleic acid, purine and pyrimidine bases, nucleosides and nucleotides, members of the B vitamin group, the amino acids found in casein hydrolysate, succinate, fumarate, pyruvate, citrate, or orotic acid. He found (Slotnick, 1958), however, that actinomycin D suppressed the assimilation of ammonia and completely inhibited the formation of certain inducible enzymes. As a result of his studies he concluded that actinomycin D interfered at some point in reactions leading to protein synthesis. DiPaolo and co-workers (1957) found that calcium pantothenate failed to reverse the action of actinomycin in prolonging the life of mice bearing Krebs-2 ascites cells or sarcoma 180. Likewise, Farber and co-workers (1958) found no reversal by pantothenate in tumor-bearing mice. Raven and Hess (1959) have recently reported that actinomycin C and actinomycin D inhibited the growth of *Neurospora crassa*, and that this inhibition was partially prevented by p-aminobenzoic acid, tyrosine, and phenylalanine. Also recently Mascitelli-Coriandoli (1959) found that actinomycin C caused a decrease in the quantity of pyridoxal phosphate, but not of pyridoxine, in the livers of rats. He assumed therefore that the antibiotics interfered with the conversion of the various forms of Vitamin B_6 into pyridoxal-5-phosphate. Therefore, no clear picture of the mechanism of action of the actinomycins is yet available.

Although puromycin has been available for several years, little work related to its mechanism of action has been reported. Bortle and Olson (1955) reported that the inhibition of the growth of *Tetrahymena puriformis* was specifically reversed by guanylic acid. Guanine hydrochloride and guanosine were ineffective, as were also adenine, uracil, cytosine, and the corresponding ribonucleosides and ribonucleotides. Succinic acid, citric acid, malic acid, lactic acid, and ribose were also ineffective. Hewitt and co-workers (1954) studied the reversal of the inhibition of *Trypanosoma equiperdum* by the aminonucleoside of puromycin and found that a number of purines and purine derivatives were effective in reversing the inhibition. On the other hand, a number of other substituted purines were ineffective. No studies in mammalian systems have been reported.

Reilly and her associates (1958) reported that the inhibition of *E. coli* by mitomycin C could be prevented by either guanine or xanthine, but that these purines did not prevent the formation of abnormally long filaments by the cultures. Mitomycin C also caused the inhibition of the formation of 5-amino-4-imidazolecarboxamide by resting B96 cells. These results suggest that this antibiotic blocks the biosynthesis of purines at some stage prior to the formation of 5-amino-4-imidazolecarboxamide. It was reported that there was some evidence which suggested that the block was at a point different from that at which inhibition by azaserine occurs. Two groups of Japanese workers (Shiba *et al.*, 1959, Sekiguchi *et al.*, 1959) have reported that mitomycin C inhibits the synthesis of desoxyribonucleic acids to a much greater extent than it inhibits the synthesis of ribonucleic acids or proteins.

The investigation of properties of actinobolin showed that this material could complex with metal ions probably by formation of chelates; therefore, it was of interest to determine if metal ions could prevent the inhibition of bacteria by this antibiotic. Using the paper disc on agar method, Burchenal and co-workers (1959) found that ferric chloride, aluminum chloride, magnesium sulfate, and ammonium molybdate would prevent the inhibition of *Streptococcus faecalis* by actinobolin. The following metal salts, however, did not prevent the inhibition: ferrous sulfate, calcium gluconate, cadmium chloride, copper acetate, cobalt chloride, zinc acetate, and barium chloride. However, when ferric chloride or aluminum chloride was injected intraperitoneally into mice at the maximum tolerated dose, the salts did not prevent the antileukemic effect of the actinobolin. These results raise some question as to whether the primary action of actinobolin in preventing the development of leukemia is due to the formation of chelates with certain essential metals. The last-mentioned workers found that sublines of L1210 leukemia that were resistant to amethopterin, to 6-mercaptopurine, to both amethopterin and 6-mercaptopurine, or to glyoxal bis (guanylhydrazone) were still sensitive to actinobolin. Also, a subline of L1210 leukemia that was resistant to actinobolin was still sensitive to amethopterin, 6-mercaptopurine, DON, fluorinated pyrimidines, and methylglyoxal bis (guanylhydrazone). These results would indicate that the antibiotic is working by some mode other than those of the other antileukemic agents that were tested.

Fumagillin, sarkomycin, actidione, carzinophilin, and streptovitacins A and B are other antibiotics that have been found to have some anticancer activity, but no studies of the mechanisms of action of these antibiotics have been reported. It is perhaps of significance that the information that has been obtained concerning the mechanisms of action of several of the anticancer antibiotics suggests that these agents interfere either directly or indirectly with the biosynthesis of nucleotides or nucleic acids. This fact is of particular interest, since a number of other types of known anticancer agents, such as the antifolic-acid compounds, the antipurines, alkylating agents, and radiation affect the biosynthesis of nucleic acids. It remains to be established, however, whether the interference with the synthesis of nucleotides, or nucleic acids, is the primary reason for the anticancer activity.

At the beginning of this presentation, I listed several questions which I said presented a challenge to the scientist. At this point in the presentation, I think that most of us would agree that the results that have been presented are also a challenge, because these results point out that much remains to be learned concerning the modes of action of anticancer antibiotics.

REFERENCES

Aaronson. S. 1959 Mode of action of azaserine on *Gaffkya homari*. J. Bact., 77, 548.

Anderson. E. P., Levenberg, B., and Law, L. W. 1957 Purine biosynthesis in azaserine-sensitive and -resistant lines of the mouse plasma cell neoplasm 70429. Fed. Proc., 16, 145.

Barclay, R. K., Garfinkel, E., and Phillipps, M. 1956 Effects of DON on incorporations of precursors into nucleic acids. Proc. Am. Assoc. Cancer Research, 2, 93.

Bennett, L. L., Jr., Schabel, F. M., Jr., and Skipper, H. E. 1956 Studies on the mode of action of azaserine. Arch. Biochem. and Biophys., 64, 423.

Bentley, M., and Abrams, R. 1956 Amide-N of glutamine as source of guanine amino group. Fed. Proc., 15, 218.

Bortle, L., and Oleson, J. J. 1955 Effect of puromycin on *Tetrahymena pyriformis* nutrition. Antibiotics Annual 1954-55, 770.

Buchanan, J. M. 1958 The interference of azaserine in purine biosynthesis, in "Amino Acids and Peptides with Antimetabolic Activity," Ciba Symposium, J. and A. Churchill, Ltd., London, 75-81.

Burchenal, J. H., Holmberg, E. A. D., Reilly, H. C., Hemphill, S., and Reppert, J. A. 1959 The effects of actinobolin on transplanted mouse leukemias. Antibiotics Annual 1958-59, 528.

Clarke, D. A., Reilly, H. C., and Stock, C. C. 1957 A comparative study of 6-diazo-5-oxo-L-norleucine and O-diazoacetyl-L-serine on sarcoma 180. Antibiot. and Chemotherapy, 7, 653.

Dagg, C. P., and Karnofsky, D. A. 1956 Protection against teratogenic action of azaserine and DON. Fed. Proc., 15, 238.

Davis, J. C., and Mudd, S. 1956 Cytological effects of ultraviolet radiation and azaserine on *Corynebacterium diphtheriae*. J. Gen. Microbiol., 14, 527.

DiPaolo, J. A., Moore, G. E., and Niedbala, T. F. 1957 Experimental studies with actinomycin D. Cancer Research, 17, 1127.

Eidinoff, M. L., Knoll, J. E., Marano, B., and Cheong, L. 1958 Pyrimidine studies. I. Effect of DON (6-diazo-5-oxo-L-norleucine) on incorporation of precursors into nucleic acid pyrimidines. Cancer Research, 18, 105.

Farber, S. 1958 Clinical and biological studies with actinomycins, in "Amino Acids and Peptides with Antimetabolic Activity," Ciba Symposium, J. and A. Churchill Ltd., London, 138-145.

Foley, G. E. 1956 Preliminary observations on the mechanism of action of actinomycin D in microbiologic systems. Antibiotics Annual 1955-56, 432.

Goldthwait, D. A. 1957 Purine nucleotide biosynthesis and neoplasia, in "Henry Ford Hospital International Symposium on the Leukemias: Etiology, Pathophysiology, Treatment," pp. 555-565. Edited by J. W. Rebuck, F. H. Bethell, and R. W. Monto, Academic Press, Inc., New York.

Gots, J. S., and Gollub, E. G. 1956 Purine metabolism in bacteria. IV. L-Azaserine as an inhibitor. J. Bact., 72, 858.

Greenlees, J., and LePage, G. A. 1956 Purine biosynthesis and inhibitors in ascites cell tumors. Cancer Research, 16, 808.

Hedegaard, J., Thoai, N-v., and Roche, J. 1959 The influence of histidine on the biosynthesis of purines in *Escherichia coli*. Arch. Biochem. and Biophys., 83, 183.

Henderson, J. F., LePage, G. A., and McIver, F. A. 1957 Observations on the action of azaserine in mammalian tissues. Cancer Research, 17, 609.

Hewitt, R. I., Gumble, A. R., Wallace, W. S., and Williams, J. H. 1954 Experimental chemotherapy of trypanosomiasis. IV. Reversal by purines of the *in vivo* activity of puromycin, and an amino nucleoside analog, against *Trypanosoma equiperdum*. Antibiot. and Chemotherapy, 4, 1222.

Kammen, H. O., and Hurlbert, R. B. 1959 The formation of cytidine nucleotides and RNA cytosine from orotic acid by the Novikoff tumor *in vitro*. Cancer Research, 19, 654.

Kaplan, L., and Stock, C. C. 1954 Azaserine, an inhibitor of amino acid synthesis in *Escherichia coli*. Fed. Proc., 13, 239.

Karnofsky, D. A., and Bevelander, G. 1958 Effects of DON (6-diazo-5-oxo-L-norleucine) and azaserine on the sand-dollar embryo. Proc. Soc. Exp. Biol. and Med., 97, 32.

Langan, T. A., Jr., Kaplan, N. O., and Shuster, L. 1959 Formation of the nicotinic acid analogue of diphosphopyridine nucleotide after nicotinamide administration. J. Biol. Chem., 234, 2161.

Mascitelli-Coriandoli, E. 1959 Effects of actinomycin-C upon the pyridoxal phosphate contents of the liver. Biochem. Pharmacol., 2, 79.

Maxwell, R. E., and Nickel, V. S. 1954 Filament formation in *E. coli* induced by azaserine and other antineoplastic agents. Science, 120, 270.

Maxwell, R. E., and Nickel, V. S. 1957 6-Diazo-5-oxo-L-norleucine, a new tumor-inhibitory substance. V. Microbiologic studies of mode of action. Antibiot. and Chemotherapy, 7, 81.

Nakamura, M. 1956 Amoebicidal action of azaserine. Nature, 178, 1119.

Narrod, S. A., Langan, T. A., Jr., Kaplan, N. O., and Goldin, A. 1959 Effect of azaserine (o-diazoacetyl-L-serine) on the pyridine nucleotide levels of mouse liver. Nature, 183, 1674.

Potter, M., and Law, L. W. 1957 Studies on a plasma-cell neoplasm of the mouse. I. Characterization of neoplasm 70429, including its sensitivity to various antimetabolites with the rapid development of resistance to azaserine, DON, and N-methylformamide. J. Nat. Cancer Inst., 18, 413.

Preiss, J., and Handler, P. 1958 Biosynthesis of diphosphorpyridine nucleotide. II. Enzymatic aspects. J. Biol. Chem., 233, 493.

Raven, H. M., and Hess, G. 1959 Die Wirkung der Actinomycin C und D auf *Neurospora crassa*. Z. fur Physiolog. Chem., 315, 70.

Reilly, H. C. 1954 The effect of amino acids upon the antimicrobial activity of azaserine. Proc. Am. Assoc. Cancer Research, 1, No. 2, 40.

Reilly, H. C. 1958 Some aspects of azaserine, 6-diazo-5-oxo-L-norleucine, and β-2-thienylalanine, in "Amino Acids and Peptides with Antimetabolic Activity." Ciba Symposium, J. and A. Churchill, Ltd., London, 62-74.

Reilly, H. C., Cappuccino, J. G., and Harrison, D. M. 1958 Studies on mitomycin X, A tumor-inhibiting antibiotic. Proc. Am. Assoc. Cancer Research, 2, 338.

Sekiguchi, M., and Takagi, Y. 1959 Synthesis of deoxyribonucleic acid by phage-infected *Escherichia coli* in the presence of mitomycin C. Nature, 183, 1134.

Shiba, S., Terawaki, A., Taguchi, T., and Kawamata, J. 1959 Selective inhibition of formation of deoxyribonucleic acid in *Escherichia coli* by mitomycin C. Nature, 183, 1056.

Slotnick, I. J. 1957 Studies on the mechanism of actinomycin D inhibition of *Bacillus subtilis*. I. Failure to demonstrate pantothenate relationship. Antibiot. and Chemotherapy, 7, 387.

Slotnick, I. J. 1958 Studies on the mechanism of action of actinomycin D. II. Interference with ammonia assimilation and adaptive enzyme formation. Antibiot. and Chemotherapy, 8, 476.

Srinivasan, P. R. 1959 The enzymatic synthesis of anthranilic acid from shikimic acid-5-phosphate and L-glutamine. J. Am. Chem. Soc., 81, 1772.

Thiersch, J. B. 1957a Effect of o-diazo acetyl-L-serine on rat litter. Proc. Soc. Exp. Biol. and Med., 94, 27.

Thiersch, J. B. 1957b Effect of 6-diazo-5-oxo-L-norleucine (DON) on the rat litter *in utero*. Proc. Soc. Exp. Biol. and Med., 94, 33.

Tomisek, A. J., Kelly, H. J., and Skipper, H. E. 1956 Chromatographic studies of purine metabolism. I. The effect of azaserine on purine biosynthesis in *E. coli* using various C^{14}-labeled precursors. Arch. Biochem. and Biophys., 64, 437.

Van der Meulen, P. Y. F., and Bassham, J. A. 1959 Study of inhibition of azaserine and diazo-oxo-norleucine (DON) on the algae *Scenedesmus* and *Chlorella*. J. Am. Chem. Soc., 81, 2233.

SCREENING OF ANTINEOPLASTIC AGENTS
WITH *NEUROSPORA CRASSA**

Robert Fuerst
Department of Biology, Microbiological Research
Texas Woman's University, Denton, Texas

The author of this presentation has not participated in the screening program of biological beers. He has, however, analyzed chemically defined compounds and microbial systems employed in the testing of both potential and established chemotherapeutic agents.

In March of last year, the author attended the conference on Screening Procedures for Experimental Cancer Chemotherapy of the New York Academy of Sciences. It became apparent during this meeting that many of the speakers attributed great importance to the use of *Neurospora* in test systems for potential antineoplastic agents.

A rather excellent report of evaluation of such a test system was given by one of the participants in the program who evaluated *Neurospora* as a test organism compared to other organisms and systems commonly used. The report stated that a combination of *Streptococcus faecalis* and *Neurospora crassa* would detect 93% of all tumor-active compounds in the test series. Since there were some antitumor compounds that were "missed" by all microorganisms, the 93% value indeed indicates a good presumptive test with a surprisingly great degree of accuracy. Evidently the data obtained with *Neurospora* was more consistant with actual human tumor data than most of the experimental animal work.

A substantial part of what was said at the meetings was published later on in the Annals of the Academy (Foley, *et al*. 1958).

The author was rather pleased to hear about this excellent work on *Neurospora*, but was appalled that no reference was made to the published reports on the use of *Neurospora* as a test organism for potential antineoplastic agents which he and his associates have published since 1955, and which have been presented to a number of scientific meetings. (Fuerst, Hsu, and Somers, 1956; Fuerst, Somers, and Hsu, 1956; Somers, Fuerst, and Hsu, 1957.)

The author has never considered the use of *Neurospora* superior to any other test system. He believes that the use of several systems has the advantage in that an active compound is less likely to be overlooked. The use of microorganisms in testing has already led to the development of some of the best drug therapies that are available for cancer today, and so no apology need be made for using a microorganism to test the metabolism of any drug. This is quite apparent from the excellent publication "Antimetabolites and Cancer" (Rhoads *et al*., 1955) and many of the recent reviews on cancer chemotherapy.

In the investigations to be reported, there were three test systems used: One system employs a mouse tumor in C57 black mice. Until recently this tumor was

* This work was supported by the National Institutes of Health, Research Grant CY3853 (C2).

adenocarcinoma E-0771, until a change to BW-10,232, which was obtained from the Roscoe B. Jackson Memorial Laboratory, was forced on the investigator. The second system employs *N. crassa Em5256A* grown in liquid culture, and the third system, the so-called "plate-well-method," also uses *Neurospora* on agar medium. (Fuerst and Skellenger, 1958).

Typical result obtained with *Neurospora* occurred in tests with some folic acid analogs. Growth of the organism was measured as dry mycelial weight after 65 hr incubation at 25 C in 25 ml of minimal medium in 125-ml Erlenmeyer flasks. Both aminopterin and amethopterin are good inhibitors, as was indicated by the inhibition curves obtained for these compounds. Teropterin did not inhibit in liquid culture although it is inhibitory when tested with the "plate-well-method."

At first, a large group of compounds was tested. These compounds synthesized and generously made available by Dr. Robins, belong to the 4-aminopyrazolo (3,4-d)pyrimidine compounds, abbreviated as 4-APP. 4-APP itself is an analog of adenine. The other compounds are derivatives of this general structure. (Robbins, 1956; Hsu, Robins, and Cheng, 1956.)

Many substitution groups on 4-APP were investigated. In all, 166 derivatives of 4-APP were tested. The data obtained indicated that the 4-APP itself was the best inhibitor in this group of compounds and that any side chains would only weaken its activity. (Somers, Fuerst, and Hsu, 1957.) The inhibition decrease could be correlated with the type of side chains present. The resulting inhibition series may be shown as follows: "Highest inhibition...4-APP>some chlorophenyl derivatives >1 phenyl derivative>3 methyl>1 methyl>some chlorophenyl derivatives >others...least inhibitory."

Very recently, two 4-APP derivatives were tested and found to be more inhibitory to *Neurospora* than 4-APP itself. These compounds are: 1-methyl-4-octylaminopyrazolo (3,4-d) pyrimidine and 4-heptylamino-1-methypyrazolo (3,4-d) pyrimidine. These findings have not yet been explained and are being further investigated at the present time. It is certain, however, that they will fit into the general concept of structure-activity relationships which have been so clearly demonstrated with the 4-APP group of compounds when tested with *Neurospora*. Thayer (1958) made similar observations when he tested p-diethylamino-styryl and its group of compounds.

Derivatives of 4-APP may be divided roughly into inhibitors, weak inhibitors, and relief compounds based on their effects on *Neurospora* growth. A fourth group consists of neutral compounds that neither inhibit nor relieve. Table 1 is shown in order to demonstrate some of the compounds that give relief of 4-APP inhibition. The criterion used for comparison of the relative effect of these relief compounds depends on the use of 0.2 mM 4-APP as the inhibition standard. The first part of Table 1 lists 11 4-APP derivatives that relieved the 4-APP effect. Most of them are 4-amino substituted compounds. The second part of the table lists 16 other chemicals that also relieved 0.2 mM 4-APP inhibition of *Neurospora* growth. These compounds are listed here in order to demonstrate the relative ineffectiveness of 4-APP derivatives in relieving 4-APP inhibition. Thiamine, deoxyadenosine, and five other compounds gave better relief than the most powerful 4-APP derivative, namely, 6-methyl 4-APP. Nevertheless, in spite of weaker activity, the 11 4-APP derivatives shown in the table appeared to be anti-4-APP agents. Side groups found in these compounds were identical with those producing the effects of intramolecular antagonism between these relief groups and inhibitor groups present on the same molecule and demonstrated in other experiments. Relief data suggest two possible potential pathways of 4-APP action in *Neurospora*, those of thiamine synthesis, and desoxyribose nucleic acid synthesis.

Also studied in liquid culture were 13 quinoxalines and 20 isonitrosomalonitriles and related compounds, several of which are being further investigated. 2-amino-6-purinethiol and many of its derivatives tested gave very little or no inhibition of *Neurospora*. This is true to experience since 6-MP also inhibits *Neurospora* only slightly. Perhaps *Neurospora* can use 6-MP in its normal metabolism.

Table 1. 4-APP Relief

Relief Compounds	mM of compound required to give 50, 25, or 10% relief of 0.2 mM 4-APP inhibition of Neurospora Em5256A		
	50%	25%	10%
6-Methyl-4-aminopyrazolo(3,4-d)pyrimidine	0.137	0.054	
4-Benzylamino-6-methylpyrazolo(3,4-d) pyrimidine	0.143	0.057	
4-n-Butylamino-6-methylpyrazolo(3,4-d) pyrimidine	0.230	0.090	
4-Dimethylaminopyrazolo(3,4-d)pyrimidine	0.236	0.112	
4-Methylaminopyrazolo(3,4-d)pyrimidine	0.258	0.078	
4-n-Butylaminopyrazolo(3,4-d)pyrimidine	0.301	0.138	
4-Isopropylaminopyrazolo(3,4-d)pyrimidine	0.400	0.100	
4-Ethylaminopyrazolo(3,4-d)pyrimidine	not reached	0.192	
4-Methylamino-6-methylpyrazolo(3,4-d) pyrimidine	not reached	0.350	
4-Amino-1-(o-chlorophenyl)-6-methylpyrazolo (3,4-d)pyrimidine	not reached	0.350	
4-Ethylamino-6-methylpyrazolo(3,4-d) pyrimidine	not reached	0.410	
Thiamine	0.024	0.011	
Pyridoxine	0.035	0.015	
Adenine	0.038	0.018	
2,6 Di(diethylamino)purine	0.045	0.030	
Pyridoxamine	0.063	0.017	
Thiamine pyrophosphate chloride	0.085	0.018	
Deoxyadenosine	0.115	0.035	
Deoxyadenylic acid	0.16	0.04	
Caffeine	not reached	0.08	
Desoxyribose nucleic acid	not reached	0.12	
Ribonucleic acid	not reached	0.28	
Hypoxanthine		not reached	0.10
Adenylic acid		not reached	0.11
2,4,5,6-Tetra-aminopyrimidine sulfate		not reached	0.13
Cytidylic Acid		not reached	0.16
2,4,6-Triamino pyrimidine		not reached	0.27

Table 2 shows a list of important inhibitors of *Neurospora* that were studied extensively. Many of the most active clinically-used antineoplastic agents are included in this table. The compound found to give best inhibition in lowest doses used was 6-diazo-5-oxonorleucine, commonly abbreviated as DON. DON, which is supposed to lose its activity quickly, was generously supplied to this laboratory through the courtesy of Parke, Davis & Company. Some DON was received about three years ago, and a new supply was shipped again recently. A comparison of the two batches of DON was made by testing their effect on *Neurospora* growth. The results obtained were not expected. The new batch is somewhat better than the old one, but the old supply seemed to have gained in potency during the time it was stored in the deepfreeze, and even for a few days in the refrigerator. Perhaps this is not very significant, but it shows one thing—that DON is a very potent inhibitor of *Neurospora*.

The second method used employs 4% minimal agar medium in petri plates. It was referred to earlier in this report as the "plate-well-method." Don was also tested by this method. In essence, these results are very analogous to those obtained using the flask method. Of course, the reported 50% inhibition is identical with the LD_{50} dose used by other investigators.

A new group of compounds, the thiophenes, prepared by Dr. R. W. Higgins at Texas Woman's University, proved to be of interest when first studied in liquid

Table 2. Inhibition of *Neurospora crassa*
Em 5256A by the Flask Method

Inhibitor	mM of compound required to give 50% and 25% inhibition as compared to normal weight	
	50%	25%
6-diazo-5-oxonorleucine	0.000016	0.000006
5-fluorouridine	0.00096	0.00047
oxythiamine chloride	0.0016	0.0006
azaserine	0.0026	0.0007
amethopterin	0.0065	0.0035
1-methyl-3-nitro-1-nitrosoguanidine	0.007	0.003
aminopterin	0.011	0.006
8-azaguanine	0.025	0.015
2,6 diamino purine sulfate	0.030	0.015
nitrogen mustard	0.046	0.019
4-aminopyrazolo(3,4-d)pyrimidine	0.047	0.030
8-aza-2,6-diamino-purine sulfate	0.056	0.300
8-azaadenine	0.100	0.038
actinobolin	0.180	0.066
4-hydroxy-6-aminopyrazolo(3,4-d)-pyrimidine	0.270	0.123
1-methyl-4-n-butylaminopyrazolo-(3,4-d)pyrimidine	0.280	0.460
2-amino-6-(pentylthio)purine	0.49	0.44
5-fluorouracil	0.62	0.09
2,5-bis(1)-aziridinyl)3,6-bis-(2-methoxyethoxy)-p-benzoquinone	0.94	0.46

culture for *Neurospora* inhibition. It was found that these compounds had a definite inhibitory capacity towards *N. crassa.* Furthermore, upon injecting these compounds into C57 black mice, it was discovered that they had a marked effect causing, in some instances, paralysis and death. When tested against some adenocarcinomas of these mice, an increase of survival time for some of the injected animals was noted; and while injections were given, a slowing down of tumor growth was apparent. These tests were not discouraging and thus these tests were extended to some more of the *Neurospora* -inhibiting thiophenes. Beta-(2-thienyl)-DL alanine is an analog of phenylalanine. Biesele and Jacquez (1954) reported polyploidization and multinucleation of sarcoma cells in C57 black mice with the formation of so-called giant cells, brought about by beta-(2-thienyl)-DL-alanine, which is a thiophene.

With both test systems, the compounds to be tested were prepared by diluting with propylene glycol and sterilizing with ethylene oxide. Ethylene oxide was used instead of any form of heat-sterilization in order to avoid any decomposition of the compound. Results of these tests indicated that 4-(2-thienyl) butanoic acid was the best inhibitor tested (Fuerst and Higgins, 1960). 2-Heptylthiophene and 2-(2-methylpropyl)thiophene also inhibited well. The latter two are inhibitors that also colonize *Neurospora*. This colonial growth of *Neurospora* is not unusual and has been investigated extensively at his laboratory. Gammexane, sodium desoxycholate, L-sorbose, and duponol were some of the other chemical substances giving similar results. Experiments are now under way to combine the essential features of the molecules of the best inhibitors, and to make all possible hybrids. Perhaps a compound with more activity can be produced.

Data obtained with the "plate-well-method" on the thiophene compounds presented interesting relationships. Actual results and theoretical values, calculated from the results, taking into consideration the molecular weight of the compound and assuming the sp gr=1, were recorded. 2-Ethyl-5-propylthiophene, did not reach 50% inhibition, whereas 4-(2-thienyl) butanoic acid reached 50% inhibition in surprisingly low concentration. The results seemed to indicate that, in general, as the molecular weight of the compounds increases, the inhibitory capacity seems to increase.

At low levels casein hydrolysate relieves the inhibition of some thiophenes, but at high concentrations, it becomes toxic itself. An extensive study was undertaken of the interaction of inhibitors of *N. crassa.* Only some general findings are represented here. The data will be published elsewhere. This study is, with many reservations, analogous to combination chemotherapy in the mammalian system. Systems of this type have also been discussed by other investigators. (Mantel, 1958.)

When 4-APP and aminopterin were tested together in many different concentrations of each compound, it was found that 4-APP acted as relief agent and did reverse the inhibition effect of aminopterin. This is shown in Table 3. Some inhibitors like 6-methyl-2-thiouracil and purine did not affect 4-APP inhibition. Some inhibitors like 2,4,6-triamino pyrimidine did relieve 4-APP inhibition. Most of the inhibitors tested increased 4-APP inhibition, such as 2-amino purine,

2,4,5-triamino-6-hydroxypyrimidine sulfate, and others. In one experiment, DON was tested with 4-APP in different concentrations. The results of this experiment are shown in Table 4 and are self-explanatory.

Enhancement of inhibition was observed between DON and azaserine, between 4-APP and azaserine, and between DON and 8-azaadenine. 8-Azaadenine with azaserine showed no additive effect of inhibition. At higher concentrations, 8-azaadenine even relieved some of the inhibition of azaserine. Most drugs acted independently, evidently affecting different enzyme systems in the organism. Combination chemotherapy is certainly of greatest importance, at least until the real super-chemotherapeutic agent is found. None of these avenues of approach should be overlooked.

The author believes that the use of *Neurospora* as a tool for study of new chemical substances is important—not more important than the use of other microorganisms or tissue-culture techniques, or animal-tumor experimentations—but important nevertheless, at least until everything is known about the one perfect test system—the *human* tumor.

Table 3. Interaction of Inhibitors of *Neurospora crassa*
Em 5256A in Liquid Minimal Medium

Compounds tested	% of growth as compared to normal in mM conc. of the compound			
	0.05 mM	0.1 mM	0.2 mM	0.4 mM
	%	%	%	%
4-APP as relief agent				
aminopterin	4.9			
amethopterin	4.9			
4-APP	45.0	35.2	20.5	13.7
4-APP and 0.05 mM aminopterin in each	36.2	22.5	20.5	13.7
4-APP and 0.05 mM amethopterin in each	18.6	18.6	17.4	13.7

Table 4. Interaction of 4-Aminopyrazolo(3,4-d)pyrimidine(4-APP) and 6-Diazo-5-oxonorleucine (DON) as Tested with *Neurospora crassa* Em5256A after 65 Hours Incubation

25 mM 4-APP	DON 0.0025 mM		
	0	0.1	1.0
ml	ml	ml	ml
0	78.5	56.4	21.8
0.2	20.4	14.8	8.2
0.8	15.2	9.8	2.4

REFERENCES

Biesele, John J. and Jacquez, John A. 1954 Mitotic effects of certain amino acid analogs in tissue culture. Annals of the New York Academy of Sciences, 58, 1276-1287.

Foley, G. E., McCarthy, R. E., Binns, V. M., Snell, E. E., Guirard, B. M., Kidder, G. W., Dewey, V. C., and Thayer, P. S. 1958 A comparative study of the use of microorganisms in the screening of potential antitumor agents. Annals of the New York Academy of Sciences, 76, 412-441.

Fuerst, R., and Higgins, R. W. The biology of some thiophene compounds. In Preparation.

Fuerst, R., Hsu, T. C., and Somers, C. D. 1956 *Neurospora crassa* as an assay organism for antimetabolites. Bact. Proc., 32.

Fuerst, R., and Skellenger, W. M. 1958 A *Neurospora* plate method for testing antimetabolites. Antibiotics & Chemotherapy, 8, 76-80.

Fuerst, R., Somers, C. E., and Hsu, T. C. 1956 Studies on 4-aminopyrazolo(3,4-d)pyrimidine: growth inhibition and relief in *Neurospora crassa* . J. Bact., 72, 387-393.

Hsu, T. C., Robins, R. K., and Cheng, C. C. 1956 Studies on 4-APP: antineoplastic action *in vitro*. Science, 123, 848-849.

Rhoads, C. P., Ed. 1955 Antimetabolites and Cancer: A Symposium. Annals of the New York Academy of Sciences, 89-105.

Robins, R. K. 1956 Potential purine antagonists. I. Synthesis of some 4,6-substituted pyrazolo(3,4-d)pyrimidines. J. Am. Chem. Soc., 78, 784-790.

Somers, C. E., Fuerst, R., and Hsu, T. C. 1957 Studies on 4-aminopyrazolo(3,4-d) pyrimidine: structural relationships among inhibiting and relieving agents in *Neurospora* .
Antibiotics & Chemotherapy, 7, 363-373.

HALLUCINOGENIC FUNGI

Dr. Leon R. Kneebone, presiding

Panelist: Dr. Sam I. Stein

METHODS FOR THE PRODUCTION OF CERTAIN HALLUCINOGENIC AGARICS

(ABSTRACT)

Leon R. Kneebone
Pennsylvania State University

Since research on commercial mushrooms had been conducted continuously for the past thirty years at the Pennsylvania Agricultural Experiment Station, requests were received to provide sporocarps of a number of species of hallucinogenic mushrooms.

Initially, the conventional methods of commercial mushroom production were used, namely, (1) the preparation of a compost consisting chiefly of horse manure, (2) the preparation of spawn of each species of interest on hand, (3) the planting of the spawn in the pasteurized compost, (4) the application of a layer of sterilized loam after the spawn had grown through the compost, and finally, (5) managing the crop to obtain a maximum harvest. Satisfactory quantities of *Psilocybe cubensis, Panaeolus sphinctrinus,* and *P. venenosus* were obtained.

Further efforts resulted in the production of *Copelandia caerulescens* on uncased horse manure compost. *Psi. aztecorum* was produced on an uncased compost consisting of supplemented sugar cane bagasse.

Psi. mexicana and *Psi. cubensis* produced sporocarps on potato-dextrose-yeast extract agar slants after several weeks of incubation and the latter also fruited on rye grain.

SOME BIOCHEMICAL AND PHYSIOLOGICAL CORRELATIONS DEVELOPED FROM CLINICAL OBSERVATIONS WITH VARIOUS TOXIC MUSHROOMS AND MEDICINAL PRODUCTS

Sam I. Stein

Bertram and Roberta Stein Neuropsychiatric Research Program, Inc., Chicago, Ill.

Perhaps, first, I should establish a rationale for having a neuropsychiatrist on the program of industrial microbiologists. Individuals in your category must focus a sharp scientific eye on very small things, and the other eye might be focused on the idea of practical application. In fact, initially, it may take the fullest or complete focus of your attention to a small biological sphere or area before the mind's eye turns to classify the observation in its correct place in the larger biological scheme. Contemporary psychiatry would do well to (or must) pass through a similar "micro" orientation regarding the foundations of its material.

Terminology is always important in science. The term psychiatry is misleading. It implies an emphasis on psyche or mind. But the mind in the neurological or physiological sense is only another variant in the fundamental functioning of a neural system. True, the mind has a special significance or importance in that its specific content is coded into the human individual as its life period progresses. But for the psychiatrist to place a main emphasis on the quality and quantity of the mind-content makes him only a competitor with the social scientist, the educator, the psychologist, the clergyman, or the sociologist. To presume that the properties of the organic and physiological substrate of the mind in the nervous system are always constant is to overlook the more significant feature of potential instability in the entire sequence in the complex of mind functioning. Grundfest in "Evolution of Conduction" states, "The rapidly expanding organizational and functional complexities of the nervous system arising from mere increase in their number accompany, and in turn make possible, the expanding capacities of animals in different evolutionary stages to cope with their environment. One may hope optimistically that *Homo sapiens* of future generations will be endowed with a more highly developed and better functioning nervous system than are his current progenitors." To most of you who are not active in my field of work I may not have the problem with psychiatric preconception. I hope that I made my point clear that in psychiatry an emphasis of observation is needed upon man's internal environment, i.e., the mind's organic and physiological substrate, how this latter integrates with the fundamental functions of the nervous system, and in turn upon the functional interdependence of the nervous system with other systems of the body, most significantly with the metabolic.

Now I should like to approach my observations of mushrooms with the following clinical analogy: If a drop of dilute atropine is placed in the conjunctival sac, the pupil of the eye dilates within a few minutes. Here is a situation in clinical practice where the ophthalmologist produces a relatively local effect with one specific chemical, hopefully upon the one specific nerve which anatomically is so located as to make this technique feasible. I said, "hopefully upon the one nerve" because the atropine is also absorbed into the local tissue fluids from whence it enters the systemic circulation. Even the few drops needed to dilate the pupils, although applied to a serous membrane type of tissue rather than by the usual routes of medicinal administration, produce in some people undesirable systemic effects (palpitation, dry mouth, disturbed intestinal activity). Diluted atropine is now being replaced by even more dilute products capable of dilating the pupil but not as active upon other tissues if absorbed systemically. Please note how many observations can be derived from, and how many variables must be considered in this one situation. Most of these situations or most of these observations are pertinent to the clinical study of all neurophysiologically significant pharmacological materials. Among these variables of consistent consequence are the factors of dosage, avenues of absorption, local effects, systemic effects, undesirable side effects, time required for induction of effects, and the variability of response of the individual human. In psychiatry, in addition, the subjective feeling-tone arising in a complex situation, such as having some "burning" drops placed in the eyes at a doctor's office by personnel of variable attitudes, is of consequence in the measurement of drug effects, since the subject's stressor mechanism may become activated in the situation to produce complicating effects from the sympathetic-autonomic apparatus. Perhaps in some instances the tension-reaction, that is, the anxiety, is so great that the atropine is almost unnecessary to dilate the pupils. An additional item within the feature of control in this type of clinical test observation is the observer's ability to be objective, and here again semantically correct in his reporting. Returning to the subject with his pupils now dilated, one could describe him as "wide-eyed" or "wild-eyed." The former expression intends to describe the status of the pupils only, whereas the latter implies in addition an observation of the affective, attitudinal, or mental state of the subject. From this example of a specific, fairly well-known chemical applied by a specific route, keeping in mind some of the tangible and seemingly intangible variables to be measured, let us turn to a situation where a human individual has taken a bite of food—let us say meat or perhaps mushroom. Now we have a condition where a number of specific chemicals or prechemicals are brought into the organism by another and natural route. The variables to be considered may not be more numerous but each variable becomes more complex, inasmuch as the time factor in securing effects depends upon almost the entire complex of metabolism in general, upon the special process of some particular chemical, the additional complication of the interaction between the metabolites formed from the bite of food, and upon interaction between the newly processed metabolites present with those in the body's pool of such substances. Fortunately, much is known about the metabolic process. Perhaps most of it was learned from the usually species-fixed metabolism of the subhuman animal, but much of it applies to man. Whether this is also true regarding the human nervous system is controversial. In my opinion, in this most important area, namely, the controlling effects and the needs of the nervous system, the neurally-affected metabolism in man contains either primary or potentially significant differences and/or variations of pathways from those of the infrahuman form. Such variations may become so extreme at times that they might be considered under the term errors of metabolism or metabolic errors.

All of the preceding has been presented in the hope that it will serve as a background to my discussion of mushrooms. Actually, my statements here are mainly concerned with finding a method to study and to understand the effects of some mushrooms as observed in the human. There is an extensive literature rapidly accumulating on the history of the mushroom's place in medicinal practices. This history dates as far back as there are intelligible records. Some of the reports and claims that have been expressed of effects of mushrooms on man's mind and/or behavior are so unusual that they appear imaginative. This is a separate topic and I do not have time for it here. Besides, my own interest in neuropharmacological mushroom effects did not arise from an outside source of stimulation but mainly by chance in 1949 when I personally became involved in an unusual experience that occurred after a meal which included apparently cultured mushrooms. A subsequent study of the literature verified my inference that the phenomenon which I had experienced and observed was most probably precipitated by the mushrooms in that particular meal. But I shall not pursue further the concepts or speculations I had developed about mushrooms from that experience. In addition to formal graduate study in medicine, I continued to apply the scientific method in my medical practice. Following the experience, mentioned above, I gradually shifted my research to the investigation of mushrooms. This is clinical research of the human which probably involves the most complex complexes of neuroid substance and of metabolism. I said, "I gradually shifted my research" because prior to approximately 1952, it would have been considered irrational or complete heresy to imply that symptomatic adjustments in psychiatric material could be secured even by standard drugs or chemicals. Prior to that time, to have made any favorable psychiatric claims for toxic mushrooms might have cost one his membership in organized psychiatry if not even his medical license.

Through a sequence of correspondence and personal contacts which started late in 1951 with Dr. A. H. Smith of Michigan, I met Dr. Rolf Singer in 1953 at the Chicago Museum of Natural History. Dr. Singer verified my inferences as to the neuropharmacological effects of mushrooms, particularly the genus, *Panaeolus*. It took several years before I was able to develop an organization to sponsor such a study. In 1955 our group became active, and soon thereafter, I discovered that another group consisting of the two Wassons and Dr. Roger Heim, the mycologist, were also investigating similar effects of mushrooms among the Mexicans, which had come to their attention as the lately deceased Mrs. Wasson, a physician, was pursuing her avocational interest in mushrooms.

My group made contact with Dr. Ralph Kneebone, who, as you may know, does research in the growing of mushrooms at Pennsylvania State University. Upon consulting with Drs. Smith and Singer again in 1957 we decided to explore *Panaeolus venenosus*, since this mushroom was the most likely contaminant of the cultured variety. It was known to have caused unusual central nervous system effects in instances when it was accidentally ingested. Happily, Dr. Kneebone succeeded in growing a sizable amount of *P. venenosus* upon his first try. Dr. Singer was despatched to Mexico to secure the significant mushrooms there which we intended to use comparatively in our studies of *Panaeolus*. In addition, Drs. Kneebone and

Singer were able to grow the Mexican varieties of the genus, *Psilocybe*, at Penn State so that we were assured of a constant and sufficient supply of this material.

I am prepared to report in part, three separate items of observation derived from the use of our mushroom material. The first item is one already to be found in the literature and consists of my taking approximately 5 g of oven-dried *Psi. cubensis*, fried in butter. For whatever it may be worth scientifically, a tendency has developed in the neurotropic drug investigations, for the investigator to sample his own soup, so to speak. I had already tried various amounts of *Psilocybe* material before. In my adventure with *Psi. cubensis* on a particular Sunday in December, 1957, I was seeking to determine whether an effect would or could be produced comparable to the one I had experienced in 1949, the one which came my way serendipitously. First, I want to report that the essential theme of my *Psi. cubensis* experience is one in which I was as close to feeling dead as I ever want to be, and actually for a period of nearly six hours I felt sicker than at any other time in my life. Unfortunately, very few objective measurements could be made on myself since I was completely unprepared for what had evolved and transpired. I can tell you that unhappily for myself, my mind remained clear, but I believe that whatever hallucinating or disorganization of mind that is alleged to occur as a result of the *Psilocybe* substance was taking place in all other parts of my nervous system and other tissues. My skin was so anesthetized that I was unable to feel my own pulse. Literally, it appeared as if both the sympathetic and parasympathetic parts of my autonomic apparatus were doing battle at various times. Presumably, the sympathetic component emerged victor, if the saucer-like size of my pupils were to be used as a main criterion. I believe I have personally tested every drug that has been deposited in my office by the pharmaceutical salesmen. Certainly I have tested some of those being used in neuropsychiatric investigation such as mescaline and lysergic acid diethylamide, which are other so-called hallucinogens. Here I experienced the standard effects I had observed or which had been reported by others. In view of my *Psi. cubensis* observations, I proceeded cautiously, or more judiciously, in the experimental application of our whole mushroom material. The dried products were granulated and weighed in ½ g amounts. I decided on this dosage on the basis of weighing such an amount of whole, dried Mexican mushrooms, as was reported to produce a pharmacological effect among the natives in Mexico. Of the three items to be described here, in contrast to the first, namely, my *Psi. cubensis* experience with its relatively random approach, the following two, which are separate case studies, might be viewed as systematized, if they are considered in the perspective of clinical psychiatric practices.

The first of these cases might be summarized as follows: My patient, J. H., age 27, had been under my care since April 13, 1958. Because of homosexual thoughts, rapid pulse (112), and a moderate hypertension (165/90), he was on the following medication (May 20) designed to control his symptoms:

Reserpine - conc. sol., 16 drops qid
Iproniazid - 25 mg tablet, ½ tablet bid
Amphetamine - 5 mg tablet, ½ tablet
Amytal - ½ grain bid

Draw-A-Person	Draw-A-Person	Draw-A-Person	Draw-A-Person
April 16, 1958	June 3, 1958	April 16, 1958	June 3, 1958
(before treatment)	(after treatment)	(before treatment)	(after treatment)

Fig. 1.   Fig. 2.

On May 4, the subject had reported a clearing of his homosexual thoughts; and by May 6, he could not restore these thoughts even if he tried. On May 6 his clinical readings were: blood pressure 120/70, pulse 72, and weight 126. The Draw-A-Person test was used almost daily to validate his subjective claims. In such a test the patient is asked to draw a man and a woman. You will note in Figs. 1 and 2 that the drawings made before treatment show a rather effeminate looking man and an essentially sexless woman. The sketches made after treatment were of a virile man and a feminine woman. Although homosexual thoughts were inactivated, the positive heterosexual thoughts were asserting themselves only slightly. The subject was from Ireland via Canada, and his visa time was becoming short. He had read of my work with mushrooms, and had originally come to Chicago in the hope that mushroom effects would help. He volunteered spontaneously to try the crude material in hope of expediting the effect he was seeking. On May 20, while still on the medication listed, ½ g *P. venenosus* was given without significant effect. At intervals of two or three days more experiments were attempted. One gram produced positive results. One and one-half grams produced strong favorable effects. One and one-half grams of *Psi. caerulescens*, was also tried, but a different, relatively hallucinogenic, and much less pleasant effect was secured or reported. The *P. venenosus* effect was only minimally hallucinogenic and only in the 1 ½ g portions. There was no doubt from these observations that the *P. venenosus* contained a psychoneurophysiologic ingredient. Its effect was clearly different from that of *Psilocybe*. Its effect was mainly stimulatory (euphoriant) and no undesirable side effects were reported or noted. The subject asked for more of the effect which was secured with 1½ g of the *Panaeolus*.

I have rushed through the findings in this case since I believe the material of the succeeding study not only is more extensive but the subject was a nonpatient and his findings were not complicated by any premedication. In fact, the subject to whom I now refer is Assistant Professor Gerhard Closs of the University of Chicago who is collaborating in our research activity and who offered to be the figurative "guinea pig" since he wanted to experience personally the effects of the chemicals he is seeking to isolate and identify. I had invited him to give his various observations and inferences personally, but unfortunately this has been prevented by earlier commitments.

As these case studies proceed, I have been making the following observations:

a. Weight of subject
b. Oral temperature at variable intervals
c. Pulse, blood pressure, and sweat reactions at 10 to 15 minute intervals
d. Estimated pupillary size and reflexes at frequent intervals
e. Patient's subjective feeling-tone statement requested (almost as often as the pulse reading)
f. My observation of the subject's reaction is continuous since I never leave the subject after he partakes of the mushrooms until the evidence indicates that the experimental process has stopped.

Time will not permit covering all the steps of even one experiment, since the observations cover from 2- to 5-hr periods. However, some of this material has been in press since March, and I hope that the journal will be publishing it soon.

My summary of the findings to the present time reads as follows: There were noticeable similarities and differences in the effects of *P. venenosus* and *Psi. caerulescens* which appeared to have no correlation with the quantity of raw mushroom ingested. In the experiments conducted so far, the subjective and objective effects developed within about the same time interval after ingestion, 25 to 30 minutes, with one exception: namely, yawning, which was observed here and elsewhere only as occurring with the *Psilocybe* material, usually within 5 to 10 minutes and continuing until other symptoms and signs present themselves. Reported symptoms or subjective effects which occured with *Panaeolus* only or mainly were (a) a mild or moderate relaxing (tranquil) type of inebriation, (b) disturbed equilibrium, (c) either diminished motivation or blocking in the psychic (thought) process, and (d) paresthesias. Features which obtained with *Psilocybe* only or mainly were (a) the aforementioned yawning in the induction period, (b) "burning" sensation in the esophageal and stomach areas, (c) a feeling of extreme tiredness, and (d) a feeling-tone of anxiety bordering on collapse associated with a sense of strong intoxication.

The factors occurring in common were (a) the extreme reduction in the pulse rate, slightly more with *Psilocybe*, (b) the steadiness of the systolic and diastolic components of the blood pressure with both varieties of mushrooms, except for a

very brief period of drop in both diastole and systole with both *Panaeolus* and
Psilocybe during the period of strong intoxication, (c) dilatation of the pupils, (d)
a drop in body temperature, seemingly more marked with *Panaeolus*, but produc-
tive of much more subjective effect (feeling of cold) during the period of Psilocybe
intoxication, and (e) sweating. With *Panaeolus* there was a general drop in the total
personality integration, but the internal sensation was one of feeling pleasant and
relaxed; whereas with *Psilocybe*, there was actually disorganization, associated
with the feeling of panic and disagreeable intoxication. A most significant feature
is that Dr. Closs at no time perceived or reported effects that could be categorized
as hallucinatory. However, in my first subject a quasi hallucinosis developed, but
more with the thought processes than with colors. In my own *Psi. cubensis* experience, I had observed strange color effects and disturbed stereognosis.

My formulation as to the meaning of these observations reads as follows:
Obviously, it is unwise and perhaps unnecessary to try to correlate clinical observations where unknown chemicals or chemical complexes are ingested in combination. Since I have had considerable experience with psychoneuropharmacological
(psychoneurotropic) material and also with the literature of biochemistry, I offer
the following tentative opinion: The relaxing subjective effects, the hypothermia,
and the diminished pulse found with *Panaeolus* resemble the effects attributed to
serotonin. But the inebriation and the pupillary mydriasis probably cannot be explained as effects of serotonin.

Some of the effects observed here (and elsewhere) with *Psilocybe* are probably
due to the serotonin-related psilocybin already isolated from the *Psi. mexicana*
(Heim) which substance probably also obtains in the *Psi. caerulescens*, the species
used in our experiments. Again however, it seems unlikely that psilocybin itself
will be found to have caused the signs of mydriasis and of cerebral or cortical
intoxication. Where might one turn for some reasonable explanation of this variety
of effects? Is there an orienting concept that one might develop to serve as a guide

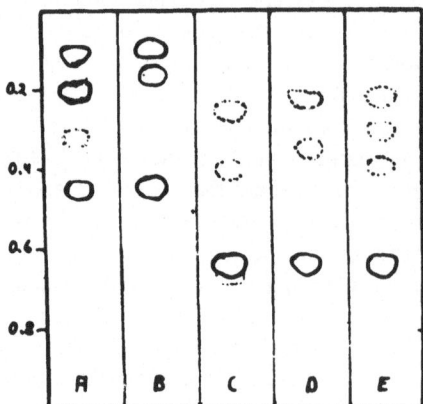

Fig. 3. Paper chromatograms of A) P. venenosus;
B) P. sphrinctrinus; C) Psi. mexicana; D) Psi.
cubensis, and E) Psi. caerulescens. Solvent: iso-
propanol - water 9:1; detecting reagent: Ehrlich's
reagent.

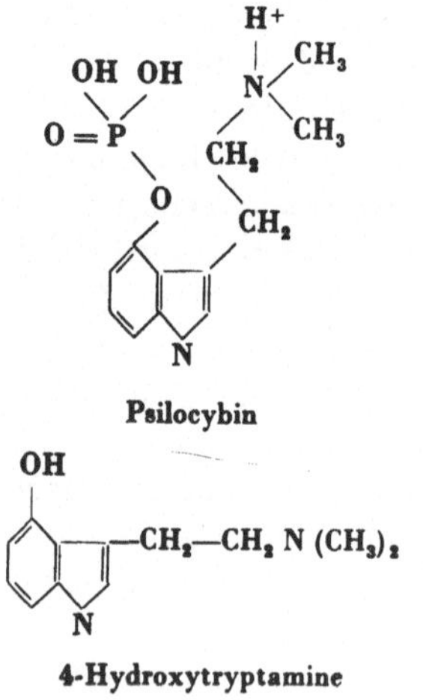

Psilocybin

4-Hydroxytryptamine

Fig. 4.

to the present information, and to what may it evolve? Our *Panaeolus* mushroom as yet has not been chemically analyzed. Figure 3 shows Dr. Closs's paper chromatographic spread. I shall not attempt to interpret Dr. Closs's findings but will state his summarizing paragraph of our combined paper which is in press:

"The extracts of *P. venenosus*, *P. sphinctrinus*, *Psi. mexicana*, *Psi. cubensis*, and *Psi. caerulescens* have been compared by paper chromatography. It was found that neither of the *Panaeoli* contains psilocybin, the main constituent of the three *Psilocybe* species. The extract from *P. venenosus* has been chromatographed, and two of the three major constituents were purified and obtained crystalline. One of these compounds could be shown to possess the same chromophore as psilocybin, a 4-oxygenated indole system, and seems most likely to be the active compound of the mushroom."

Since our material has not been isolated and identified in a final chemical way, I shall proceed with my tentative formulation of trying to explain clinical results by turning to the known chemical structure of psilocybin (Fig. 4) which when hydrolyzed yields 4-OH-tryptamine instead of 5-OH-tryptamine which is serotonin. Before proceeding further, I should like to bring to your attention some chemical and clinical observations with tryptophan. I have reference to the L-form of this substance, which is active in human metabolism, and which I am using clinically in a variety of combinations with B vitamins to good advantage. Tryptophan is an essential amino acid which may be derived from the metabolism of a variety of animal and vegetable proteins eaten by man. Allegedly, the metabolism of L-tryptophan yields the following:

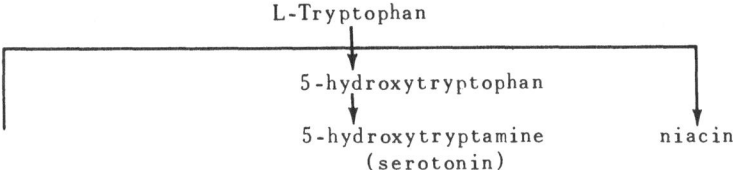

L-Tryptophan

5-hydroxytryptophan

5-hydroxytryptamine
(serotonin)   niacin

Here we have serotonin (5-hydroxytryptamine) which chemically is closely related to 4-hydroxytryptamine. The quieting effects derived from serotonin which result seemingly from its stimulation of the parasympathetic-autonomic system have been extensively studied, and essentially consist clinically of lowered temperature, narrowed pupils, diminished pulse rate, decreased blood pressure, and subjectively less tension or anxiety. This is actually the opposite clinical picture to euphoria. However, when one turns to the other prominent metabolite of L-tryptophan, namely, niacin or nicotinic acid, entirely different inferences might be drawn (Fig. 5). A considerable literature exists to inform you that a deficiency of this substance or its precursor, tryptophan, such as has occurred in large populations on this planet has been the basis for wide-spread pellagra, a condition characterized by mental changes, such as anxiety, hallucinations, depression, and disorientation. In my clinical observations with niacin, either as it might be derived in exaggerated amounts from L-tryptophan or where applied in its isolated state, a variety of results occur which suggest that many tissues are affected by the niacin. Often among these findings are those of improved energy-coefficient and

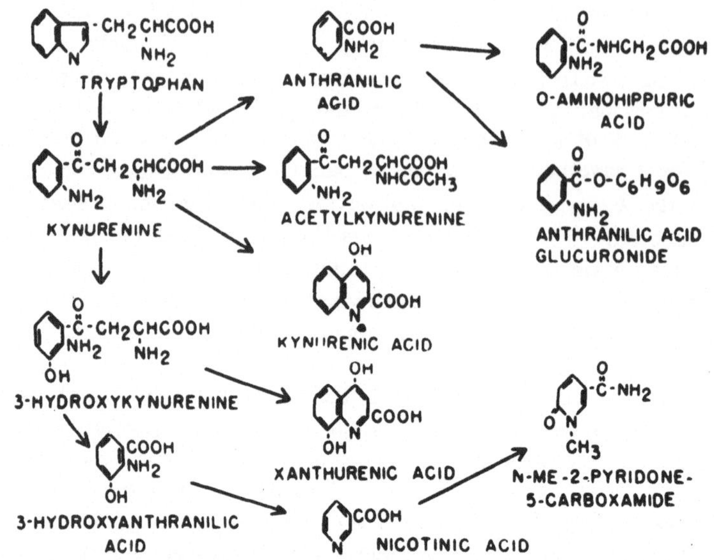

Fig. 5. Abbreviated diagram of the pathway from tryptophan to niacin and its chief metabolite, N-methyl-2-pyridone-5-carboxamide, showing the interrelationships of the various metabolites. This figure has been published previously.

mood modified upwards to the level of almost mania in some. It is my impression that the niacin factor has a huge enterprise in the body. Whether its excitatory effect is accomplished through its known action in the Krebs cycle, or as an amine oxidase inhibitor, I am not certain. However, I am more impressed with its alleged role in the formation of diphosphopyridine nucleotide (DPN) and triphosphopyridine nucleotide (TPN), which substances join the general metabolism to cellular energy. It is reported that niacin is involved in 35 important, separate metabolic steps. The above line of reasoning inevitably leads one to consider the probability that effects in or on enzyme systems will probably be found as the basis of the mushroom observations.

Besides the dosage being important in these various substances as to the effects produced from individual to individual, must we not also take into consideration the various possible pathways of such a substance as niacin which each human may have inherited or which exist in us as potentials to assert themselves under differing metabolic conditions? Is this a part of what is meant by the biochemical individual differences that are encountered in this class of research which involves nervous system and metabolism? My work with L-tryptophan has shown that many adjustments of nervous system functioning can be secured through its correct manipulation. I believe that somewhere in its metabolism there exists close relationships to the effects observed with whole mushrooms or isolated compounds secured from them. With specific reference to mushrooms, probably several chemical compounds, working alone or in combination, will be found to be psychoneurophysiologically effective in each significant mushroom, and not only those related to tryptophan as I have been suggesting. Useful explanations will be uncovered for such unusual phenomena which have been alleged or reported, and which I believe have occurred in situations where so-called "sacred" mushroom material has been ingested, such as extrasensory perception, psi-activity and hallucinosis. To me, it is understandable how unsophisticated or primitive people came to label this material "sacred."

In closing, I should like to prognosticate that conceivably man's total maturity of neural functioning, including his mood, motivation, energy, even quality of thought, and some currently unsolved medical conditions, eventually may be significantly determined by manipulating combinations of food, and/or its ingredients, and/or other biological vegetations which are in the process of being explored. The term, psychodietetics, used recently at a nutrition symposium, seems very apt. In that context, the quality and quantity of what one would write in a paper to be presented to a society might even depend considerably on whether he was ingesting candy-bars, pretzels, L-tryptophan, or mushrooms while writing.

DETERIORATION OF PROTEINS

Dr. Bernard Wolnak, presiding

Panelists: Dr. Leland A. Underkofler
Robert L. Charles
Dr. William T. Roddy
Dr. Theone C. Cordon

INTRODUCTION

Bernard Wolnak

Chemlab, Inc., Franklin Park, Ill.

When in the spring of this year (1959) Dr. Klens asked me to organize this symposium I agreed to do so, thinking that in this particular field there would be no difficulty in obtaining a panel.

It was not long before I found out that I was completely wrong. In retrospect this situation can be explained: first, the amount of work going on in this field is not as widespread as one might believe; secondly, in many laboratories the work which is being done is not being published because of company policies; and thirdly, much work appears to be done on the prevention of the microbial decomposition of only a small group of economically important proteins.

In this introduction the theme shall be what I strongly believe is the comparatively inadequate total effort being made in this important field today.

Thus we all know that microbes possess, produce, and secrete proteolytic enzymes. The occurrence of enzymes and microbes is widespread, all microbes producing these enzymes to a greater or lesser extent. But how many of you have compared the recent literature which exists with respect to enzymes derived from plant and animal sources with that concerned with the microbial enzymes?

In the past decade we have seen some wonderful advances take place in the general field of microbiology and enzymology; and since the enzyme is basically a protein molecule, if we add to this the tremendous advances taking place in protein chemistry, we see some very fundamental and brilliant basic work taking place with these large, sometimes difficult, but always fascinating molecules.

Without belaboring the point I should like to make a list of the data which are available on the enzymes pepsin and trypsin, and to compare these data with the data which are available for comparable microbial enzymes.

121

CRYSTALLINITY AND PURITY

Many of the animal enzymes have been crystallized and the absolute purity of the preparations established.

An intensive study of the literature has revealed that only a few microbial enzymes have been crystallized and fewer still proteolytic microbial enzymes have been crystallized. Thus Herbert and Pursent reported on a crystalline bacterial catalase in 1948 and more recently the Japanese workers have reported upon a crystalline proteinase derived from an *Aspergillus* strain.

The multitudinous reports on microbial enzyme activity available in the literature are in the main based upon materials of variable purity: purities ranging from simple cellular extracts, through crude preparations, to materials of reasonable purity. It also appears that these data reported previously are valid only for the particular preparations described and that these data might have little similarity to data obtained with preparations of higher purity and with crystalline materials.

Since many of the following properties are valid only for pure or crystalline materials it can be seen that there is a large void in our basic know-how. Thus for pepsin and trypsin we know the:

1. Molecular weight from
    a) osmotic pressure
    b) diffusion coefficient
    c) sedimentation data
    d) chemical composition
2. Analytical and specific amino acid composition
3. Arrangement of the amino acids
4. Crystal structure from x-ray studies
5. UV spectra
6. Effect of reaction with other reagents upon activity
7. Specificity with regard to rate and to substrate
8. Electrophoresis patterns
9. Isoelectric points

With the following, purity is a lesser, but important criterion and here many studies have been made.

1. Effect of pH change on activity
2. Effect of heat on activity
3. Effect of presence of metallic ions as activators or as poisons
4. Reaction rates
5. Effect of products produced upon reaction rates
6. In microbial enzymes the effect of variation of conditions of growth upon the amount and type of enzymatic activity
7. Effect of oxidizing agents, alkylating agents, and sulfhydryl group reagents

It is therefore surprising and disturbing that comparatively small amounts of this type of work are directed toward enzymes and/or proteins derived from microbiological sources. The reason for this lack of "basic research" is not readily apparent and in this discussion perhaps is not pertinent.

However this is the *Society for Industrial Microbiology* and we must therefore be interested in the industrial and practical aspects of these enzymes: perhaps in recent years—especially in the past two decades—we have placed too much emphasis upon the practical. It occurs to me that at this time it might be worth while to carefully analyze this growing mountain of practical know-how and to see if we have not reached the point of diminishing returns. Should we not begin to build a foundation of knowledge which might be called "basic"? It may be very trite to say that it is important that we have basic research, but at this stage only good, practical, and important economic results can come from this approach.

It does not seem that it will be difficult to prepare pure crystalline enzymes from these sources. Once these are at hand, all of the data previously mentioned, as well as some not mentioned, can be obtained. Thus we will be able to discuss the values for and the significance of their molecular weights, amino acid contents and ratios, isoelectric points, pH stabilities, temperature stabilities and abilities to decompose certain substrates.

Several areas which we have not mentioned previously and about which we know comparatively little are these:
1. The relationship of the morphology of a species or group to proteolytic activity, and type of enzyme produced.
2. The genetic relationships of these organisms vis-a-vis enzymes, has not yet been reported on in detail, although I'm sure it is being worked upon.
3. The new and useful techniques of chromotography have yet to yield their values in this field.
4. We know comparatively little concerning protein biosynthesis in relation to enzyme synthesis.

The emphasis in the past then has been almost completely on practical approaches; this is often something with which one cannot argue. However, I want to make one more point in this regard. The large research laboratories of the several companies in this field have not published to any significant degree in this area since they think such a move might be of some advantage to a competitor. We find the situation therefore in which each group has built up its own *know-how* and *art* and has been completely dependent upon its own efforts to produce marketable items.

That this situation exists is in part an explanation of the scarcity of basic know-how being revealed. This situation, of course, prevails in other technical and scientific fields, but this does not lessen the weaknesses of this method of operation to a greater or lesser degree.

I am sure that on my say-so and recommendation these various groups are not going to fall all over themselves in a rush to reveal and publish their existing data. However, for what it is worth, I can say that present procedures help no one, least of all the company concerned. I realize also that it is difficult for me to say to the president or research director of a successful multimillion dollar company that he should do more fundamental research and more practical research, and that following the filing of patent applications, he should publish such data. I have not the time here to give the many convincing arguments that the lack of work directed to "the need to know" and the process of working in utmost secrecy is a self-defeating one, but I am hopeful that such discussions as might be developed will lead to some improvements in the situation.

What do we know about our subject? Some of the areas of general enzyme preparation, isolation, and uses are either very well known or are better known by members of our panel and I will leave it to them to discuss. Suffice it to say here, that many proteolytic enzyme concentrates are available and these find use in such areas as cereal foods, prevention of chill haze in beer, tenderizing meats, bakery bread and crackers, dry cleaning, acids, and others.

There has been some study of the proteolysis of milk proteins by fungal enzymes, especially in relation to the function of these enzymes in the ripening of highly flavored cheeses. This work, while highly important, suffers from the drawbacks outlined previously; we have here an extremely fertile field for more fundamental research which could lead to dividends in a hurry.

At the other end of the scale we have been interested in preventing the proteolytic activity of the microbial enzymes on such important industrial proteins as collagen (leather) and keratin (wool).

In this field we know practically nothing of the mechanism of decomposition of these proteins by enzymes whether they be secreted by microbes or by the larva of a higher organism such as a moth or carpet beetle. At this stage we can not say for sure whether the decomposition is caused by the larval enzymes, the enzymes secreted by the larval gut, or both. Suffice it to say here that some well-planned research could pay handsome yields.

A discourse on where basic research should and could go is not my purpose; I wish merely to emphasize, not to belabor the point.

I am hopeful that with our increasing research efforts in the not too distant future we can see progress being made in this field. I shall be able to judge well by comparing notes with the person who organizes the next symposium.

PROTEOLYSIS BY MICROBIAL PROTEINASES

Leland A. Underkofler and Robert L. Charles

Miles Chem. Co. (Takamine Plant), Div. of Miles Laboratories, Inc.
Clifton, N.J.

All microorganisms which utilize proteins as nutrient materials produce proteolytic enzymes necessary for hydrolysis of the proteins. It therefore follows that there must be a large number of microbial proteinases produced by various species of microorganisms, including many bacteria, molds, and yeasts. Only a very limited number of these numerous microbial proteinases have been investigated at all, and those available as commercial products are few indeed. However, these latter have become of great practical and economic importance.

The earliest known practical application of microbial proteinases was the koji of the Japanese. The term koji first referred specifically to the product resulting from growing the mold *Aspergillus oryzae* on cereal grains. In current usage it embraces almost any organism grown in surface culture on a solid cereal material.

Records show that the mold koji was introduced into Japan from China over 1700 years ago. The proteolytic activity of the koji has been for centuries and still is used in manufacturing sake, soy sauce, miso, and other Japanese beverage and food products. The enzymatic nature and mode of action of koji was not recognized until the latter half of the nineteenth century after Japan opened her doors to Western civilization.

Dr. Jokichi Takamine introduced the use of koji into the United States about 1890. Moreover, he also developed the procedure of growing the mold on wheat bran, followed by extraction of the enzymes from the resulting bran koji with water and precipitation of the enzymes from the aqueous extract with alcohol. This marked the beginning of the commercial application of concentrated fungal enzymes. Boidin and Effront, in France, about twenty years later were the pioneers in producing bacterial enzymes.

At present only a relatively small number of commercial applications for microbial proteinases have developed—but all of these are important. The number of applications is increasing through research, and the field will undoubtedly be much expanded in the future.

COMMERCIAL MICROBIAL PROTEINASES

The commercially available microbial proteinases, differing in their specificity and conditions for action, are obtained from bacterial or fungal sources. The amounts of desired enzymes produced vary markedly between species and even between strains of the same species. Hence, isolation and cultivation of a strain having the potential for production of maximum proteolytic activity when cultured in an optimum environment is necessary. Two principal methods for cultivation are used today, the older one involving modifications of Dr. Takamine's original koji process, and the newer one, submerged culture procedures.

In the koji process the microorganisms are cultivated on the surface of a moist, solid substrate. Wheat bran has come to be recognized as the most satisfactory basic substrate for this method, although other fibrous materials can be employed. Additives may be used to stimulate proteinase production such as soybean meal, beet pulp, nutrient salts, and acid or buffer to regulate the pH. The moistened bran medium may be steamed or sterilized under pressure, cooled, and inoculated with the chosen culture. The inoculated bran medium can then be spread on trays and incubated in chambers where the temperature and humidity are controlled within certain limits by forced circulation of air until maximum enzyme production has occurred.

As is well known to all microbiologists, the submerged culture method was first extensively used for production of penicillin and other antibiotics, although it had prior limited use in such fermentations as for production of sorbose and gluconic acid. Submerged culture methods for production of microbial enzymes were introduced only about ten years ago, but are now extensively employed.

Enzymes for commercial sale are obtained from koji by extraction with water, followed by filtration or centrifugation to obtain clear solutions. From submerged culture processes, the microbial cells are filtered from the beer. The solutions may then be concentrated by vacuum evaporation. For commercial products, suitable preservatives and stabilizers may be added to the solutions to permit sale and use in this liquid form. Most commercial enzymes, however, are solid products, generally precipitated from the aqueous solutions by addition of a solvent such as acetone or aliphatic alcohol, followed by filtration or centrifugation, and drying of the precipitate at low temperature, at atmospheric pressure, or under vacuum. The dry microbial proteinase powders may be sold as undiluted concentrate on an enzyme potency basis, or, for most applications, may be diluted to an established standard potency with acceptable diluents.

Table 1. Commercial Uses for Microbial Proteinases

Industry and Use	Bacterial	Fungal
Baking— crackers, bread		X
Cereal breakfast foods	X	
Meat-tenderizer	X	X
Brewing— chillproofing	X	X
Animal feeds — ingredient	X	X
Animal feeds — fish solubles	X	
Protein hydrolysates	X	
Leather— bating hides	X	X
Leather— unhairing hides	X	
Dry Cleaning— spot removal	X	
Textiles— desizing	X	
Photographic— film stripping	X	
Medicine— wound débridement	X	
Medicine— relief of inflammation, bruises, blood clots	X	X

Use is made of microbial proteinase in cracker doughs because such treatment results in (1) reduction of mixing time, (2) uniform rolling of dough, without stickiness or "bucking" at edges, (3) increase of spread, (4) reduction in the amount of shortening required to give the desired tenderness, (5) exceptionally even bake, (6) uniform and improved browning, (7) more open grain, and (8) enhancement of flavor and taste.

A majority of bread bakers now employ fungal proteinase which markedly decreases mixing time of the doughs by slight degradation of the wheat flour gluten. In addition, the fungal proteinase increases the extensibility of the doughs, thus ensuring proper machinability. Secondary benefits are improvement of grain, texture, and compressibility of the bread.

Cereal foods are treated with proteolytic enzymes to modify the protein to obtain better processing, including improved product handling, increased drying capacity, and lower power requirements.

Proteolytic enzymes are used for tenderizing meats and animal casings for processed meats. The commercial products contain papain or bromelin as active agents, but since different proteolytic enzymes preferentially attack different meat tissues, combination of plant, bacterial, and fungal proteinases have an advantage over any single enzyme for meat-tenderizing.

To prevent development of undesirable haze in beer and ale when these beverages are cooled, proteolytic enzymes are added during the finishing operation to "chillproof" these beverages. Chillproofing agents contain pepsin, papain, bromelin, fungal, and bacterial proteinases in various combinations, which digest enough of the protein to prevent formation of haze.

Recently a great deal of attention has been given to addition of enzymes to animal feeds for poultry, swine, cattle, and sheep. Addition of suitable enzymes, usually of bacterial origin, has been shown to increase rate of gain in weight, and decrease the amount of feed required per unit of weight gain for young animals. A great deal of research remains to ascertain clearly with which enzymes and with what rations greatest growth response may be obtained, whether enzyme supplementation will increase egg or milk production in mature animals, and the like.

Protein hydrolyzates for condiments and special diets, and for animal feeds, are obtained by extensive enzymatic hydrolysis of plant, meat and fish, and milk proteins. Enzymatic processing has the advantage over acid or alkaline hydrolysis of proteins in the simple equipment employed and the lack of destruction or racemization of amino acids. Examples of the use of bacterial proteinases are in the fishing industry where inedible and scrap fish are processed for oil, meal, and solubles. The conventional process has been to cook the fish, and press to remove the oil. The residual solids are dried to produce fish meal and the aqueous phase is evaporated to about 50% solids to obtain fish solubles. The fish meal and fish solubles are valued livestock feeds. Addition of proteolytic enzymes to the aqueous phase prior to evaporation increases evaporation efficiency by reducing viscosity, and addition of enzyme to the concentrated fish solubles containing 50% solids results in a product which will not congeal in cold weather.

The leather industry makes use of microbial proteinases in bating hides, these having replaced the rather obnoxious method once employed of soaking the hides in lime pits with dog dung as a means of bating. Bacterial proteinases have also been suggested for unhairing hides for leather making. Research actively in progress at the Eastern Regional Research Laboratory, by the Tanners' Council, and other laboratories in this country and in Japan, has shown bacterial enzymes to be very effective in unhairing cattlehides. Such enzymatic unhairing may be expected to replace the slower and less efficient lime-sulfide soaking methods which have been in common use.

In the drycleaning industry, solvents will not remove proteinaceous stains of milk, egg, blood, and the like from clothing. Products containing bacterial proteinases are used throughout the drycleaning industry to solubilize such stains without damaging the fabrics. A somewhat similar application is the use of bacterial proteinases for degumming and desizing textiles.

Another application of bacterial proteinases is for recovering silver from photographic film by enzyme digestion and solubilization of the gelatin emulsion coating.

Pharmaceutical and clinical applications for bacterial proteinases, especially of the streptokinase-streptodornase type, include débridement of wounds, and injection or buccal administration to relieve inflammation, bruises, and blood clots.

ACTION OF MICROBIAL PROTEINASES

Modes of Action

All proteinases, whether they are of animal, plant, or microbial origin are, of course, characterized by their ability to catalyze hydrolytic cleavage of peptide linkages between amino acids. The term endopeptidase is used to designate an enzyme which attacks internal peptide bonds in the high-molecular-weight polypeptide protein molecules, thus forming peptides of lower molecular weight. The term exopeptidase denotes an enzyme which attacks terminal peptide bonds adjacent to free polar groups thus resulting in the liberation of free amino acids. The exopeptidases may be further classified as aminopeptidases or carboxypeptidases, depending upon the free polar groups (amino or carboxyl groups, respectively) required for their action.

The common commercially available proteolytic enzymes are mixtures of several proteinases, some being endopeptidases and others exopeptidases. Hence, their proteolytic action is extremely complicated, both because of the complexity of the high-molecular-weight protein structures to be hydrolyzed which may contain some 20 different amino acids in various sequence, and the multiple nature of the enzyme preparations themselves.

However, even in these commercial enzyme mixtures, it is possible to demonstrate the presence and the difference in action of the endopeptidases and the exopeptidases present. An example is with gelatin, which is commonly used as a substrate for measuring proteolytic activity. One method for enzyme assay measures viscosity reduction of a gelatin solution when it is treated with a proteinase, and another method measures the amino acids liberated by proteinase action by means of the formol titration procedure. The viscosity reduction method is primarily an indication of endopeptidase activity. The formol titration method, of course, measures total amino acids liberated by the exopeptidase activity, but this in turn will be partially dependent upon endopeptidase activity making available increased numbers of terminal polar groups for the exopeptidase reactions.

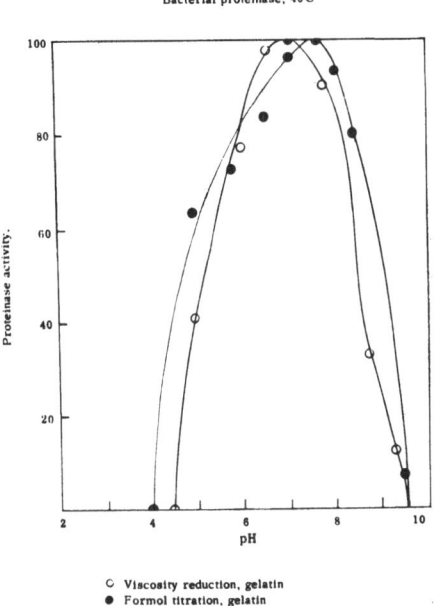

Fig. 1. Effect of pH on activity of a commercial
bacterial proteinase.

As shown in Fig. 1, the pH activity curves for a commercial bacterial proteinase by both the viscometric and formol titration procedures are somewhat similar, but not identical. By the viscometric procedure there is no activity below pH 4.5. A slight reduction in viscosity occurs at this pH within two to three minutes after introduction of the enzyme, but no further viscosity reduction occurs on longer incubation.

The pH optimum is at 7.2, and activity ceases at pH 9.6. As measured by formol titration some activity is shown at pH 4.5, the optimum is pH 7.8, and all measurable activity ceases at pH 9.6. In general usage, it is recommended that bacterial proteinase be used over the narrow range of about pH 6.5 to 8.

As shown in Fig. 2 the commercial fungal proteinases will tolerate and work effectively over a wider pH range. By both the viscometric and formol titration procedures there is no activity at pH 4.0, but very marked activity still is shown at pH 10.5. The optimum is pH 6.5 for the formol titration method, and pH 7.8 for the viscosity reduction procedure.

Specificity

Most individual proteinases which have been isolated in pure form have been found to be quite specific with regard to which peptide linkages they can split. For example, crystalline trypsin hydrolyzes links involving the carboxyl groups of the basic amino acids lysine and arginine. The specificities of other crystal-

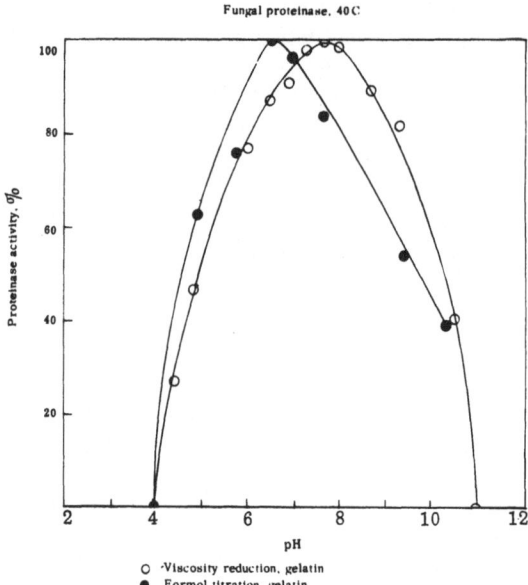

Fig. 2. Effect of pH on activity of a commercial fungal
proteinase.

line proteinases such as pepsin, chymotrypsin, papain, carboxypeptidase, leucine, and aminopeptidase have been investigated, but in most cases not completely elucidated.

In the case of microbial proteinases, few have been extensively purified or crystallized, and few investigations of the specificity of these enzymes have been undertaken. Certain Japanese workers have crystallized a proteinase of *A. oryzae*, and others a crystalline protease from a strain of *Bacillus subtilis*. European workers have also crystallized a proteinase from *B. subtilis*. However, no reports have been made on the enzyme specificity of these enzymes.

An interesting crystalline proteinase has been obtained by Yoshida from the black mold designated as *A. saitoi*. This enzyme is markedly different from all previous microbial proteinases in that its optimum pH for activity is pH 2.5 - 3.0, and for stability pH 4.0. It is rapidly inactivated at room temperature at pH levels above 4, although the presence of casein stabilizes the enzyme up to pH 6. Yoshida made an extensive investigation of this enzyme, with respect to its thermal inactivation at the various pH levels of 1.5, 2.5, 3.0, 3.5, and 4.0, the effect of 42 chemicals as possible inhibitors, and its specificity for certain substrates. While the investigation of specificity was not exhaustive, the enzyme was found to hydrolyze links involving the carboxyl groups of L-arginine and L-glutamic acid rapidly, and of L-leucine and L-alanine slowly; glycyl-L-aspartic acid was hydrolyzed slowly, but not glycyl-glycine, or glycyl-L-leucine.

From a practical standpoint, since the specificity of the proteinase enzyme mixtures constituting commercial enzyme products is unknown, for any particular application it is necessary to select the appropriate proteinase mixture, or combination of enzymes. Usually this can only be determined by trial and error methods. It is by such empirical methods that most of the present-day uses have developed. With proper selection of enzymes, with appropriate conditions of time, temperature, and pH, either limited proteolysis or complete hydrolysis of most proteins can be brought about.

Conditions for Action

Like all enzymes, microbial proteinases are sensitive to conditions of temperature, pH, and available substrate. They may be completely denatured and inactivated by exposure to high temperature or to strongly acid or alkaline conditions. The optimum temperature for action of a microbial proteinase, as for most enzymes, is usually defined as the temperature at which the most rapid enzymatic proteolysis occurs. This optimum may vary depending upon the pH, substrate, presence or absence of activators or inhibitors, and other factors. At best, the optimum temperature is a compromise, since increasing temperature simultaneously increases the rate of enzymatic catalysis, and the rate of enzyme inactivation. The rate of inactivation of an enzyme by heat again depends upon the pH, substrate, presence or absence of activators or inhibitors, and other factors. Typical heat inactivation curves at 50 C and 60 C for a commercial bacterial proteinase are shown in Fig. 3. The enzyme suffers no measurable loss of activity at 40 C in 60 min. From the graphs, at 50 C, with gelatin substrate at pH 7.0, the enzyme loses half its activity in about 51 min, whereas at 60 C, half the activity is lost in less than 15 min. At 60 C with casein substrate at pH 8.0, half the activity is lost in 19 min.

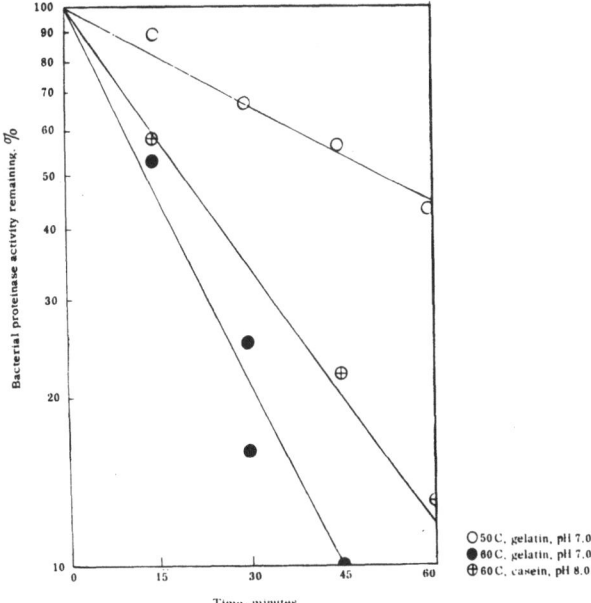

Fig. 3. Effect of temperature and time of heating on
inactivation of a commercial bacterial proteinase.

Such information is quite important for commercial applications. Although maximum enzyme stability is 40 C or below, the optimum temperature for this enzyme is usually given as 50 C because at this temperature the rate of proteolysis is about twice that at 40 C and the rate of thermal inactivation is not excessive. For some uses the temperature of 60 C may be permissible, that is, where very rapid but limited proteolysis is desired.

It is readily apparent from our discussion that knowledge concerning the degradation of proteins by microbial proteinases is extremely fragmentary. The numerous microorganisms which utilize proteins as nutrient materials produce a multiplicity of proteinases, as a group and individually. Only by the isolation of each of these, a very laborious and difficult task, and a thorough study of each, will it be possible to pinpoint the specificity and optimum conditions of action for each.

In the meantime, until that desirable but distant goal is achieved, empirical, trial-and-error methods will continue to be used in developing many useful applications for microbial proteinases in many industries.

MICROBIAL DETERIORATION OF SKIN PROTEINS

William T. Roddy

Tanners' Council Research Laboratory, University of Cincinnati, Cincinnati 21, Ohio

INTRODUCTION

A freshly flayed skin or hide at the packing plant or slaughterhouse contains many bacteria. The bacteria vary in their morphology, their biological characteristics, and their action upon the skin or hide. The proteolytic type bacteria present are the organisms which must be killed or inactivated to prevent skin decomposition. Elimination of manure, blood, and excess fatty flesh from the skin or hide also aids in the control of bacterial activity. In recent years the addition of bacteriostatic agents to the sodium chloride cure salt or the brines used for curing hides has helped materially in controlling bacterial decomposition. However, there are still areas where the best curing practice is not used and skins and hides in various stages of damage are constantly being sent to us by tanners for examination and identification of the source of the damage.

In addition to bacterial attack the skins and hides are also subjected to autolytic and hydrolytic breakdown when not properly cured and stored. All three retrograde actions may proceed simultaneously. The skins may be properly cured and during shipping may be exposed to poor conditions at which time the bacteria and molds inactivated before shipment become active and cause damage.

The autolytic and hydrolytic action on the skin proteins along with control of the type of microorganisms has been used for many years in some areas of the world to unhair sheepskins and other types of skins. This practice is called "sweating" where the skins are exposed at high humidities and optimum temperatures to microorganisms. The disadvantage to the "sweating" practice is that unwanted strains of bacteria contaminate the "sweating" chambers on occasion and cause considerable damage until they are eliminated.

DETERIORATION DURING CURE AND STORAGE

Skins and hides shipped into the United States from restricted areas before being put into process must first be soaked for 24 hr in a 1 to 10,000 solution of sodium bifluoride or in a 1 to 7,500 solution of sodium silicofluoride to prevent spread of rinderpest or foot-and-mouth disease.

In practice, where most of the skins and hides are laid down, flesh up and covered with cure salt, many alterations in the skin proteins can occur. Conditions of impurities in the salt, insufficient salt, the presence of manure and blood, as well as previously used salt and storage temperatures above 70 F found in some establishments result in ideal conditions for bacterial invasion of skins and hides.

When bacteria invade the epidermal area of skin or hide the epidermis is separated from the underlying derma and hair slip occurs, the hair and epidermis being easy to pull off. To the tanner this hair slip is a sign of poor cure. It is possible to have hair slip and no bacteria present, but in general the bacteria are usually present when hair slip occurs. These organisms, usually of the coccus variety, may or may not harm the underlying collagen.

Frey and Stuart (1936) used fresh sodium chloride admixed with used cure salt in a 40-skin experiment at a tannery. Bacteriostatic agents were added to the treated lots to prevent bacterial invasion. The calfskins were held at room temperature for 90 days, the temperature ranging from 61 F to 82 F. The calfskins treated with salt alone after curing were judged to be almost ruined, yet they made acceptable finished leather with no observable decrease in tensile strength. The calfskins cured with the salt containing the bacteriostatic agents were clean and free of odor after cure but the finished leather from the treated calfskins was no better than the calfskins treated with salt alone.

It was suggested by Frey and Stuart that at the end of the cure period the autolytic, microbial, and other deteriorative actions had proceeded in the skins treated with the dirty salt no further than modification brought about on the globular proteins in the pretanning operations of soaking, liming, and bating.

The work of Frey and Stuart indicates that the bacteria make their inroad into the skin or hide by solubilization of the globular proteins of the skin. In cross section it has been observed by Roddy (1942) that the globular proteins are located in the interspaces between the fiber bundles in the corium and in the blood vessels in small amounts and present to a greater amount in the epidermis and its appendages. During the pretanning operations of soaking, liming, unhairing, and bating, the globular proteins are partially or almost completely removed by the process operations according to Tancous (1955). In her work she isolated the skin proteins by fractionation (of the total nitrogen of the skin), and histologically by examination of sections stained to trace the fate of the globular proteins.

The globular proteins may be removed by action of bacteria which in general do not harm the collagen during cure and storage, but there is no assurance that the bacteria will be as specific in action as obtained in the experiment by Frey and Stuart. Therefore it is becoming a practice in many skin and hide storage areas to add a bacteriostatic agent to the cure salt. Kritzinger and van Zyl (1954) have advocated the use of sodium silicofluoride as an additive to cure salt for the purpose and it is being used in many areas throughout the world to retard harmful bacterial growth on skins and hides. It is interesting to note that this bacteriostatic agent is a required treatment on hides shipped into the United States from restricted areas where it serves as a viricidal agent.

The presence of mold growth is also observed on stock not properly cured or stored. The black mold, *Aspergillus niger* and the green mold, *Pencillium glaucum* are noticed on undercured stock on occasion. These molds also grow on and in pickled skins during storage and on leather. These organisms, unlike bacteria, are observed by the naked eye and fungicides are used to prevent their growth. The main known damage they cause in the leather is the discoloration which they produce and tanners add fungicides to prevent their occurrence.

When the moisture content of leather, not treated with a fungicide, becomes greater than 15%, mold growth occurs. Therefore, the presence of a fungicide (p-nitrophenol) is required in all military leathers.

UNHAIRING OF CATTLEHIDES

In regular tannery practice the hair is removed by liming the hides in saturated lime solution containing 10% excess on the stock weight and accelerators such as sulfides and amines. In such a system there is little or no bacterial activity. On occasion, if the hides do not contain sufficient lime, bacteria may still be active in the hides as they enter the tan liquors.

After the hides are unhaired the tanner bates the stock to further remove unwanted proteins remaining in the skin. The bates are a mixture of an enzyme and a deliming salt. The enzyme used has been that of the tryptic type. During World War II, because of the shortage of pancreatic material, the suppliers of bating materials found it necessary to produce bacterial- and fungal-enzyme bating materials. These bates were effective but did not necessarily incorporate the same action as the pancreatic bates. Pfannmuller (1956) in his chapter in the "Chemistry and Technology of Leather" indicated that the exact mechanism of the action was not known but that some alteration (peptizing) of the collagen takes place.

It is possible to unhair cattlehides by the use of microorganisms but variable results occur and the use of enzymes has been suggested for better control. This work is now under way at the University of Cincinnati, using the pattern established by Dr. Cordon of the Eastern Utilization Research Laboratory of the USDA. The project is one where the use of enzymes in the tannery for unhairing will be undertaken.

In the past year at the Tanners' Council Laboratory in using a preparation of the whole raw pancreas which had been activated, desiccated, and defatted,* Lindroth (1959) was able to unhair salt-cured cattlehide which had been soaked in water previous to being treated in the enzyme system. In his work he histologically traced the penetration of the enzyme into the hide by its action on the elastic tissue. The enzyme system caused a removal of elastic tissue and when it had penetrated to the base of the hair pockets, it was then possible to unhair the hide. When the hide was allowed to remain in the system for extended periods the collagen also was noticeably altered.

LEATHER

All of us are very familiar with the ability of leather items to become contaminated with mold growth. Bacteria may be found on the finished leather and also yeast organisms growing in some of the ingredients used in the finishes, but seemingly once the collagen is tanned the bacteria or molds while present do not attack the tanned collagen.

There are numerous reports in the literature on leather items showing very bad damage thought to be due to the presence of mold and its activity on the leather. The mold growth will remove grease which is used to lubricate the leather, alter the ability of the fiber to hold water, and ultimately cause a decrease in strength of the leather.

To determine the influence of mold growth on the leather, Roddy and Jansing (1954) studied the changes in tensile, stitch tear, and burst strength, and in the chemical composition of leather exposed up to 16 weeks to 100% relative humidity

* Viokase Powder. P-145-D. Produced by the Viobin Corporation, Monticello, Ill.

at temperatures of 20 C, 30 C, and 40 C. Half of the specimens were kept free of mold throughout the aging period by soaking the specimens in a "merthiolate" solution prior to aging. Based on the results obtained on vegetable-tanned side leather there was a close similarity of the strength losses by the most severe treatment, (exposure at 100% humidity at 40 C) irrespective of mold growth. Uninhibited mold growth caused substantial losses in grease and a considerable increase in acidity. In the absence of mold, the same environment causes only modest changes.

SUMMARY

The inclusion of a bacteriostatic agent or a fungicide definitely controls or retards microbial action but does not prevent all hydrolysis. However, the tanner has found through research and experience that he can upgrade his leather quality by use of bacteriostatic agents and fungicides as a precautionary measure. Considering the chemistry involved in the processing of salt-cured skin which is at a pH of 6, after immersion in an unhairing bath at a pH of 12.5-12.8 for three to five days, bating and deliming at pH levels of 8 to 10, immersion in a pickle liquor of salt and acid at a pH of 2-2.5, usually overnight, and then tanning to a pH of 3-3.5, the skin proteins are subjected to very severe chemical action. Any measures which will hold the hydrolysis of the skin proteins to a minimum, exclusive of the tanner's treatment, will aid in making a better product.

REFERENCES

Frey, R. W., and Stuart, L. S. 1936 J. Am. Leather Chemists Assoc., 31, 254.

Kritzinger, C. C., and van Zyl, J. H. M. 1954 J. Am. Leather Chemists' Assoc., 49, 207.

Lindroth, E. 1959 Thesis, University of Cincinnati.

Pfannmuller, J. 1956 In chapter 11, "Chemistry and Technology of Leather", Reinhold Publishing Corp.

Roddy, W. T. 1942 J. Am. Leather Chemists' Assoc., 37, 410.

Roddy, W. T., and Jansing, J. 1954 J. Am. Leather Chemists' Assoc., 49, 793.

Tancous, J. 1955 J. Am. Leather Chemists' Assoc., 50, 278.

SOME ASPECTS OF PROTEIN DEGRADATION IN LEATHER PROCESSING

Theone C. Cordon

Eastern Regional Research Laboratory,[*] Philadelphia 18, Pennsylvania

INTRODUCTION

Animal hides and skins have been used by man before recorded time. These prehistoric men must have found that freshly flayed skins will decompose rapidly if not protected. Means of accomplishing this were found, chiefly by drying and treatment with salt, and these methods, with refinements, are still in use today.

The skin of an animal is a highly complex structure. Figure 1 is a diagrammatic representation of a cross section of a steerhide and the complexity is clearly seen. During the processing into leather many of these components are removed and those remaining are more or less altered. When a hide or skin is in a condition ready for tanning, it consists primarily of the fibrous protein collagen. However, residues of such other components as elastin, reticulin, muscle fibers, mucopolysaccharides, fat, and some cellular material still remain.

Some of the degradative changes which take place during curing and depilation will be discussed.

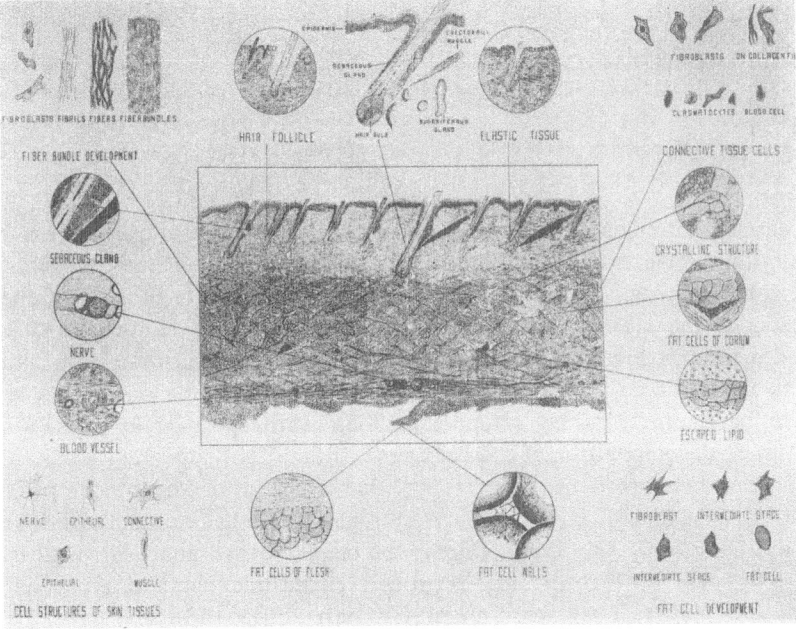

Fig. 1. Diagram of a cross section of steerhide showing its complexity. (Courtesy of William Roddy, University of Cincinnati.)

[*] Eastern Utilization Research and Development Division, Agricultural Research Service, United States Department of Agriculture.

Table 1. Salt Tolerance and Collagen Solubilizing Activity of Representative Single Cultures (from Everett and Cordon, 1956)

Cultures	Growth without Salt	Growth with 4M Salt(23%)	Activity* in 15% Salt Broth + Substrate		
			12% Gelatin	Limed+ Collagen	Unlimed‡ Collagen
Uninoculated controls	-	-	0	10-20%	3-9%
Gram-pos. rods:					
No. 12	Yes	No	+++	62	8
No. 11	Yes	Yes	+++	32	5
No. 87	No	Yes	++	34	6
Gram-neg. rods:					
No. 21	Yes	Yes	+++	41	5
No. 22	Yes	Yes	+++	36	3
No. 20	No	Yes	++	20	4
Gram-pos. cocci:					
No. 28	No	Yes	++	98	5
No. 78	No	Yes	++	31	7
No. 59	Yes	Yes	+	25	7
Gram-var. red cocci:					
No. 112	No	Yes	+	20	6
Gram-var. irreg. rods:					
No. 104	No	Yes	++	20	10

*Gelatin, complete liquefaction: +++ in 4 weeks; ++ 4 to 8 weeks; + over 8 weeks. Collagen, solubilization: hydroxyproline in solution as per cent of total found.

+Center layer of cowhide soaked in saturated lime liquor; chemically sterilized.

‡Center layer of cowhide carefully purified without liming; chemically sterilized.

CURING

Treatment of hides with common salt is the usual procedure for curing. Salt reacts with the moisture in the hide and produces a brine which dissolves some of the globular proteins along with some of the constituents of blood. This process also reduces the moisture level of the hide from 60-65% to about 45-50%. This lowered moisture, together with the high salt content, prevents the growth of most bacteria and other microorganisms. Some halophilic bacteria are able to develop, however, particularly during extended periods of storage and may cause staining, hair slip, or decomposition of the collagen.

We were interested in determining what types of bacteria were present on salt-cured hide and what effect pure cultures would have on collagen (Everett, A. L. and Cordon, T. C., 1956). During the course of the study, several hundred cultures of bacteria were isolated from salt-cured hides, brines, etc., and grown on high-salt media. They were tested for their ability to liquefy gelatin and the positive ones were used for studies on decomposition of collagen.

Collagen was prepared in essentially unaltered condition by splitting off the grain and flesh layers and using only the center portion. This is known as the corium split. Small pieces of this material were sterilized by the method of Maxwell (1945). Five to ten gram lots of this material were treated in succession in closed containers, with 100 ml of each of the following solutions: (1) 0.025 M sodium metabisulfite for 17 hr; (2) 0.3 M hydrogen peroxide for 7 hr; (3) sterile 0.1 M sodium bicarbonate for 17 hr; (4) two changes of sterile 10% salt. The bisulfite solution was at pH 2.0 and the first three solutions contained 10% salt to prevent swelling.

Small pieces of the sterilized collagen were placed in tubes containing a complete broth medium and incubated at 30 C for several weeks to check sterility and then inoculated with test cultures. After two to three months of incubation the solid material was filtered off, both fractions were hydrolyzed by autoclaving in 6N HCl in sealed tubes, and hydroxyproline determined by the method of Neuman and Logan (1950). Since this amino acid is unique to collagen this is an effective measure of collagen breakdown.

Table 1 shows the results with some representative cultures. Several of them required salt for growth, all but one could grow in a 23% salt solution, all attacked gelatin more or less strongly, and most made some attack on limed collagen; several attacked it very extensively. None of these pure cultures was able to attack unlimed collagen.

Results with mixtures of these pure cultures are given in Table 2. Here again, limed collagen was attacked by most of the combinations but only a few were able

Table 2. Enhanced Activity Shown by Various Mixed Cultures
(from Everett and Cordon, 1956)

Composition of Culture Mixtures	Total No. of strains	Activity* in 15% Salt Broth + Substrate	
		Limed[+] Collagen	Unlimed[‡] Collagen
Uninoculated controls	0	10-20%	3-9%
A Gram-pos. rods	19	47	10
B Gram-neg. rods	5	32	5
C Gram-pos. cocci	30	40	5
D Gram-var. red cocci	10	15	8
E Gram-var. irreg. rods	3	42	8
Group Combinations			
A + B	24	65	8
A + C	49	29	-
A + D	29	22	-
A + E	22	41	11
B + C	35	37	5
B + D	15	21	6
B + E	8	91	11
C + D	40	24	-
C + E	33	62	14
D + E	13	58	8
A + B + C + D + E	67	93	15
Selected Mixtures			
No. 19 (B) + E	4	88	36
No. 104 (E) + B	6	93	40

*Collagen solubilization: hydroxyproline in solution as per cent of total found.
[+]Center layer of cowhide soaked in saturated lime liquor; chemically sterilized.
[‡]Center layer of cowhide carefully purified without liming; chemically sterilized.

to degrade the unlimed collagen. These results were surprising in view of the degradation which sometimes takes place in salt-cured hides. It appears that suitable combinations of organisms under favorable growth conditions may be required to effect this type of action. The effect of mild hydrolysis of collagen by liming on its susceptibility to microbial attack is clearly shown.

UNHAIRING

The almost universal method of removing the hair from animal hides consists of a treatment with saturated lime containing a "sharpening" agent, such as sodium sulfide, arsenic sulfide, sodium sulfhydrate, or amines. We have just seen how treatment of collagen with lime renders it susceptible to microbial attack. Hormann and Schubert (1958) have shown that treatment of collagen with lime causes some solubilization of a component which is rich in hydroxyproline. Fifteen percent of the nonammoniacal nitrogen in the solubilized portion is hydroxyproline, compared with only 6.9% in whole collagen. It appears that a hydroxyproline-rich portion of the molecule must have been attacked. This lime-treated collagen was subjected to the action of trypsin and compared with unlimed material. Trypsin was able to attack the limed but not the unlimed material.

Grassmann and Hannig (1958) have studied the effect of the denaturation of collagen by mild heat on its susceptibility to attack by trypsin and chymotrypsin. Both of these enzymes were able to attack the heated collagen but not the unheated. Here again a mild denaturation has resulted in making collagen susceptible to the action of biological agents.

In order to alleviate waste disposal, shorten the processing time, and produce a better leather, we have been investigating the depilatory action of enzymes on hides and skins. We found very soon that freshly flayed hides were not nearly so readily unhaired by enzymes as were salt-cured hides or hides which had been soaked in dilute salt solution. Beuchler and Lollar (1949) have shown that salt alone has a depilatory action under certain conditions. If fresh hide is allowed to stand for several days in a 2-10% salt solution containing a disinfectant to prevent microbial growth, the entire epidermis becomes loosened and can be removed intact. Unfortunately, it is not loose enough for commercial unhairing. Figure 2 shows a piece of intact epidermis being removed and Fig. 3 shows the intact epidermis as it appears under the microscope. Note that the epidermal lining continues around the hair roots and that the sebaceous or oil glands are part of this system. Parts of the sweat glands can often be seen in such preparations.

We were very much interested in learning how salt made hides more susceptible to enzyme action and employed histological techniques to help elucidate this phenomenon. Burton, Reed, and Flint (1953) have postulated that the binding material which holds the epidermis to the dermis, sometimes called the basement membrane, is a mucopolysaccharide. We cut thin sections from untreated hide pieces or those soaked in water or salt solutions, and stained them with the periodic acid-Schiff stain which visualizes carbohydrate material. Figure 4 consists of photomicrographs of some of these sections. The section of untreated hide shows an intensely stained band at the juncture of the dermis and epidermis. After soaking in water some of the background staining material is removed, but the intense band

Fig. 2. Removal of intact epidermis from steerhide after soaking in dilute salt water.

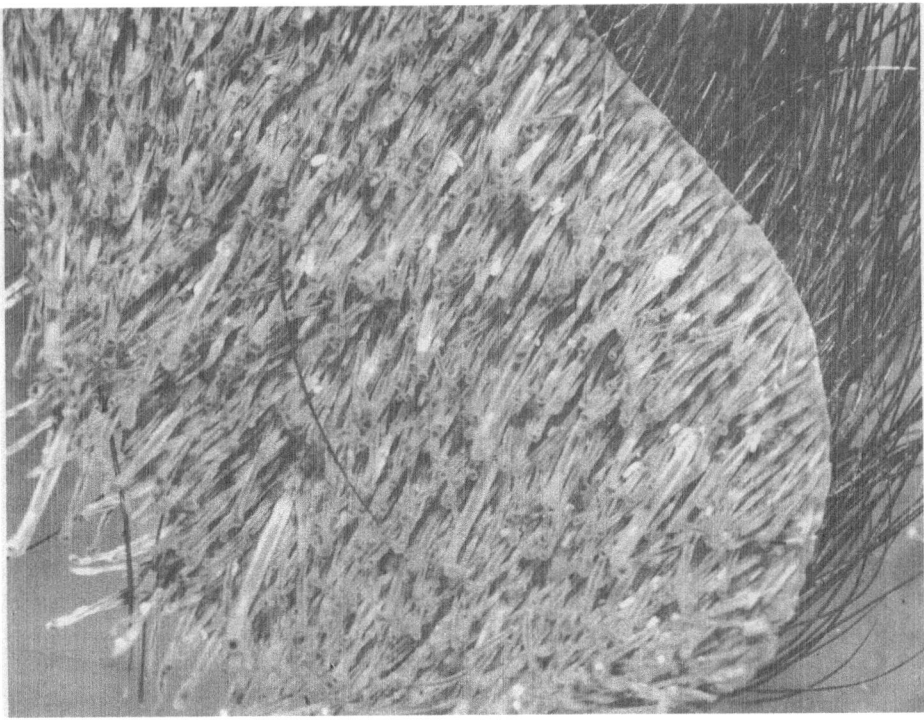

Fig. 3. Intact epidermis from steerhide showing details of its structure. (From Everett and Cordon, 1959.)

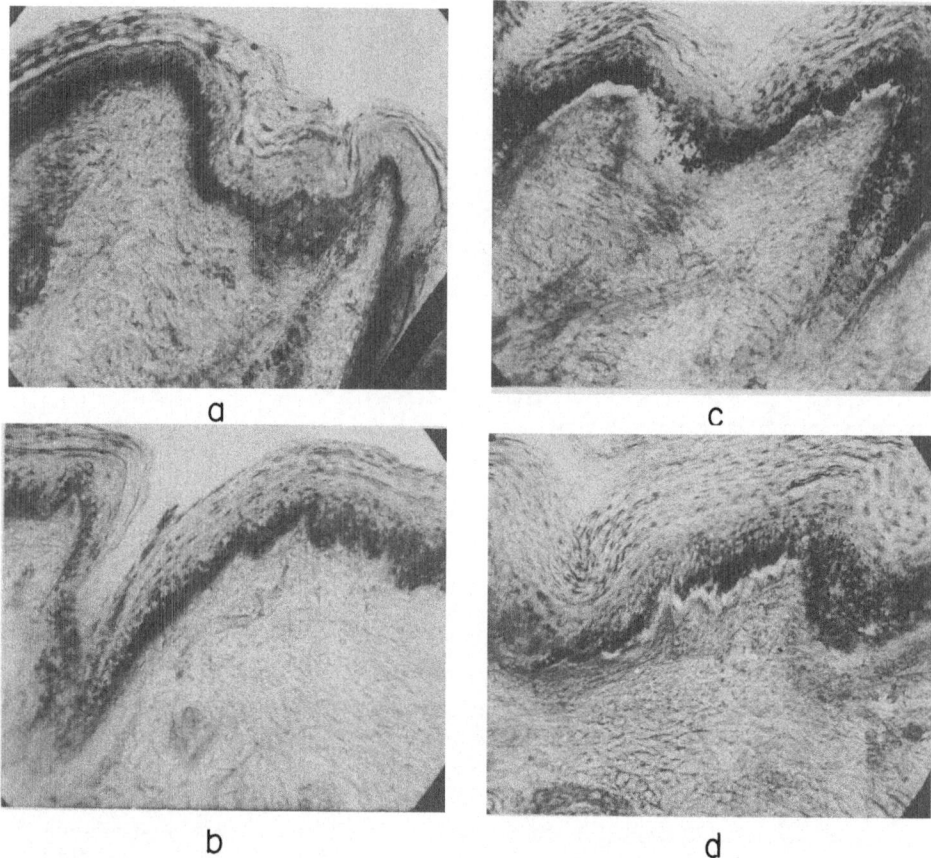

Fig. 4. Cross sections of fresh hide, PAS stain. X200. a) untreated; b) soaked in water; c) soaked in 4.8% salt water; d) soaked in 9.1% salt water. (From Everett and Cordon, 1958.)

still remains. After soaking in dilute salt the intensely stained layer is much more diffuse and the epidermis has separated from the dermis in some places. After treatment with depilatory enzymes this carbohydrate-staining material is completely removed. We believe that salt treatment causes a peptization or partial degradation of the bonding layer and thus conditions it for enzyme action.

As another approach to uncover the nature of the material which holds the epidermis to the dermis, we have tried to correlate the depilatory action of various enzymes with assays on known substrates (Cordon, T. C. *et al*., 1958). The data in Table 3 illustrate the results. Assay values for action on casein (PV), gelatin (formol titration), and starch (DV), are tabulated so that any relation to hair loosening could be observed. For several enzymes we have also measured gelatinase activity as a function of lowering the viscosity and keratinase activity on scoured wool. None of these activities showed a direct correlation with hair-loosening action, but in general enzymes showing good depilatory action also have high proteolytic activity.

Table 3.   Relation of Hair Loosening to Other Activities of Some
Selected Enzyme Preparations (from Cordon *et al.*, 1959)

Enzyme	PV(a) Units/g	Formol Titration(b) meq-N/g	DV(c) Units/g	Concentration Used g/100 ml	Hair Looseness P(d) R(e)
HT Concentrate(f)	190,000	46.0	95,000	0.1	7-84
HT Proteolytic(f)	331,000	41.8	9,200	0.06	7-83
L-56D(g)	100,000	35.5	4,500	0.19	7-84
Protease 15 Concentrate(h)	162,000	26.7	2,200	0.12	8-77
4511-3(i)	46,300	19.1	17,000	0.41	9-76
Papain	24,800	10.0	<300	0.77	3-97
Bromelin	46,200	46.3	<400	0.41	4-91
Pancreatin 3 X USP	66,800	36.6	16,300	0.28	10-25
Trypsin 4X USP Pancreatin	66,500	38.6	3,200	0.29	10-0

(a) Action on casein.
(b) Action on gelatin.
(c) Starch-dextrinizing power.
(d) Number of pulls with scraper. See Fig. 1.
(e) Estimation of percent of hair removed.
(f) Bacterial enzymes from the Takamine Laboratories.
(g) Bacterial enzyme from Pabst Laboratories.
(h) Bacterial enzyme from Röhm & Haas Company.
(i) Bacterial enzyme from Wallerstein Company.

The fibrous protein elastin has been suggested as being implicated in bonding the epidermis to the dermis and with this in mind we have assayed a number of depilatory enzymes for elastase activity. Figure 5 shows photographs of cross sections of hide after enzyme treatment and staining with the Fullmer-Lillie elastin stain. The elastin fibers are stained black and run roughly parallel to the surface of the hide. In some cases enzyme treatment completely removes these fibers.

Some of the enzymes which showed strong depilatory action also show strong elastase activity but here again the relationship to hair loosening is only casual and no direct correlation could be made. It appears that a mucopolysaccharide material may be involved in cementing the epidermis to the dermis; at least such a material is present before but not after the action of depilatory enzymes. Such enzymes also degrade elastin but no direct relation to hair loosening has been proved.

SUMMARY

It is concluded that during the processing of hides and skins in preparation for tanning many degradative processes take place. Some of these are necessary for the removal of unwanted constituents and some are required to condition the hide for tanning.

Some of these changes are undoubtedly brought about by halophilic bacteria existing in the salt-cured hides during storage. This represents a complex problem of microbial ecology. For it has been shown that undenatured collagen is very

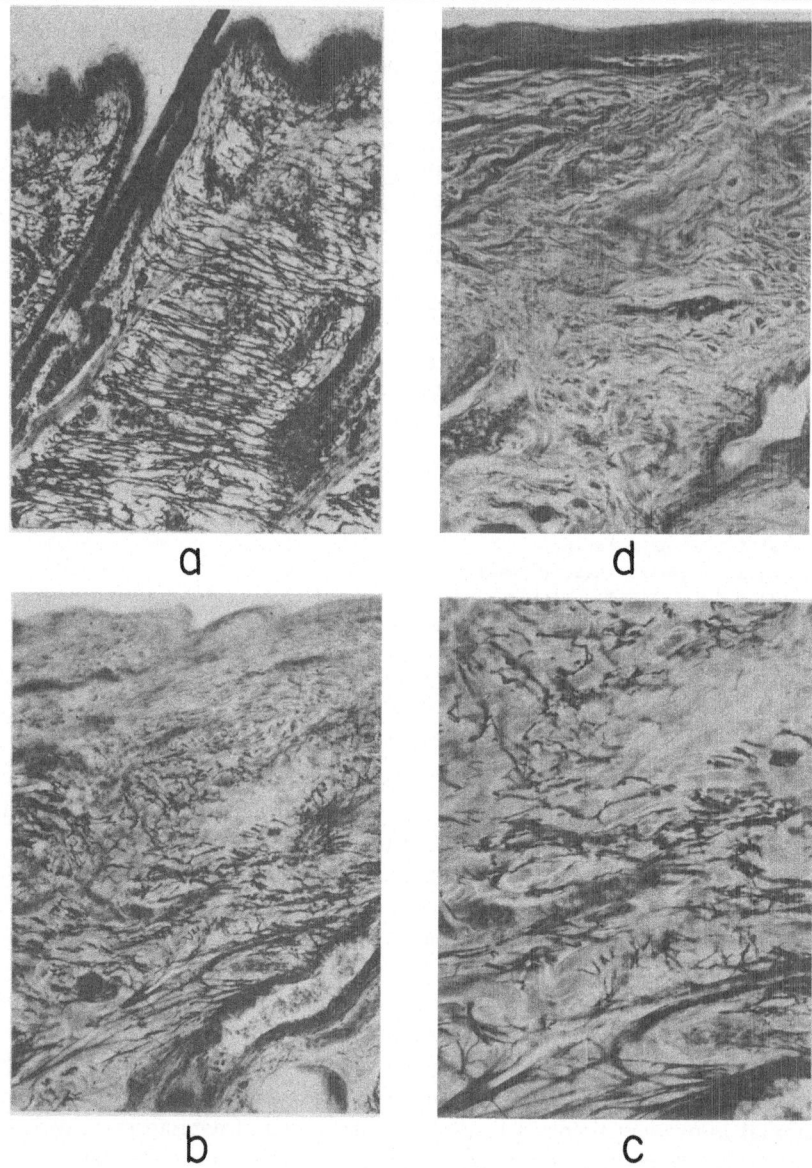

Fig. 5. Cross section of steerhide, stained for elastin by Fullmer-Lillie method; magnificauon 100 ×
for a, b, and d, 200× for c. a) untreated control; b and c) treated with papain solution for about 7 hr.
Illustrates incomplete removal of elastin. d) treated with papain solution overnight. Illustrates almost complete removal of elastin. (From Everett et al., 1959.)

resistant to attack by single pure cultures, while degradation could be obtained
with certain mixtures. However, collagen that has been denatured by treatment
with lime, is susceptible to attack even by single cultures. Similarly, trypsin and
chymotrypsin can degrade heat-denatured or lime-treated collagen but not native
collagen.

REFERENCES

Beuchler, P. R., and Lollar, R. M. 1949 Some sulfur distribution studies on cattle hair and epidermis. J. Am. Leather Chem. Assoc., 44, 359-370.

Burton, D., Reed, R., and Flint, F. O. 1953 The unhairing of hides and skins without lime and sulfide. The use of mucolytic enzymes. J. Am. Leather Trades' Chemists, 37, 82-87.

Cordon, T. C., Jones, H. W., Clarke, I. D., and Naghski, J. 1958 Microbial and other enzymes as depilatory agents. Appl. Microbiol., 6, 293-297.

Cordon, T. C., Windus, W., Clark, I. D., and Naghski, J. 1959 A progress report on the enzyme depilation of cattle hides. J. Am. Leather Chem. Assoc., 54, 122-139.

Everett, A. L., and Cordon, T. C. 1956 Resistance of purified collagen to degradation by salt-tolerant bacteria in pure culture. J. Am. Leather Chem. Assoc., 51, 59-55.

Everett, A. L., and Cordon, T. C. 1958 The use of salt to assist enzymatic unhairing of fresh hides, with histochemical observations. J. Am. Leather Chem. Assoc., 53, 548-567.

Everett, A. L., Cordon, T. C., Kravitz, E., and Naghski, J. 1959 *In situ* histological evaluation of elastase activity. Stain Technology, 34 (6), 325-330.

Grassmann, W., and Hannig, K. Cited in Hormann and Schubert, Das Leder 9.

Hormann, H., and Schubert, B. 1958 Chemical changes of hide substance under liming and bating conditions. Das Leder 9, 265-274.

Maxwell, M. E. 1945 Fellmongering investigations. IV Bacteria responsible for the loosening of wool on sheepskins. C.S.I.R. Australia, Bull. No. 184, 89-116.

Neuman, R. E., and Logan, M. A. 1950 The determination of hydroxyproline. J. Biol. Chem. 184, 299-306.

DETERIORATION OF WOOD BY LOWER FUNGI

Dr. Catherine G. Duncan, presiding

Panelists: Dr. S. P. Meyers
Dr. E. S. Reynolds
Dr. Elwyn T. Reese

INTRODUCTION

C. G. Duncan

The title of our symposium might well imply that we are to discuss the role of the Phycomycetes in the deterioration of wood, since taxonomically they are the lower fungi. However, we are to be concerned with the Ascomycetes and Fungi Imperfecti which are in the taxonomic sense, the higher fungi. Rather than "lower fungi," the term "microfungi" has been used by some. These two groups, until recently, have been considered relatively unimportant in wood deterioration since they were believed incapable of attacking cellulose as it occurs in combination with lignin in wood. Only the Basidiomycete group of higher fungi were believed to contain members which readily attacked lignocellulose.

I will first review the status of research on the terrestrial wood-inhabiting Ascomycetes and Fungi Imperfecti which cause the type of deterioration we recognize as soft rot. Dr. Samuel Meyers will then discuss the importance of representatives of these two groups that deteriorate wood in marine habitats.

DETERIORATION OF WOOD BY TERRESTRIAL ASCOMYCETES AND FUNGI IMPERFECTI

C. G. Duncan

Forest Products Laboratory,[*] Forest Service, U.S. Department of Agriculture

It has been known for some time that certain Ascomycetes and Fungi Imperfecti are capable of attacking cellulose and nonlignified materials. For instance:

1. A great many can attack cellulose plant fibers such as cotton, causing serious destruction of fabrics.

2. Others may grow through wood, living on stored food materials, such as starches and sugars in the ray cells, and thereby stain and discolor the wood.

3. Others not only utilize the food material within the ray cells but also the pectin and hemicelluloses of the walls. When the cells are dissolved, porosity is increased, and the wood becomes more absorptive of water and preservatives. It is of interest to note here that certain bacteria also are capable of dissolving the ray cells in logs stored in ponds for the purpose of preventing stains and decay (Ellwood and Ecklund, 1959). Increased absorptiveness may cause considerable drying and paint-maintenance problems. Increased porosity in wood can, on the other hand, be of value to the wood preserver. In experiments directed toward improving the penetration of preservatives, chemicals, such as sodium fluoride, have been sprayed on logs to stimulate the growth of molds, particularly Trichoderma (Lindgren, 1952).

4. A few bark-inhabiting Ascomycetes are known to decay small branches and twigs. Their attack on the lignified walls of tracheids, vessels, and fibers, however, has been regarded as unimportant.

Since the work of Hartig in 1874, important deterioration of cellulose, as it occurs with lignin in the tracheary cells and fibers of wood, has been considered to be the province of a relatively small group of basidiomycetous fungi, namely the Agaricales, and perhaps a few of the larger Ascomycetes, such as *Daldinia* and *Xylaria*. Descriptions of the manner in which these Basidiomycetes attack the wood have been characteristically similar. That is, they penetrate the cell wall transversely, through pits or by forming their own channels, called bore holes, and proliferate in the cell lumina.

However, even before a definite relationship was established between the Basidiomycetes and decay, Schacht in 1863, Wiesner in 1864, and Dippel in 1898, observed fungal hyphae that did not characteristically penetrate the wall transversely but grew longitudinally within the walls. The fungi responsible for this type of cell-wall attack remained unknown for many years.

Bailey in 1913 called further attention to such fungi, and noted that they form cavities within the thick, secondary walls of pine. Subsequently, he and Vestal

[*]Maintained at Madison, Wis., in cooperation with the University of Wisconsin.

(1937) frequently observed these cavities in tracheary cells and fibers during their anatomical comparisons of a wide range of woods from diverse environments. Their observations, published in 1937, characterized these fungi by their ability to grow within the secondary wall, where they enzymatically dissolved the wall substance to form cavities that were oriented either helically around or parallel to the long axis of the cell. The arrangement of the cavities suggested to them that hydrolysis proceeded along planes determined by the structural orientation of the cellulose. D. H. Linder thought that the responsible fungi might be Pyrenomycetes, or imperfect stages of this group, since he had observed a similar type of attack in maple by a *Brachysporium* species. A few years later, Barghoorn and Linder (1944) showed that wood in marine habitats was attacked in a like manner, and that Fungi Imperfecti and Ascomycetes isolated from the wood were capable of attacking cellulosic substrates in pure culture.

No serious attempt, however, was made to investigate this type of decay until about ten years ago when the problem arose as to the cause of deterioration in industrial cooling towers where water flows over wooden slats. This deterioration was first assumed, at least by us in the United States, to be entirely of a chemical nature. The Basidiomycetes were not suspected since it was known that these fungi could not develop in wood that is submerged or thoroughly soaked in water, where the oxygen supply under such conditions is below their minimum needs. Findlay and Savory (1950, 1954) at the Forest Products Research Laboratory in England were among those investigating deterioration in cooling towers. They found that fungi growing within cell walls were invariably present near the surface of deteriorated slats in cooling towers. They also began to find this type of attack in wood that was in contact with wet soil, and not uncommonly in wood long exposed to the atmosphere even though it was wet for only short periods of time. Savory was soon able to demonstrate in laboratory tests (1954a, 1954b, 1955) that some of the Ascomycetes, notably *Chaetomium*, and a few Fungi Imperfecti isolated from cooling towers and wood in contact with wet soil were capable under suitable conditions of causing decay of hardwood species. Since the surface of the wood, especially where it had been wet for a number of years, was typically softened, the term "soft rot" was applied. This term has since come into general use for referring to the type of decay produced by fungi that grow within the walls of wood. (This is in contrast then to the terms "white rot" and "brown rot" caused by the Basidiomycetes.)

The discovery that some of the Ascomycetes and Fungi Imperfecti were capable of causing decay, at least under special environmental conditions of extreme wetness or of frequent dryness--where basidiomycetous fungi cannot survive--aroused considerable interest. It was soon recognized that they might also be responsible for some failures of treated wood in the ground. Anyone who has attempted to isolate fungi from decaying wood, finds many Fungi Imperfecti. Heretofore, these have been considered harmless casual associates of the wood-destroying Basidiomycetes. At the Forest Products Laboratory in Madison we have been attempting to determine the prevalence and practical significance of these nonbasidiomycetous fungus inhabitants of wood. In this connection, considerable attention has been given to some of their physiological characteristics. A detailed report of this investigation will be published in a few months (Duncan, 1960), but the results of this study will be briefly summarized under the separate topics of the discussion to follow.

ENVIRONMENTS IN WHICH SOFT ROT HAS BEEN OBSERVED

Seemingly, soft rot can be found almost anywhere one chooses to look now that we know what it is. Dr. Bailey in 1937 found the decay we now call soft rot in many hardwoods and softwoods, over 100 species, that came from many different places. We are, however, more apt to find soft rot in places where nature has provided suitable conditions for the more rapid growth of the fungi such as (1) very wet wood and (2) wood subjected to either constant leaching or only occasionally wet-- such as is typical near the surface of exposed items. Leaching not only removes extractives contributing to decay resistance but also may loosen the lignocellulose bond.

A cooling tower adequately provides these conditions and thus soft rot in cooling towers has become an economic problem. Soft rot is also found in boats and piling in fresh water, in wood in contact with wet soil such as the belowground portion of fence posts, test stakes, and telephone poles, and in greenhouse benches. It is found in warm, humid places such as humidity rooms and the tropics. Soft rot is not uncommon, though it develops slowly, on wood exposed to the atmosphere where it is only occasionally wet but, nevertheless, may be freely leached. Such wood occurs, for example, in the tops of fence posts and telephone poles, in old rail fences, and in unpainted siding of houses and barns. It probably is usually associated with most "weathered wood."

WOOD SPECIES ATTACKED BY SOFT ROT FUNGI

No actual survey has been made but it appears that soft rot is more prevalent in hardwoods than in softwoods, although both are subject to attack. In nature, no wood apparently is entirely resistant to attack, since very resistant woods such as redwood, greenheart, locust, yew, oak, and many of the durable woods of the tropics have been found to be soft-rotted. Pines, gums, poplar, hickory, beech, etc. are woods that have been personally observed to be quite susceptible to soft-rotting.

MACROSCOPIC APPEARANCE OF SOFT-ROTTED WOOD

In nature, the surface of wood submerged in water, or subjected to excessively wet conditions for a number of years, becomes darkened and softened if attacked by soft-rot fungi. When still wet, the softened surface can easily be scraped away to reveal relatively sound wood. When the wood is dried it usually retains its shape and may appear normal, though it breaks with a brash fracture. In wood which has been subjected to very wet conditions involving leaching, small cracks develop across the grain in the darkened surfaces and give an appearance similar to that of lightly charred wood.

MICROSCOPIC APPEARANCE OF SOFT-ROTTED WOOD

In its early stages, soft rot is difficult to detect without the aid of a microscope. Under the microscope it has a characteristic appearance and is so unlike the white or brown rot decay of Basidiomycetes that it can be recognized quite easily. In the type of deterioration known as soft rot, the hyphae ramify within the cell wall and make cavities that run longitudinally in the direction of the cellulose fibrils. These cavities in longitudinal sections of the wood appear as spirals or lie parallel to the long axis of the vessels in hardwoods. In polarized light (Aaron and Wilson, 1955), the hyphae can be seen to lie within cavities having pointed ends. [Roelofsen (1956) offers an explanation of corrosion figures.] These cavities are confined to the less-lignified secondary walls of tracheary cells and fibers, and are more conspicuous in the summer- than in the springwood, especially in softwoods. [Electron microscope studies made by Meirs (1955,1957).] In later stages, the hyphae may become enlarged and thick walled. In cross section, the cavities appear as holes that equal or exceed the diameter of the hyphae. Some of the soft-rot fungi dissolve the ray cells, grow abundantly in the lumen, while others penetrate the wall transversely, forming bore holes, a characteristic of basidiomycetous attack. In this connection it might be mentioned that we have found that one of the Basidiomycetes, *Poria nigrescens*, which decays the drier redwood structural members in cooling towers, may attack the wall in a manner similar to soft-rot fungi.

GENERA OF FUNGI IMPERFECTI AND ASCOMYCETES CAUSING SOFT ROT

Savory, in England, has observed that several species of *Chaetomium* can cause considerable rot in beech, but little attack of Scotch pine sapwood. Likewise he has found that some Fungi Imperfecti species of *Trichurus, Bispora, Stysanus, Stemphylium,* and *Coniothyrium,* also can decay beech but to a lesser degree than *Chaetomium* . Since only *Chaetomium* seemed to cause soft rot with an acceptable degree of reproducibility, it has been favored by Savory to represent the soft-rot fungi in his studies.

At Madison, we have isolated more than 150 fungi from soft-rotted wood, the majority of which have produced soft rot under laboratory conditions. Several of these appear to have a greater capacity, under our test conditions, to attack wood than *Chaetomium*. The majority of fungi we have isolated are Fungi Imperfecti, but their taxonomic place within this group has been difficult to determine since only vegetative stages were available in most cases. Certain of the organisms undoubtedly represent new species.

It appears that most of the Fungi Imperfecti that produced a soft-rot condition in wood under laboratory conditions are in the order Moniliales. These have tentatively been placed in such general as: *Acremonium, Alternaria, Bisporomyces, Cephalosporium, Chalaropsis, Cylindrocarpon, Diplococcium, Haplochalara, Helicosporium, Helminthosporium, Hormiscium, Hymenella, Nematogonium, Phialophora, Pullularia, Sporocybe, Stysanus, Torula,* and *Trichurus.* A few of the Fungi Imperfecti, which were among the most destructive, are in three other orders: Sphaeropsidales, tentatively of the genera *Coniothyrium, Cytospora, Cytosporella* , and *Phoma*; Melanconiales, represented by *Pestalozzia;* and Mycelia Sterilia, represented by *Sclerotium* . *Fusarium, Penicillum,* and *Trichoderma* of the order Moniliales frequently occurred in all isolation attempts, but these isolates have not been found as yet to attack the secondary walls of wood and, therefore, could not be classed as soft-rotters.

Among the isolates that were capable of causing soft rot, an occasional Ascomycete also was found. There were three species of *Chaetomium* (*C. cochliodes, C. funicolum,* and *C. globosum*) and one species each of *Xylaria* and*Orbicula. Chaetomium* is seldom isolated from cooling towers in the United States, but we commonly isolate it from soft-rotted wood in soil contact.

PRODUCTION OF SOFT ROT UNDER LABORATORY CONDITIONS

It was known from our isolation work that the majority of fungi isolated were cellulolytic since they grew well on a mineral vitamin agar with filter paper as the carbon source. Therefore, the assessment of the destructive capacity of the fungi from a practical standpoint was considered to be the capacity of the fungi to attack cellulose as it occurs with lignin in wood. Various-sized wood blocks or veneer strips of sweetgum, beech, or pine sapwood, and redwood heartwood were subjected to pure cultures of approximately 100 of the fungi isolated from soft-rotted wood. The results of these tests indicated that a majority of the isolates were capable of causing substantial decay in the sweetgum and beech, and about 50% of them were also capable of attacking pine although not as readily as gum. Leached pine, however, was attacked by about as many isolates as attacked the hardwoods, although generally not as severely. (We have indicated that in nature one is more apt to find soft rot in hardwoods than in softwoods.) None of the isolates decayed either normal or leached redwood heartwood unless it was buried in wet soil for a long time.

Several factors contributed to an increase in the amount of soft rot:
1. Minerals and vitamins, applied either to the wood or in an agar substrate, or additional organic matter in the soil, generally increased soft-rot attack.
2. A mineral-vitamin agar provided a more favorable condition for soft-rotting than wet soil probably because it provided better moisture conditions in the test block and nutrient ingredients necessary to cellulolytic activity.
3. Leaching the wood in hot or cold water alone led to a somewhat greater attack of pine sapwood but not of redwood. The presence of chlorine in the leach water, however, increased the susceptibility of both pine and redwood to soft-rot attack.

The chlorination effect would indicate that the presence of chlorine in the water of redwood cooling towers may accelerate attack of the wood by soft-rot organisms. Possible explanations for this accelerated attack are that chlorine contributes to the removal of toxic extractives or changes the lignocellulose complex in wood to make the cellulose more susceptible. Whether the elimination of chlorine would substantially reduce the soft-rot problem in cooling towers is not yet known. Since redwood buried in wet soil for a long time was decayed slowly, it is probable that over a longer period the leaching by water alone would lead to gradual soft rot.

The soil-block technique (ASTM, 1956) which has been used so successfully with the basidiomycetous fungi was not satisfactory for testing soft-rot isolates. This failure was attributed to the drier soil and the use of a feeder strip of wood between the test wood and soil, both of which reduced the wetness of the sample to be tested.

A shake culture technique appeared to have promise for rapid appraisal of soft-rot potentials of fungi on susceptible hardwoods, but not on coniferous woods.

There was a close correlation between losses in weight and decrease in bending tolerance [determined on mandrels (Scheffer and Duncan, 1944)] of infected wood; consequently, either appeared to be suitable as a measure of the relative amounts of attack by the soft-rot fungi. Since the use of bending radii necessitates particular care in the selection of test material, however, weight determinations generally seemed preferable.

Armstrong and Savory (1959) indicated that in beech decayed by *Chaetomium globosum* there was (1) a sharp decrease in resistance to impact loading (measured by total work, or the energy absorbed/total fracture in bending) before significant weight losses occurred in the wood, similar to decay by Basidiomycetes, (2), a gradual reduction in the ultimate bending strength, but it was not significant until definite weight losses were evident, and (3) the effect on impact resistance was more like that of white than of brown rot caused by Basidiomycetes.

In appraising a potential soft-rotter, microscopical examination of the wood proved to be a necessary supplement to the other observations, especially in cases where there was little weight loss.

PHYSIOLOGICAL CHARACTERISTICS OF THE SOFT-ROT FUNGI

We have made several preliminary studies to determine the general nature of the soft-rot isolates with respect to commonly investigated physiological characteristics. The four preliminary physiological studies were directed at (1) temperature relations, (2) pH relations, (3) tolerance to toxic materials, and (4) oxidase activities of the isolates. Thirty-two of the isolates, selected to represent the various groups, were used in the studies.

Temperature relations

Twelve per cent of the isolates showed a temperature optimum of 22 C; 41% of 28 C (also the optimum for most wood-destroying Basiodiomycetes); 41% of 34 C; and 6% of 38 C. From these data and those gathered by Humphrey and Siggers (1933) for 64 basidiomycetous fungi, it was apparent that the temperature optima tend to be considerably higher among the soft-rot fungi than among the wood-destroying Basidiomycetes.

Among the soft-rot isolates, there were also many more from cooling towers than from soil that had relatively high temperature maxima. Thus, a great many of the cooling-tower fungi may represent an ecological group with a greater tolerance of higher temperatures as well as a greater capacity for rapid growth at temperatures above 30 C. This may be significant in view of the higher temperatures of water that passes through cooling towers.

pH relations

As to pH relations, our results indicated that about 40% of the isolates failed to grow at pH 3. All grew at pH 4 and 5, however, and apparently reached their maximum rate of growth at pH 6. The optimum tended to extend broadly into levels above pH 6; about half of the isolates maintained their maximum rate of growth at pH 7 and a fourth of them at pH 8. A pH of 9 had a pronounced retarding effect on all the isolates, but growth was not prevented.

When the results are compared with what is known about pH relations of Basidiomycetes, it can be said that both the soft-rot isolates and the basidiomycetous wood-destroyers commonly have a growth optimum that includes pH 6, but only among the soft-rot fungi does the optimum commonly extend to a much higher pH. Most of the Basidiomycete isolates seem to be inhibited by a mildly alkaline substrate [pH between 7 and 8 (Wolpert, 1924)].

Relative tolerances to some wood-preserving chemicals

Savory (1955) and Scholles (1957) have indicated that *Chaetomium globosum* was considerably more tolerant to some preservatives than standard test Basidiomycetes. There are also some indications in nature that the soft-rot fungi might be more tolerant to preservatives than Basidiomycetes. For instance, we find soft-rot fungi on the surface of telephone poles and test stakes impregnated with retentions of preservatives that inhibit decay by Basidiomycetes.

Results of our toxicity tests, made to determine the minimum concentration of a given preservative in malt agar needed to inhibit growth of the fungus, indicated that (1) there was a considerable range of tolerances among the soft-rot isolates for most of nine preservatives tested, (2) at least one-fourth, and generally considerably more, of the soft-rot isolates were more tolerant of sodium fluoride, sodium arsenate, zinc chloride, and possibly creosote than were nine Basidiomycetes, and (3) none to only a few isolates were more tolerant of mercuric chloride, copper sulfate, or sodium pentachlorophenate than the most tolerant Basidiomycete. Relative tolerances to sodium borate were not clear, because the lowest concentration used (0.5%) inhibited most fungi.

Oxidase production

A knowledge of the oxidase-producing capacity of a fungus gives us some indication of their rot-producing potential. It has also been a valuable taxonomic tool, along with microscopic and macroscopic characters, for separating the Basidiomycetes into the white-rot and brown-rot groups (Davidson et al., 1938). In one of the several tests (Bavendamm, 1928) used to determine this capacity, the white-rot but not the brown-rot fungi are capable of oxidizing tannic or gallic acid in the medium, causing a brown diffusion zone. From our studies, the indications are that some of the Ascomycetes and Fungi Imperfecti which produce soft rot also react positively to the oxidase test, while others do not. We have not yet determined whether there is any relationship between the oxidase-producing capacity of the soft-rot fungi and the visible characteristics of the rot produced by them. There are indications that wood with soft rot may undergo chemical changes characteristic of both the brown- and white-rotted wood.

CHEMICAL NATURE OF SOFT-ROTTED WOOD

Recently Savory and Pinion (1958) have made a chemical analysis of beech-wood decayed by *Chaetomium globosum* (*C. globosum* is oxidase-negative and therefore like a brown-rot Basidiomycete). They found that the carbohydrates were virtually depleted while the lignin, although it decreased steadily, was not markedly attacked. In this respect, the wood was similar to that with brown-rot, and unlike white-rotted wood—in which much of the lignin as well as the carbohydrates are prominently utilized. The alkali solubility of the wood, on the other hand, increased slowly, which is a characteristic of wood undergoing white rot. Brown rot typically causes a rapid increase in alkali solubility, indicative of an accumulation of primary degradation products of cellulose.

REFERENCES

American Society for Testing Materials 1956 ASTM D1413-56T. Tentative Method of Testing Wood Preservatives by Laboratory Soil-Block Cultures. Supplement to book of ASTM Standards, Part 4, p. 142-155.

Aaron, J. R., and Wilson, K. 1955 Soft rotting in timber. The use of the polarizing microscope. Wood 20, 186-189.

Armstrong, F. H., and Savory, J. G. 1959 The influence of fungal decay on the properties of timber. Effect of progressive decay by the soft-rot fungus, *Chaetomium globosum*, on the strength of beech. Holzforschung 13, 84-89.

Bailey, I. W. 1913 The preservative treatment of wood, I. The validity of certain theories concerning the penetration of gases and preservatives into wood. Forestry Quarterly 11, 5-11.

Bailey, I. W., and Vestal, M. R. 1937 The significance of certain wood-destroying fungi in the study of the enzymatic hydrolysis of cellulose. Jour. Arnold Arboretum 18, 193-205.

Barghoorn, E. S., and Linder, D. H. 1944 Marine fungi: their taxonomy and biology. Farlowia 1, 395-467.

Bavendamm, W. 1928 Über das Vorkommen und den Nachweis von Oxydasen bei holzzerstörenden Pilzen. Zuschr. Pflanzenkrank. u. Pflanzenshutz 38, 257-276.

Davidson, R. W., Campbell, W. A., and Blaisdell, D. J. 1938 Differentiation of wood-decaying fungi by their reactions on gallic and tannic acid medium. J. Agr. Research 57, 683-695.

Dippel, L. 1898 Das Mikroskop und die Anwendung des Mikroskopes. II. Teil. Anwendung des Mikroskopes auf die Histologie der Gewachse. Braunschweig, F. Vieweg u. Sohn.

Duncan, C. G. 1960 Wood-attacking capacities and physiology of soft-rot fungi. Forest Products Laboratory Report No. 2173. 1-28.

Ellwood, E. L., and Ecklund, B. A. 1959 Bacterial attack of pine logs in pond storage. Forest Products Journal 9, 283-292.

Findlay, W. P. K., and Savory, J. G. 1950 Breakdown of timber in water-cooling towers. Int. Bot. Congr. Proceedings 7, 315-316.

Findlay, W. P. K, and Savory, J. G. 1954 Moderfäule. Die Zersetzung von Holz durch niedere Pilze. Holz als Roh-und Werk-stoff 12, 293-296.

Hartig, R. 1878 Die Zersetzungsercheinungen des Holzes der Nadelbäume und der Eiche. Berlin.

Humphrey, C. G., and Siggers, P. V. 1933 Temperature relations of wood-destroying fungi. Jour. Agri. Res. 47, 997-1008.

Lindgren, R. M. 1952 Permeability of southern pine as affected by mold and other fungus infection. Proc. Amer. Wood-Preservers' Assoc. 48, 158-168.

Meir, H. 1955 Über den Zellwand Abbau durch Holzvermorschungs-pilze und die submikroskopische Struktur von Fichten Tracheiden und Birkenholzfasern. Holz als Roh-u. Werkstoff 13, 323-328.

Meir, H. 1957 Discussion of the cell wall organization of tracheids and fibers. Holz-forschung 11, 41-46

Roelofsen, P. A. 1956 Eine mögliche Erklärung der typischen Korrosionsfiguren der Holzfasern bei Moderfäule. Holz als Roh-und Werkstoff 6, 208-210.

Savory, J. G. 1954a Damage to wood caused by microorganisms. Journ. Appl. Bacteriology (London) 17, 2-3-218.

Savory, J. G. 1954b Breakdown of timber by Ascomycetes and Fungi Imperfecti. Annals of Appl. Biology 41, 336-347.

Savory, J. G. 1955 The role of the microfungi in the decomposition of wood. Rec. Brit. Wood Pres. Assoc. 5, 3-19.

Savory, J. G., and Pinion, L. C. 1958 Chemical aspects of decay of beechwood by *Chaetomium globosum*. Holzforschung 12, 99-103.

Schacht, H. 1863 Jahr. fur Wiss. Botanik 3, 442-483.

Scheffer, T. C., and Duncan, C. G. 1944 Breaking radius of discolored wood in aircraft veneers. Forest Pathology Special Release No. 22.

Scholles, W. 1957 Über die pilz-und insektenwidrigen Eigenschaften von Naphthensäuren und Metallnaphthenaten als Wirkstoffe in Holzschutzmitteln. Holz als Roh-und Werkstoff 15, 128-137.

Wiesner, J. 1864 Über die Zerstörung der Hölzer an der Atmosphare. Sitzungsber d. k. Akad. d. Wiss. Wien. 49, 61-94.

Wolpert, F. S. 1924 Studies in the physiology of the fungi. XVII. The growth of certain wood-destroying fungi in relation to the H-ion concentration of the media. Ann. Missouri Bot. Gard. 11, 43-97.

CELLULOLYTIC ACTIVITY OF LIGNICOLOUS MARINE ASCOMYCETES AND DEUTEROMYCETES[*]

Samuel P. Meyers and Ernest S. Reynolds

The Marine Laboratory, University of Miami, Miami, Florida[**]

The stimulus of marine mycological research in recent years has initiated the development of aspects of this discipline of direct and considerable interest to workers in marine biology as well as to those in fields concerned with deterioration and microbial physiology. The subject selected for this symposium--the cellulolytic activity of wood-inhabiting marine fungi--has been the focus of much attention since the early studies of Barghoorn and Linder (1944) wherein the occurrence of an abundant fungal biota on submerged wood was first recognized. Since 1944, the amount of literature dealing with marine mycology has increased greatly (Johnson and Meyers, 1957), and has noted the existence of a vast nutritional and ecological array of fungi in the sea. These vary from fastidious stenohaline representatives of the Myxomycetes and Phycomycetes, many of which are parasites on marine plants and animals, to members of the Ascomycetes and Deuteromycetes, a large number of which attack submerged wood and appear to be active in various decomposition processes. With the exception of the Basidiomycetes, it is quite probable that most, if not all, of the large or main fungal groups are represented in the sea.

In our laboratory as well as elsewhere in the United States and in various other parts of the world, marine mycologists are examining the nature of the wood-inhabiting marine mycota, with ever increasing emphasis on the physiological activities of this population, its contribution to marine productivity, and the possible relationship of these microorganisms to associated wood-attacking animals. Considerable speculation has been raised regarding the latter problem, resulting in much controversy as to whether woodboring animals, including species of *Teredo* and *Limnoria*, have an actual nutritional dependency on marine fungi or their metabolic products (Becker, *et al.*, 1957; Ray and Stuntz, 1958; Kohlmeyer, 1958; Kamp, *et al.*, 1959). This problem is far from resolved and only more intensive basic work on the physiology and nutritional requirements of the lignicolous marine fungi and the wood-destroying animals will clarify many of the present conflicting views.

Since 1952, with the extensive cooperation of scientific and naval personnel, The Marine Laboratory has been studying the occurrence of wood-inhabiting marine fungi at various marine sites throughout the United States, Canada, Alaska, Nova Scotia, Newfoundland, the Caribbean Sea, and the Canal Zone. A report of this fungal distribution, especially in the northern marine areas will be presented in a subsequent publication. However, certain noteworthy ecological aspects can be mentioned here.

[*] These studies were supported by ONR Project No. 103-305, Microbiology Branch, Office of Naval Research.
[**] Contribution No. 267 from The Marine Laboratory, University of Miami.

Differences in the dominant mycota between northern and southern environments are apparent, especially among the Deuteromycetes. Various genera and species extremely prevalent in warmer oceanic areas have not been found in colder marine environments. Our studies of fungal attack at northern sites including Kodiak, Alaska, and Halifax, Nova Scotia, have indicated the abundance of an active ascomycetous and deuteromycetous mycota during periods of water temperature of 0 C to 15 C. Often, reproduction by species of *Lulworthia* and *Ceriosporopsis* occurred on wood submerged during periods of water temperature of less than 5 C. The presence of this significant fungal population in arctic and northern marine areas necessitates a consideration of their diverse degradative and synthetic processes in subsequent investigations of the biota of these localities.

The use of special "traps" composed of panels of yellow pine and basswood along with the development of methodology, including extremely effective incubation techniques (Meyers and Reynolds, 1958), has permitted us to isolate and subsequently secure in pure culture a large and diverse number of ascomycetous and deuteromycetous genera and species. Incubation of the infested wood panels has indicated considerable fungal attack upon the panels in addition to that found at the initial examination, and has facilitated a more complete evaluation of the *total* lignicolous mycota, as well as establishing the early infestation of the wood. It is possible that the results obtained by Ray and Stuntz (1958), wherein fungal attack on submerged wood could not be demonstrated, may be attributed to their failure to incubate the freshly collected wood. Studies in our laboratory on many thousands of panels submerged from Alaska to Panama, and the results of investigators throughout the world, have shown repeatedly that a diverse marine mycota colonizes and attacks submerged wood. Frequently, vigorous degradation of the wood surfaces by reproductive structures of these fungi can be observed quite readily immediately following removal of the test panel.

Further anatomical evidence of the degradative activities of these fungi, in addition to the presence of surface and imbedded reproductive structures, includes: (1) softening and disintegration of the outer wood tissues, often to a depth of several millimeters, (2) proliferation of the fungal hyphae throughout the wood, including ramification within the lumina of the tracheids and the wood rays (Fig. 1), and, (3) direct penetration of the fungi through the walls of the wood elements (Fig. 2).

In the paper given before this society last year, we presented evidence for the occurrence of active cellulolytic mechanisms in various species of marine fungi. Cell-free filtrates from fungi comprising the significant part of the mycota on submerged wood, incubated under specified conditions on such cellulosic materials as carboxymethylcellulose, powdered cellulose, and balsa wood, have produced from 0.075 to 1.5 mg reducing sugars per milliliter. These organisms develop readily in a sea-water medium containing 0.1% yeast extract and balsa wood strips (Fig. 3). Wood and wood products incorporated in culture media have a stimulative effect on perithecial production in species of Ascomycetes studied (Meyers and Reynolds, 1959b).

In current work, Manila twine, a lignocellulose material comparable to wood in proximate analysis (Hessler and Merola, 1949), has been used to evaluate the

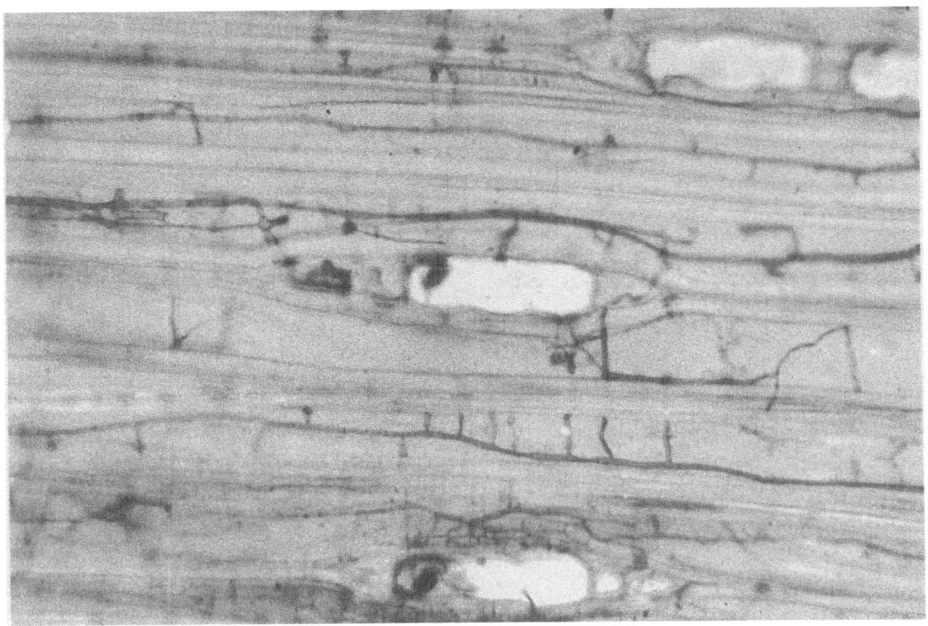

Fig. 1. Ramification of fungal hyphae within the lumina of yellow pine.

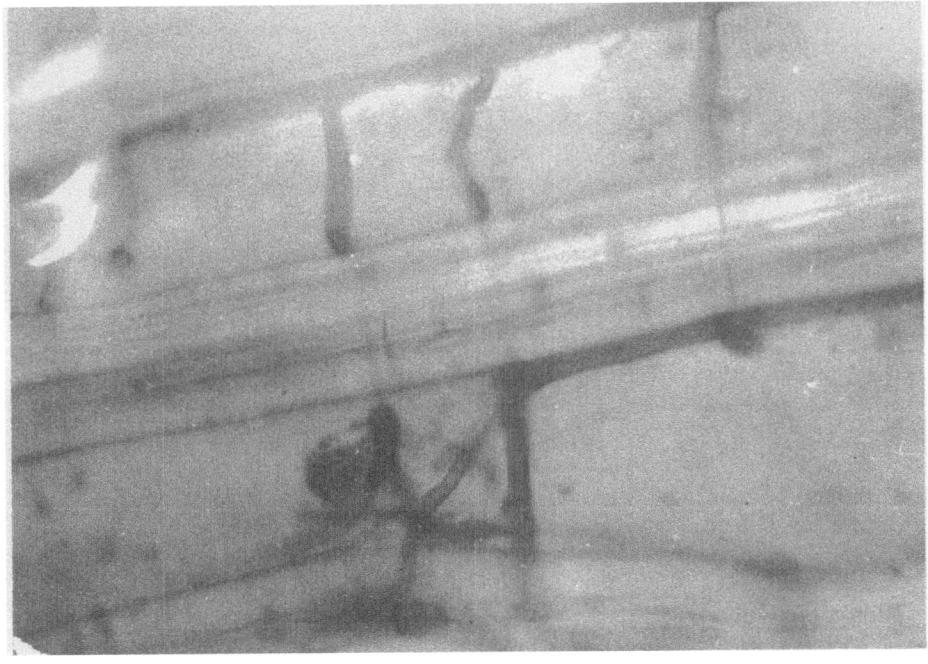

Fig. 2. Direct penetration of fungal hyphae through the wood element walls. (Note the constriction of the hyphae as they pass through the wall itself.)

Fig. 3. Infestation of balsa wood strips in yeast-extract sea-water vessels. Uninoculated flask is in center.

Fig. 4. Fernbach flasks containing five coils of Manila twine submerged on yeast-extract sea-water broth. The flask on the left has been inoculated with *Culcitalna achrospora*.

enzymatic activity of the marine fungi in their degradation of vascular tissue resulting in a loss of strength of the substrate. We have selected this material since it can be readily standardized and tested, can be autoclaved in a nutrient medium without the formation of compounds toxic to fungal growth, and lastly, but of no minor importance, is quite susceptible to fungal attack and supports vigorous reproduction. Another consideration, to be applied in our future work, is the availability of individual fibers of Manila for detailed studies of the mode of fungal attack and subsequent degradation of the plant tissues.

In the exposure of the Manila twine to uniclonal fungal attack, five coils, each 10 ft long, were placed in 2-L Fernbach flasks containing 1100 ml of 0.1% yeast-extract sea-water broth (Fig. 4). The Manila was 100% grade F, non-Davao (Tagaon) fiber, taken from standard lots from the Plymouth Cordage Company, Plymouth, Massachusetts.

Prior to inoculation, the flasks containing the Manila yeast-extract medium were sterilized at 121 C for 20 min. The inoculum for the culture vessels consisted of 5 to 10 ml of spore and mycelial suspensions from individual stock cultures, prepared by flooding the slant cultures with sterile sea water. The fungi used in these tests are listed in Table 1.

Inoculated flasks were incubated at 20 to 25 C for 23 to 78 days, and were harvested when vigorous fungal growth was evident on the twine and throughout the liquid. Each culture vessel was examined for the extent of vegetative growth and reproduction, cellulolytic activity of the cell-free filtrate, and residual reducing sugars present, and tensile strength measurements of the infested twine.

In certain tests, the coils of twine on removal were numbered 1 to 5 to indicate their relative position in the culture vessel, i.e., No. 1 coil was at the surface of

Table 1. A List of the Species of Ascomycetes and
Deuteromycetes Used

Fungus	Reference No.
Ascomycetes	
Torpedospora radiata Meyers	F-187
Lulworthia floridana Meyers	F-190
Ceriosporopsis halima Linder	F-189
Antennospora quadricornuta (Cribb & Cribb) Johnson	F-186
Lignincola laevis Höhnk	F-191
Peritrichospora integra Linder	F-192
Peritrichospora sp.	F-99
Halosphaeriopsis sp.	F-115
Deuteromycetes	
Culcitalna achraspora	F-118
Humicola alopallonella	F-123
Piricauda arcticoceanorum Moore	F-73
Helicoma sp.	F-150, 151
Alternaria sp.	F-109
Cremasteria cymatilis	F-63
Unidentified marine sp.	F-29
Pestalotia sp.	F-260

the liquid while the Nos. 3-5 coils, and occasionally No. 2, were completely sub-merged. This method was developed to ascertain if maximum fungal growth on the twine, as determined visually, was correlated with low tensile strength read-ings. The coils were dried at 37 C for 8 hr or overnight and shipped to the Plymouth Cordage Company for testing.

A Model Q Scott-Horizontal Testing Machine, using the 500 lb scale, was used to determine the tensile strength of the twines. The latter is expressed here as the lb "pull" recorded at the actual breaking point of the fiber. Fifty feet of twine were used for each test, providing material for ten breaks, i.e., two breaks on each of the five 10-ft coils. A significant loss of strength, expressed as "residual tensile strength," is indicated whenever the breaking point of the test twine is below 200 lb.

The filtrate from each Manila vessel was collected separately and filtered through an 03 Selas filter candle to insure a completely cell-free medium. The analytical procedure for the determination of cellulolytic activity, including the amounts and treatment of the cellulosic preparations, Walseth cellulose (Walseth, 1952) and Carboxymethylcellulose, CMC 50 T, has been described previously (Meyers and Reynolds, 1959a, 1959c). The Nelson-Somogyi method (Somogyi, 1952) was used to determine the reducing sugars, RS, of the fungal filtrates and the RS produced during the enzymatic tests. The tests with Walseth cellulose, using 100 mg amounts, were run for a 24 hr period, while those with 1.0% CMC were run for 1 and 3 hr. All tests were at 50 C, and except for specific pH studies, at pH 6.0. The latter were adjusted with M/15 Sorensen's buffer made isotonic with sea water by the addition of NaCl. The amount of glucose in the hydrolysates and in the fungal filtrates was measured with the Glucostat Micromethod.[*]

TENSILE STRENGTH TESTS OF MANILA TWINE

A grand average tensile strength reading of 260 lb was obtained for the four uninoculated sets, or controls, subjected to sterilization in the yeast-extract sea-water medium for 20 min at 121 C. Further analyses were made of the effect of the autoclaving process *alone* (without soaking of the fiber in the nutrient medium) upon Manila. Fifty breaks were made with unsterilized twine and 50 with sterilized twine. This material was dried and tested in the same manner as that used for the inoculated twines. Measurements of tensile strength, expressed as the average of 50 breaks, indicated that this sterilization process, as well as that noted above, had *no apparent affect* on the physical properties of the Manila. Furthermore, no significant difference in residual tensile strength was observed between twine soaked in the yeast-extract sea-water medium for as long as 40 days and material removed and tested immediately after sterilization in the nutrient medium. The grand average tensile strengths of samples of twine attacked by species of Asco-mycetes and Deuteromycetes are presented in Figs. 5 and 6 respectively. The age of individual cultures, in days, is noted in parentheses.

All of the fungi reduced the strength of the twine to less than 200 lb, thus in-dicating considerable degradation of the fiber. The cordage in the 23-day culture

[*] Worthington Biochemical Corp., Freehold, New Jersey.

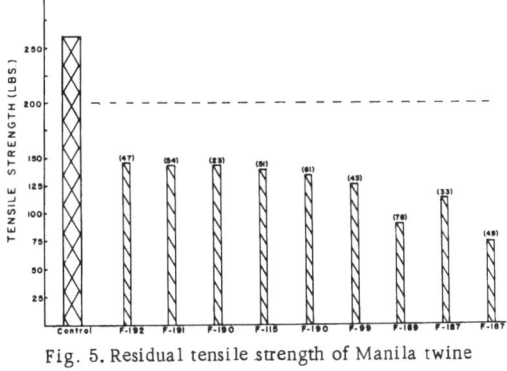

Fig. 5. Residual tensile strength of Manila twine
attacked by species of Ascomycetes. Control
twine at left.

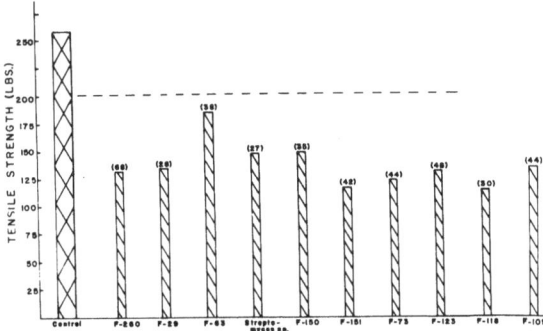

Fig. 6. Residual tensile strength of Manila twine attacked
by species of Deuteromycetes and a species of Streptomyces. Control twine at left.

of F-190 showed nearly the same grand average tensile strength as did that in
the 61-day culture of the fungus. However, in the latter vessel, the pH of the filtrate was 5.6 to 5.8 while that of the filtrate in the 23-day culture was adjusted to
7.5 before inoculation. It is possible that these differences in the initial pH of the
filtrates in the two tests affected growth and cellulolytic activity of the organism
allowing the 23-day culture to develop cellulase well in advance of the 61-day test.
The vigorous attack upon the twine apparent in the 49-day culture of F-187 and in
the 78-day culture of F-189 is quite noteworthy. Both of these ascomycetous
species exhibited considerably greater fungal activity, in terms of loss of tensile
strength, than any of the other Ascomycetes and Deuteromycetes examined. Only
very slight variation in lb breaking-strength was noted among the five coils attacked by F-187 in the 49-day culture. However, in the 78-day culture of F-189,
coil No. 1, which showed the most vigorous fungal growth and perithecial production, broke at 9 lb, while coil No. 5, showing only a slimelike development, had
a breaking point at 95 lb. In other species, the ratio of coil 1 to coil 5, an expression of differences in residual tensile strength, fell between 1:1 and 1:2, thus
indicating greater degradation in the surface, or partly exposed coil.

Similar to the Ascomycetes examined, the twine attacked by the deuteromycetous species, with the exception of F-63, had a breaking point of 150 lb or less.
Two nonmarine isolates, F-260 and a species of *Streptomyces*, grown in uniclonal
culture for comparative cellulolytic and strength determinations, also exhibited
a reduction in the strength of the twine.

Though the fungal-infested twine was harvested when extensive growth was
indicated, variability in the nature and extent of fungal development on the twine
within individual flasks was evident. Twine at the surface of the liquid, usually
coils 1 and 2, showed vigorous mycelial growth with extensive discoloration of the
fiber (Fig. 7). Fruiting structures, when present, occurred in this area of infestation. It appears likely that maximum cellulase production and activity is most
prevalent in such areas of greatest fungal reproduction. The submerged fiber,
coils 3 to 5 and occasionally 2 to 5, exhibited a more scattered fungal attack,
gelatinous in nature, and frequently without evident discoloration. Indication of
this extreme variability in fungal attack between coils is apparent in the wide
range of maximum and minimum readings within individual strength tests (Fig.
8). The 80 lb range of the control indicates the inherent variability in the test of
the 50 ft of uninoculated twine itself. It is hardly possible to ensure uniform fungal
attack on all pieces of twine in the culture vessel, especially within a total length
of 50 ft coiled in the bottom of the flask. Nevertheless, since all of the strength
determinations of a particular fungus were made from material within a single
flask, the relative minimal tensile strength reading of the inoculated twine validly
represents the maximal attack or "cellulolytic potential" of the organism.

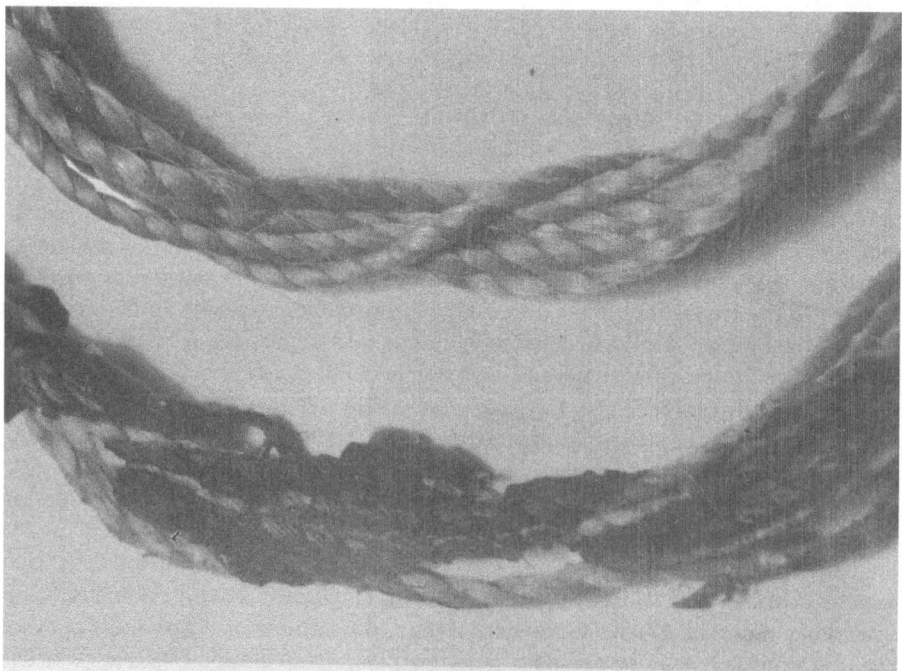

Fig. 7. An infested coil showing vigorous mycelial growth. Uninoculated coil is at top.

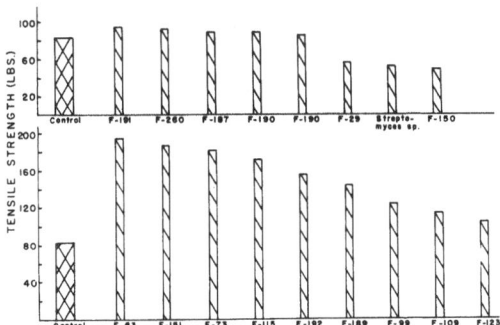

Fig. 8. Range of tensile strength readings of twine
attacked by various fungi. Control twine at left

CELLULOLYTIC DETERMINATIONS

Along with strength measurements, tests were made of the cellulolytic activity of the filtrate at the time of harvesting of the infested twine. Figure 9 presents cellulase activity of selected cell-free fungal filtrates on Walseth cellulose, while Figs.10 and 11 characterize enzymatic activity on carboxymethylcellulose, CMC 50 T. In general, there was a greater total RS production by the deuteromycetous species, especially F-109, F-123, and F-260, than by the ascomycetous isolates. The percent of glucose in the RS fraction from Walseth cellulose varied considerably, from 10 to 39% for the Ascomycetes to 17 to 100% for the Deuteromycetes examined. While it is not possible to compare directly the cellulolytic activity of the fungi on Walseth cellulose with that on CMC 50 T, in general, a high RS production on the former substrate was associated with a high RS production on the CMC 50 T. Using RS production on the latter material as a criterion of enzymatic activity, F-260 would appear to have greater activity than F-109 or F-123. However, on the Walseth cellulose, the latter two fungi showed considerably greater amounts of RS than F-260.

Selected fungal filtrates were incubated with Walseth cellulose at various pH levels, from 5.3 to 7.0. Figure 12 notes the total amount of reducing sugars (RS) produced, while Figs. 13 and 14 respectively record the milligrams of glucose in the RS fraction and the per cent of glucose represented. F-190, F-73, and F-151 all exhibited slightly greater cellulolytic activity at the lower pH levels. F-73 and F-151 showed a reduction in the per cent of glucose in the RS fraction as the pH increased, while F-190 and F-123 exhibited a gradual increase in the per cent of glucose with increasing pH, followed by a decrease in the per cent of glucose from pH 6.2 to 7.0. It is interesting that F-123 showed nearly identical RS production over the entire pH range tested. Similarly, F-187 exhibited excellent RS production from pH 5.4 to 7.0, with maximum at pH 6.0. In F-187, the per cent of glucose was identical over the entire pH range, with a slight increase at pH 6.0. The glucose content of the RS fraction in F-123 was approximately 100% (from 85 to 100%), while that of the RS fractions of the other fungi varied from 11 to 76%.

A direct comparison of enzymatic activity and residual tensile strength is not possible, especially since the recorded cellulolytic activity of the fungal filtrates

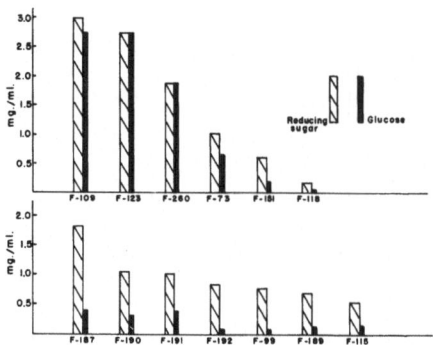

Fig. 9. Total reducing sugars and glucose produced by fungal filtrates on Walseth cellulose. Ascomycetes in lower section.

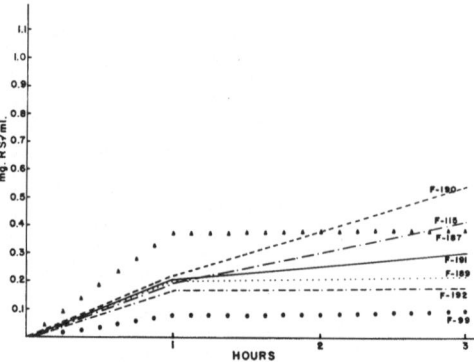

Fig. 10. Total reducing sugars (RS) produced by ascomycetous fungal filtrates on CMC-50 T for 1 and 3 hr.

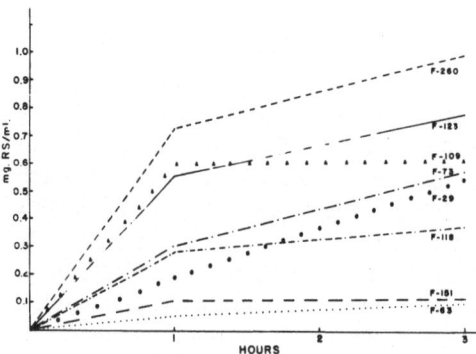

Fig. 11. Total reducing sugars (RS) produced by deuteromycetous fungal filtrates on CMC-50 T for 1 and 3 hr.

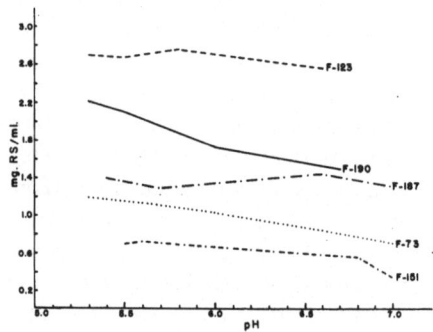

Fig. 12. Total reducing sugars (RS) produced by selected fungal filtrates on Walseth cellulose at different pH levels.

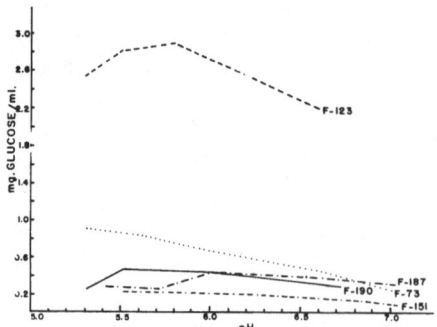

Fig. 13. Glucose content of the RS fraction produced by selected fungal filtrates on Walseth cellulose at different pH levels.

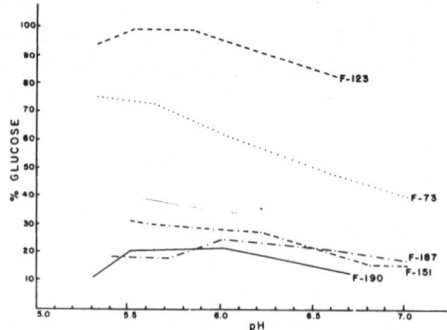

Fig. 14. Percent of glucose in the RS fraction produced by selected fungal filtrates on Walseth cellulose at different pH levels.

is specifically that of the culture medium in which the fungus had developed and excludes the role of the enzymes present within the fungal mycelia on and within the infested twine. It is conceivable that a total homogenate of the deteriorated twine, fungal mycelia, and culture medium would reveal considerably greater enzymatic activity than has been indicated in the present tests.

These tests are only the beginning of more extensive work planned to evaluate the participation of marine fungi in general biological processes, including decomposition. In particular, we wish to know more about the synthetic activities of these organisms, especially the predominant metabolites produced during growth. Since microbial deterioration is basically a multiphasic phenomenon with interrelationships and various degrees of complexities, these considerations are of paramount importance. The need for careful studies of the interrelationships among the marine organisms which infest submerged wood cannot be overemphasized. Such studies, if they are to present a *complete* picture of fouling or deterioration, must consider the contribution of the wood-inhabiting marine fungi.

SUMMARY

Enzymatic studies of pure cultures of various genera and species of wood-inhabiting marine Ascomycetes and Deuteromycetes, developing on Manila twine in vessels containing 0.1% yeast extract and sea water, have revealed extensive activity accompanied by significant loss in the tensile strength of the fiber. An increase in total reducing sugars, together with the specific production of glucose, from carboxymethylcellulose and Walseth cellulose, have been used as indices of the fungal hydrolysis of cellulose and cellulosic materials.

Loss in fiber strength varied considerably with different fungal species examined, and within individual tests correlations between maximal reproduction and strength reduction were observed. Because of the close similarity of cordage and wood in proximate analysis, except for the lower lignin content of the former, it is considered that the notable loss of strength of cordage when used as a growth substrate for the various lignicolous sea-water fungi is a positive demonstration of the destructive action of these organisms on wood fibers.

ACKNOWLEDGMENTS

The writers are indebted to Dr. Bryce Prindle, Plymouth Cordage Company, Plymouth, Massachusetts, for his invaluable cooperation and assistance in supplying the Manila twine and for his determinations of the tensile strength of the material. Similarly, we wish to acknowledge the assistance of Mr. Joe Levi and Mr. Steve Zellner of The Marine Laboratory in the analytical determinations.

REFERENCES

Barghoorn, E. S., and Linder, D. H. 1944 Marine fungi; their taxonomy and biology. Farlowia, 1, 395-467.

Becker, G., Kamp, W.-D, and Kohlmeyer, J. 1957 Zur Ernährung der Holzbohrasseln der Gattung Limnoria. Naturwiss., 44, 473-474.

Hessler, L. E., and Merole, G. V. 1949 Determination of cellulose in cotton and cordage fiber. Anal. Chem., 21, 695-698.

Johnson, T. W., Jr., and Meyers, S. P. 1957 Literature on halophilous and halolimnic fungi. Bull. Mar. Sci. Gulf and Caribbean, 7, 330-359.

Kamp, W.-D., Becker, G., and Kohlmeyer, J. 1959 Versuche über das Auffinden und den Befall von Holz durch Larven der Bohrmuschel *Teredo pedicellata* Qutrf. Zeitschrift für Angewandte Zoologie, 17, 257-283.

Kohlmeyer, J. 1958 Holzzerstörende Pilze im Meerwasser. Holz. Roh- u. Werkstoff, 16, 215-220.

Meyers, S. P., and Reynolds, E. S. 1957 Incidence of marine fungi in relation to wood-borer attack. Science, 126, 969.

Meyers, S. P., and Reynolds, E. S. 1958 A wood incubation method for the study of lignicolous marine fungi. Bull. Mar. Sci. Gulf and Caribbean, 8, 342-347.

Meyers, S. P., and Reynolds, E. S. 1959a Growth and cellulolytic activity of lignicolous Deuteromycetes from marine localities. Canad. J. Microbiol., 5, 493-503.

Meyers, S. P., and Reynolds, E. S. 1959b Effects of wood and wood products on perithecial development by lignicolous marine Ascomycetes. Mycologia, 51, 138-145.

Meyers, S. P., and Reynolds, E. S. 1959c Cellulolytic activity in lignicolous marine Ascomycetes. Bull. Mar. Sci. Gulf and Caribbean, 9, 441-455.

Ray, D. L., and Stuntz 1958 Possible relation between marine fungi and *Limnoria* attack on submerged wood. Science, 129, 93-94.

Somogyi, M. 1952 Notes on sugar determination. J. Biol. Chem., 195, 19-23.

Walseth, C. S. 1952 Occurrence of cellulases in enzyme preparations from microorganisms. Tappi, 35, 228-235.

Vishniac, H. S. 1955 Marine mycology. Trans. N.Y. Acad. Sci., 17, 352-360.

FUNGAL METABOLISM AND FUNGICIDES

Dr. Saul Rich, presiding

Panelists: Dr. Elwyn T. Reese
Dr. Mary Mandels
Dr. Robert G. Owens
Dr. Hugh D. Sisler
Dr. Bradner W. Coursen

INTRODUCTION

Saul Rich

The Connecticut Agricultural Experiment Station, New Haven, Conn.

Microbiologists wage constant battle against rots that rob man of billions of dollars yearly. These rots, caused principally by fungi, are largely combated by the use of antimicrobial chemical compounds. The search for compounds poisonous to fungi has been the primary task of many microbiologists, who not only seek the best ways to use presently available fungitoxicants, but also continue to search for more effective protecting agents.

Although such research appears to be entirely empirical, actually it is based on knowledge of how fungi perform. We label as routine the tests designed to determine the response of fungi to compounds which we hope will be fungitoxic. Two hundred years ago, the knowledge to design these tests was not available. No one knew then that rots were caused by fungi, let alone which particular fungi were involved. Rotting was caused by "miasmas" or "humors." The visible evidence of fungi, such as sporophores or mycelial wefts, were symptoms of rotting, rather than clues to the cause of rotting. The identity of the specific fungi which cause rots, and how to culture these fungi is now such common knowledge that we tend to forget how recent this knowledge is in man's history. Before it is tested adequately by scientific methods, such information is in the realm of the mysterious; afterwards it may become so commonplace that its scientific origins are forgotten.

The choice of test compounds, for the most part, has been by random selection. The great range of compounds, however, from which microbiologists can choose, is based on the tremendous growth of scientific information in the physical sciences. Not only have we had the choice of new and old synthetic compounds, but recently antimicrobial materials produced by the fungi themselves have been added to the list. From this intensive testing have come great masses of data. Now we

169

are attempting to integrate the vast amount of available information, both chemical and biological, in an effort to extrapolate and predict. What must we know about fungi that will allow us to pick out the weak links in their life processes? What must we know about fungitoxicants which will allow us to predict more efficient fungitoxic structures? Answers to these questions are the aim of the papers to be presented in this symposium.

Only within the past two decades have sufficient strides been made in defining metabolic routes and in the techniques used to study them, to make possible the kinds of papers that follow in this symposium. These studies would not have been possible 20 years ago.

The title of the symposium is "Fungal Metabolism and Fungicides." The papers presented by Owens and by Sisler and Coursen are concerned with the metabolic changes brought about in fungi by fungitoxicants. Owens describes the effect of sulfur and organic sulfur compounds on the metabolism of a fungus. Sisler and Coursen discuss the effect of an antibiotic, cycloheximide, on fungus metabolism. Reese and Mandels examine the reverse, i.e., the effect of fungal metabolism on fungitoxicants. They describe how a fungus can produce an enzyme which will detoxify an otherwise fungitoxic compound. Fungi which produce this enzyme are not poisoned by the compound, while fungi which do not secrete this enzyme are killed. Detoxification is not the only way in which the metabolism of a fungus can change chemical compounds. Byrde and Woodcock (1953) report a fungus which makes a fungitoxic derivative out of a compound which is not originally toxic. Fungi producing this enzyme are poisoned, while fungi without this enzyme are not poisoned. Rich and Horsfall (1954) have also described the relation between the production of specific enzymes by fungi and the susceptibility of these fungi to poisoning by certain compounds. Hence, fungus metabolism is not only drastically altered by fungitoxicants, but also the enzymes produced by a fungus may well decide whether that fungus will be poisoned.

It is the hope of applied microbiologists that basic studies of the kind reported here will reduce further the degree of empiricism needed to choose more effective fungitoxicants.

REFERENCES

Byrde, R. J. W., and Woodcock, D. 1953 Fungicidal activity and chemical constitution, II. Compounds related to 2,3-dichloro-1,4-naphthaquinone. Ann. Appl. Biol., 40, 675-687.

Rich, S., and Horsfall, J. G. 1954 Relation of polyphenol oxidases to fungitoxicity. Proc. Nat. Acad. Sci. U.S., 40, 139-145.

GLUCONOLACTONE AS AN INHIBITOR OF CARBOHYDRASES

E. T. Reese and Mary Mandels

Pioneering Research Division, Quartermaster Research and Engineering Center
Natick, Massachusetts

One possible means of controlling microbial decomposition is through inactivation of the extracellular enzymes involved (Siu, 1951). This possibility was investigated for cellulose decomposition (Reese and Mandels, 1957). The conclusion reached was that it requires more of the inhibitor to inactivate the degrading enzymes than it does to kill the organism.

One of the interesting compounds tested was Δ-gluconolactone. It was found to be a strong inhibitor of fungal β-glucosidases, but with no action on cellulases. This selective action was recently confirmed for the enzymes in rumen extracts (Festenstein, 1957). Gluconolactone is thus a convenient tool for studying cellulase action in the absence of cellobiase. However, the extent of inhibition of β-glucosidase by gluconolactone depends on the source of the enzyme (Ezaki, 1940; Reese and Mandels, 1957; Conchie and Levvy, 1957; Findlay et al., 1958; Jermyn, 1959).

In the present work we have extended the studies with gluconolactone to other carbohydrases, and have tested other lactones as inhibitors. We have also examined a number of fungal enzyme preparations for the presence of gluconolactonase (Brodie and Lipmann, 1955) and have found a correlation between the presence of gluconolactonase and resistance to lactone inhibition.

In these studies the carbohydrates were hydrolyzed by appropriate enzymes in the presence and absence of lactones. Substrates were usually present at 2.5 mg/ml of assay solution. Lactones were added to give inhibitor/substrate (I/S) ratios of 0.01-1.0. The assays were carried out at pH 4.8-5.4 at 40 C for 1 hr. Products were determined as reducing sugar by the dinitrosalicylic acid method (Sumner and Somers, 1944) or as glucose by glucose oxidase. Results were expressed as Ratio $\frac{\text{Inhibitor}}{\text{Substrate}}$ for 50% inhibition. The gluconolactonase assay was based on the procedure of Brodie and Lipmann (1955). The increase in decomposition over the spontaneous rate was used as a measure of the enzyme activity.

PROPERTIES OF Δ-GLUCONOLACTONE AND RELATED COMPOUNDS

Sugar lactones are inner anhydrides of the corresponding sugar acids. They may be formed from either aldonic or uronic acids. Δ-Gluconolactone is structurally similar to glucose in its pyranose form; γ-lactone to glucose in its furanose form. On hydrolysis, the lactone ring is opened, and the gluconic acid no longer retains its strong resemblance to glucose (which in solution is predominantly in the ring form).

Solutions of aldonic acids or lactones equilibrate to mixtures of the free acid and the Δ- or γ-lactone. For gluconolactone the equilibrium is in favor of the acid.

	Initial Rotation $[\alpha]_D$	Final Rotation $[\alpha]_D$
D-gluconic acid	− 6.7°	+17.5°
D-glucono-γ-lactone	+67.5°	+17.7°
D-glucono-Δ-lactone	+66.0°	+15.8°

(Pigman, 1957, p. 305)

The γ-lactones with their five-membered rings are usually the most stable. The Δ-lactones with six-membered rings tend to be much less stable. The rate of spontaneous hydrolysis to the acid is a function of the pH (Fig. 1). Hydrolysis is very rapid about pH 7.0, and least rapid below pH 5.0. The Δ-gluconolactone has a maximum half-life of about 60 min under the test conditions, the γ-glucono-lactones a maximum half-life of 180 min. Other γ-lactones (galactono-, glucurono-, arabono- ribono-, etc.) have half-lives of over 100 hr.

The length of the half-life and its dependence upon pH are important factors to consider in using the lactones as enzyme inhibitors. It is obvious that the instability of these compounds precludes their use in experiments of long duration, or as practical control measures.

INHIBITION OF ENZYMES BY LACTONES

α- and β-Glucosidases

The extensive work of Conchie and Levvy (1957, 1958) and their collaborators in England has shown that glyconolactones are remarkably specific inhibitors of glycosidases. Cellobiase of rumen was almost completely inhibited by glucono-lactone. The delta (Δ) and gamma (γ) gluconolactones were equally effective. They found, in a general way:

gluconolactone inhibits both α- and β-glucosidases
mannonolactone inhibits both α- and β-mannosidases
galactonolactone inhibits β- more than α-galactosidases
N-acetylglucosaminolactone inhibits β- more than α-N-acetylglucosamini-
   dases

Our present work on glucosidases (Table 1) indicates that β-glucosidases are much more susceptible to the action of Δ-gluconolactone than are the α-glucosidases. This difference is usually quite large—about one hundredfold. There are, however, α-glucosidases which approach β-glucosidases in their degree of susceptibility, emphasizing again the importance of source of enzyme. Considering these data with those of Conchie, it appears that, in general, the β-enzymes are more susceptible to lactone inhibition than are the α.

The γ-gluconolactone is somewhat less inhibitory than the Δ-gluconolactone (Table 2). About three to five times as much of the γ-, as of the Δ-lactone is required to give the same degree of inhibition of β-glucosidases. Lactones other than the gluconolactones showed little ability to inhibit β-glucosidases.

The source of enzyme again may be observed to have a marked effect. Arabonolactone seems to have some activity (observed also by Conchie and Levvy). Unfortunately, xylonolactone was not available. Its similarity to gluconolactone suggests its use. The English workers, however, found that xylonolactone did not

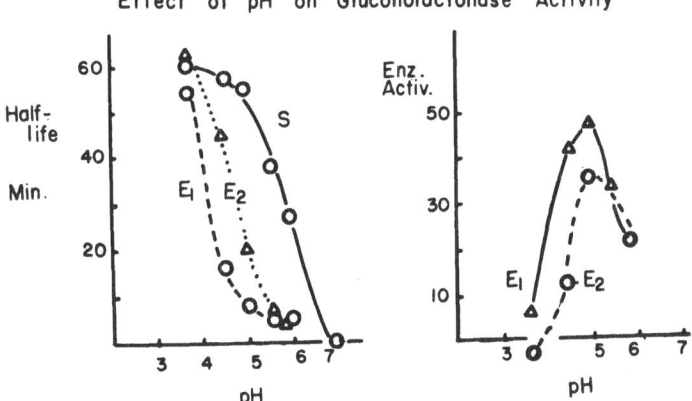

Fig. 1. Effect of pH on gluconolactonase activity. S = Spontaneous decomposition; E = Enzyme of Basidiomycete QM 806; E = Enzyme of *Pestalotiopsis westerdijkii* QM 381. Note: Enzyme activity is the difference between half-life of enzyme-treated and control (minutes).

inhibit the rumen β-glucosidases but did have a small effect on limpet β-glucosidase.

The products of a reaction often inhibit the rate of that reaction. Glucose thus is a competitive inhibitor of both α - and β-glucosidases.

$$\text{Saligenin } \beta\text{-D-glucoside + HOH} \xrightarrow{\text{E}} \text{Saligenin + Glucose}$$

We attribute gluconolactone inhibition to its similarity to glucose. Both compete with the substrate for the enzyme. The lactone is much more inhibitory, probably because it dissociates more slowly from the enzyme-lactone complex. We are at a loss to understand, however, *why* the β-glucosidases are so much more sensitive than the α-glucosidases to the action of the lactone.

Other Carbohydrases

We tested a large number of enzymes for gluconolactone inhibition. Some showed little or no inhibition at a ratio of inhibitor/substrate, R^I/S = 10. We consider an enzyme inhibited only if 50% inhibition occurs at R^I/S of 1 or less. The resistant enzymes included sucrase, β-1,4 glucanase (cellulase), β-1,4 xylanase, α-1,6 glucanase, most amylases, and most β-1,3 glucanases. The fungal sucrases resemble yeast invertase, a β-fructosidase, in their resistance to inhibition by gluconolactone. They require an R^I/S greater than 10 for inhibition, and are thus more resistant than the α-glucosidases tested (Tables 3, 1).

The inhibition of the susceptible β-1,3 glucanases (*Aspergillus sydowi, A. phoenicis*) is specific for the gluconolactones. Again, about three to five times as much of the γ-lactone is required to give the same degree of inhibition as of the Δ-lactone (data not shown). The following lactones were tested and found to be less than 1% as effective as Δ-gluconolactone: γ-galactonolactone, γ-ribonolactone, gulonolactone, arabonolactone, glucoheptonolactone.

Table 1. Ratio Gluconolactone (Δ)/Substrate (R^I/S) for 50% Inhibition of Glucosidases

Enzyme Source	QM	β-Glucosidase		α-Glucosidase
		Cellobiase	Gentiobase	Maltase
Aspergillus phoenicis	1005	0.014	0.04	2.5
Aspergillus luchuensis	873	0.005	0.01	1.0
Pestalotiopsis westerdijkii	381	0.06	0.3	10.+
β-Glucosidase No. 1 (Röhm & Haas)		0.03	0.024	10.+
Lipase (Gen. Bioch. Co.)		0.012	0.022	4.+
Poria cocos	7695	0.01	0.007	
Almond emulsin (Richtmyer)		0.025		
Trichoderma viride	6a	0.022		
Penicillium funiculosum	474			2.0
Hemicellulase				4.+
β-Amylase (Wallerstein)				4.+
Aspergillus clavatus	862			0.3; 0.6
Sclerotium rolfsii	7740			0.3
Linkage		β 1-4	β 1-6	α 1-4

Table 2. Comparative Effects of Lactones on β-Glucosidases (Salicin Hydrolysis)

Lactone	R^I/S to give 50% inhibition					
	E_1	E_2	E_3	E_4	E_5	E_6
Δ-Lactobionolactone	10.	10.	2.	10.	10.	6.
Δ-Gluconolactone	0.008	0.007	0.009	0.005	0.005	0.012
γ-Gluconolactone	0.04	0.02	0.05	0.015	0.02	0.04
γ-Mannonolactone	3.	5.	0.9	7.	10.	10.
γ-Galactonolactone	3.	5.	1.2	3.	10.	6.
γ-Gulonolactone	10.	10.	5.	10.	10.	10.
γ-Arabonolactone	1.7	1.7	6.	10.	1.2	1.6
γ-Ribonolactone	10.	10.	4.	10.	10.	10.

E_1 *Aspergillus phoenicis* QM 1005 E_4 *Poria cocos* QM 7695

E_2 *Aspergillus luchuensis* QM 873 E_5 *Aspergillus sydowi* QM 31c

E_3 Almond emulsin E_6 *Pestalotiopsis westerdijkii* 381

Table 3. Inhibition of Carbohydrases by Gluconolactone

Enzyme	Number of filtrates having 50% inhibition value at		
	R^I/S = <0.1	R^I/S = 0.1 -1.0	R^I/S = >1.0
Sucrase	0	0	9
β-Glucosidases	8	0	0
α-Glucosidases	0	2	8
β-1,4-Glucanase (cellulase)	0	0	13
β-1,3-Glucanase	2	1	27
α-1,4-Glucanase (amylase)	4	16	29
α-1,6-Glucanase (dextranase)	0	0	3
β-1,4-Xylanase	0	0	11

Table 4. Effect of Enzyme Source on Susceptibility to
Inhibition by Gluconolactone

Enzyme	Fungi having 50% inhibition at	
	R ^{I}/S = <0.1	R ^{I}/S = >1.0
β-1,3-Glucanase	*Aspergillus sydowi* 31c *Aspergillus phoenicis* 1005	*Rhizopus arrhizus* 1032 *Sporotrichum pruinosum* 826 Basidiomycete sp. 806
α-1,4-Glucanase (amylase)	*Aspergillus terreus* 82j *Mucor adventitious* 7576 *Mucor hiemalis* 7575 *Trichoderma viride* 6a	*Pestalotiopsis westerdijkii* 381 *Pestalotia bicolor* 664 Lipase (Gen. Bioch. Co.) β-Amylase (Wallerstein)

R ^{I}/S = ratio inhibitor to substrate.
Number after organism is QM Culture Collection number.

Table 5. Lactonase* Activities of Fungal Preparations

Organism or Preparation	Grown on	Lactonase	
		Δ-glucono-	γ-galactono-
Pestalotiopsis westerdijkii QM 381	Cellulose	17. μ/ml	1.4 μ/ml
Basidiomycete 806	Cellulose	17. "	0.7 "
Aspergillus luchuensis 873	Starch	3.6 "	NT
Penicillium funiculosum 474	Dextran	3. "	0.5 μ/ml
Penicillium notatum 944	Glucose	2.4 μ/mg	0.5 μ/mg
Protease, Lallouette (? *Aspergillus flavus*)	?	80. μ/vial	5.0 μ/vial
Glucose oxidase (Takamine)	?	12.5 μ/mg	1.4 μ/mg
β-Glucosidase No. 1 (Röhm and Haas)	?	1.2 "	0.5 "
No enzyme = spontaneous hydrolysis		(0.5)	(0.5)

*The values have *not* been corrected for control values indicated in last line.

Table 6. Use of Gluconolactone Inhibition to Characterize Linkage

Enzyme QM	Hydrolysis, %*			
	Cellobiose (β-1,4)		Maltose (α-1,4)	
	Alone	+ Lactone	Alone	+ Lactone
α-, β-Glucosidase mixture				
Aspergillus phoenicis 1005	20	6	22	22
Aspergillus luchuensis 873	54	13	76	74

*Substrate 2.5 mg/ml in M/20 citrate pH 4.5 40C 1 hour. Gluconolactone 0.1 mg/ml in reaction mixture.

Glucose is an inhibitor of these β-1,3 glucanases but at 100 times the concentration required for gluconolactone. Other simple sugars and glucosides are without effect. It is interesting that when glucose or gluconolactone concentrations are plotted against inhibition for various organisms, curves of similar shapes are obtained, indicating that glucose and gluconolactone may be inhibiting by the same mechanism.

The fact that some β-1,3-glucanases (and α-1,4-glucanases) are 100 times as susceptible as others to the inhibitory action of gluconolactone immediately suggests that the enzymes are acting by different mechanisms. A random-splitting enzyme might be expected to be less affected than an enzyme splitting off one glucose unit at a time. This, however, does not seem to be the explanation. In β-1,3-glucanases, we have found that of *Rhizopus arrhizus* to be a random splitter, and that of Basidiomycete 806 to act in an endwise manner. Yet both are quite resistant to the action of gluconolactone. Of course, it may be that the susceptible enzyme systems are working by a third mechanism different from either of the above. This would be compatible with the observed differences in susceptibility of the α- and β-glucosides, where glucose is the direct product of hydrolysis in both instances. In all this, we are assuming that the gluconolactone is acting because of it similarity to glucose.

An alternate explanation is that, in the two species of *Aspergillus*, the hydrolysis of β-1,3-glucan takes place in two steps: (a) a random-splitting β-1,3-glucanase, as found in *R. arrhizus* and (b) a β-glucosidase acting on the short-chain oligosaccharides. In fact, we know that these two *Aspergilli* do indeed contain very appreciable amounts of β-glucosidase. Inhibition of the β-glucosidase, known to occur, could then explain the apparent inhibition of the β-1,3-glucanase, the lower reducing values found in the presence of lactone, indicating that the reaction stopped before final conversion to glucose. This hypothesis was tested.

Laminarin (the β-1,3-glucan) was hydrolyzed by the enzyme solutions of *A. sydowi*, and of *A. phoenicis*, in the presence and in the absence of gluconolactone. The hydrolysates were spotted on paper at various times during the hydrolysis, and chromatogrammed. In all cases glucose appeared first and was the dominant reducing sugar. The dimer and trimer also were produced, but in lesser amounts. The addition of gluconolactone slowed down the hydrolysis rate, but did *not* alter the nature of the products. If the lactone had been acting by inhibiting the β-glucosidase, there should have been a decrease in glucose and an increase in dimer and trimer. Since this did not occur, this hypothesis seems to be ruled out, and we are left without a satisfactory explanation for the differences observed.

GLUCONOLACTONASES

The discovery of an enzyme which hydrolyzes gluconolactone is a recent one. Its existence had been postulated by Lipmann in 1952 (Cori and Lipmann) and demonstrated by him three years later (Brodie and Lipmann, 1955). The enzyme extracted from yeast proved to be highly specific, acting only on gluconolactone and on 6-phosphogluconolactone. It had no effect on the lactones derived from mannonic or xylonic acids, and little effect on that derived from galactonic acid. The enzyme is present also in bacteria, and in liver and kidney of rat, rabbit, guinea pig, dog, hog, cow, and monkey (Winkelman and Lehninger, 1958).

Gluconolactonases in Fungi.

Our interest in these enzymes stems from the observation that β-glucosidases of different fungi showed different degrees of inhibition by gluconolactone. Thus, for 50% inhibition of β-glucosidase, the concentration of gluconolactone required is:

0.002% for enzyme of *Trichoderma viride*, but
0.120% for enzyme of *Pestalotiopsis westerdijkii*.

The presence of a gluconolactonase in the latter could account for its apparent resistance to gluconolactone. For this reason, we devoted some attention to this rather unusual system.

Sixty-four enzyme solutions were tested for the presence of gluconolactonase. Some of these were commercial enzymes. Others were prepared by us for one reason or another. (It should be noted that the gluconolactonases found here have been secreted into the medium. No extraction of tissue was attempted.)

Two organisms stand out as active producers of gluconolactonase, Basidiomycete sp. QM 806, and *P. westerdijkii* QM 381. These produce about 17μ of enzyme per milliliter of filtrate (Table 5, and Fig. 1). It is interesting that both fungi produce the enzyme best when grown on cellulose. Little or none of the lactonase was found in solutions from cultures of these fungi grown on starch, cellobiose, or cellulose acetate. On the other hand, filtrates of other fungi grown on cellulose produce little or none of the enzyme. It is also interesting that *P. westerdijkii* QM 381 should have gluconolactonase, since its β-glucosidase is the one most resistant to inhibition by gluconolactone.

γ-Gluconolactone is much less susceptible to hydrolysis by gluconolactonase than is Δ-gluconolactone. At pH 4.8, the enzymes of Basidiomycete 806, and of *P. westerdijkii* 381 had no effect. At pH 7.2, however, these enzymes did speed up the hydrolysis (over the spontaneous rate). This substantiates the suggestion made earlier that the lactones must be reduced to a certain level of instability before the enzyme is able to attack, a change which is accomplished by raising the pH. γ-Glucuronolactone and γ-mannonolactone (not shown) resisted hydrolysis by the gluconolactonase preparations from 806 and 381. Δ-Lactobionolactone appears to be susceptible to hydrolysis, but its short half-life (29 min) makes it difficult to work with.

Gluconolactonase was found in glucose oxidase (Takamine) and in notatin (prepared by us), but not in proportion to their glucose oxidase activity. It was also found in β-glucosidase No. 1 (Röhm and Haas) and in a purified protease of fungal origin, but not in 14 other commercial preparations tested.

pH Optimum of Gluconolactonase

Most enzyme substrates with which we are familiar are highly stable. No detectable hydrolysis occurs during the investigator's lifetime. It is quite a contrast to work with something as unstable as Δ-gluconolactone. When attempting to determine the pH optimum of the enzyme, there is the added complication that the stability of the substrate is also a function of pH. To compensate for the latter, we have subtracted the half-life of the test solution from the half-life of the control (no enzyme). This gives an optimum pH value of 5.0 for the gluconolactonases of both *Pestalotiopsis* and Basidiomycete (Fig. 1). Yet it is obvious that the over-all hydrolysis rate (i.e., enzyme plus spontaneous) is greater at a pH above 5.

The effect of pH was also determined for the galactonolactonase of *Pestalotiopsis*. The maximum rate was at pH 6.9 (no higher pH values were tested). At pH 6.1 the rate was one-half maximum and at pH 5.7 the enzyme was inactive.

Other Properties of the Lactonases.

The lactonases tested were completely inactivated in 15 min at 100 C. The gluconolactonases of Basidiomycete 806, *Penicillium funiculosum* and *P. notatum* were precipitated with two volumes of cold acetone without appreciable loss in activity. The addition of versene* (4×10^{-3} M) did not alter the rate of hydrolysis of gluconolactone by the enzymes of *Pestalotiopsis* and of the Basidiomycete.

The concentration of versene is twice that used by Brodie and Lipmann (1955), who reported a requirement for Mn, Mg, or Co for activity of the gluconolactonase extract from yeast. Winkelman and Lehninger (1958) obtained inhibition of rat gluconolactonase by ethylenediaminetetraacetate which could be reversed by Mn but not by Mg, Co (II), Zn, or Fe (II). It would seem that the metal requirement varies considerably depending upon the source of enzyme.

Relationship between the Inhibitory Ability of Gluconolactone on β-Glucosidases, and the Presence of Gluconolactonase.

While it is obvious that anything tending to destroy an inhibitor will lessen its effect, it was considered desirable to see whether such an effect could be demonstrated. We have already stated that the gluconolactonase is present in *Pestalotiopsis* where the β-glucosidase is least inhibited by the lactone. We now decided to add lactonase to β-glucosidase solutions in which it was absent. At an Inhibitor/Salicin ratio of 0.012, the presence of gluconolactonase reduce gluconolactone inhibition of the β-glucosidase of almond emulsin from 77 to 40%, and of the β-glucosidase of *A. luchuensis* from 86 to 53%.

DISCUSSION AND SUMMARY

Organisms differ from each other in their resistance to killing by various chemical agents. While there are many possible reasons for the resistance of the few, the one we are concerned with here is the ability to produce enzymes which destroy the inhibitor. *Pestalotiopsis* and Basidiomycete 806 are fungi that have the ability to destroy a particular inhibitor, gluconolactone, by secreting an enzyme capable of hydrolyzing it.

Of what value is gluconolactonase to an organism? We have screened fungi only for their ability to secrete gluconolactonase. Many others may possess this enzyme internally, from which it can be obtained by extraction (as Lipmann did with yeast). The regular occurrence of the lactonase in glucose-oxidase systems indicates a role for it in the metabolism of glucose (Brodie and Lipmann, 1955):

$$2 \text{ Glucose} \xrightarrow[\substack{-2H_2 \\ +O_2}]{E_1} 2 \text{ Gluconolactone} \xrightarrow[H_2O]{E_2} 2 \text{ Gluconic Acid}$$

$$2H_2O_2 \xrightarrow{E_3} 2H_2O + O_2$$

where E_1 may be notatin, E_2 gluconolactonase, and E_3 catalase. Here are two comparatively toxic compounds (gluconolactone and hydrogen peroxide), both of rather unstable nature, the removal of which is hastened by the action of an appropriate enzyme. It seems likely that all organisms which metabolize glucose by this path-

---

* Ethylenediaminetetraacetate

way possess gluconolactonase. This lactonase has already been found in many animals, in bacteria, and in fungi.

Of what use is the information to us?

(1) The presence or absence of a lactonase explains why a β-glucosidase of one source differs from that of another in its susceptibility to gluconolactone. The information serves to point out the possibility that other unsuspected enzymes act in a similar manner to inactivate fungicides. Where there is a marked difference in the susceptibility of an enzyme, or an organism, to chemicals, the possibility of such an enzyme should be investigated. In a similar manner, an inactive compound may be converted into a toxic one. Triacetin, which is inert, is hydrolyzed to acetic acid, which inhibits the growth of certain fungi, by esterases secreted by fungi (Knight, 1957). This principle may be applicable to the development of fungicides. A chemical substitutent can be added to a toxic chemical to make it water-insoluble, or otherwise inert. Fungus growth would then produce the enzyme necessary to split off the substituent, liberating the toxic chemical and killing the organism. The assumption here is that all organisms can produce the desired enzyme.

(2) We have applied the hundredfold difference in susceptibilities of α- and β-glucosidase to gluconolactone, to the determination of linkage type in glucosides. The unknown glucoside is subjected to hydrolysis by a mixture of α- and β-glucosidases (as from *A. phoenicis*, Table 6). To one tube is added gluconolactone in a concentration known to inhibit the β-glucosidase without inhibiting the α-; to the other tube an equal amount of water. If the hydrolysis is inhibited by the gluconolactone, the linkage is β; if uninhibited, the linkage is α.

(3) Polysaccharase action can be studied without interference from contaminating oligosaccharases, since the latter are highly sensitive to inhibition by gluconolactone, while the former (most of them) are quite resistant. Festenstein (1958) has already used this procedure in his study of cellulase.

(4) The mechanism of action of polysaccharases can be investigated. Those sensitive to inhibition by gluconolactone may represent a type different from any previously studied.

(5) Glycosidases can be investigated for their specificities by subjecting them to the presence of various aldonolactonases. β-xylosidase of limpet appears to be identical with its β-glucosidase. Both activities were equally affected by gluconolactone (Conchie and Levvy, 1957). This is not true of rumen glucosidase and xylosidase.

ACKNOWLEDGMENTS

Much of the information reported has depended upon the generosity of colleagues in supplying the various lactones. Our thanks go to:

Drs. C. Bublitz and A. Lehninger   Johns Hopkins School of Medicine
Dr. H. S. Isbell   National Bureau of Standards

We also thank Dr. Lallouette for the protease used.

REFERENCES

Brodie, A. F., and Lipmann, F. 1955 Identification of a gluconolactonase. J. Biol. Chem., 212, 677-685.

Conchie, J., and Levvy, G. A. 1957 Inhibition of glycosidases by aldonolactones of corresponding configuration. Biochem. J., 65, 389-395.

Cori, O., and Lipmann, F. 1952 Primary oxidation product of enzymatic glucose-6-PO_4 oxidation. J. Biol. Chem., 194, 417-425.

Ezaki, S. 1940 Studien über carbohydrase 4. Über die β-glucosidase hemmung der α-glukoside and glukonsäure. J. Biochem. (Tokyo), 32, 107-120.

Festenstein, G. N. 1958 Cellulolytic enzymes from sheep-rumen liquor microorganisms. Biochem. J., 69, 562-567.

Festenstein, G. N. 1957 Differentiation of cellobiase and carboxymethylcellulase by glucono-1,4 lactone. Biochem. J., 65, 23 P.

Findlay, June, Levvy, G. A., and Marsh, C. A. 1958 Inhibition of glycosidases by aldonolactones of corresponding configuration. II. β-N-acetylglucosaminidase. Biochem. J., 69, 467-476.

Jermyn, M. A. 1959 Some comparative properties of β-glucosidases secreted by fungi. Austral. J. Biol. Sci., 12, 213-222.

Knight, S. G. 1957 In vitro antifungal activity of triacetin. Antibiotics and Chemotherapy, 7, 172-174.

Pigman, W. 1957 The Carbohydrates. Academic Press, N.Y.C.

Reese, E. T., and Mandels, Mary 1957 Chemical inhibition of cellulases and β-glucosidases. Research report, Microbiol. Series No. 17, 60 pages. QM Res. and Eng. Center, Natick, Mass.

Siu, R. G. H. 1951 Microbial Decomposition of Cellulose. Reinhold Publishing Co., N. Y. City, N. Y.

Sumner, J. B., and Somers, G. F. 1944 Laboratory Experiments in Biological Chemistry. Acad. Press, N. Y. City, N. Y.

Winkelman, J., and Lehninger, A. L. 1958 Aldono-, and uronolactonases of animal tissue. J. Biol. Chem., 233, 794-799.

EFFECT OF CYCLOHEXIMIDE (ACTIDIONE) ON THE METABOLISM AND GROWTH OF FUNGI

Hugh D. Sisler and Bradner W. Coursen

Department of Botany, University of Maryland, College Park, Maryland and
Department of Biology, Lawrence College, Appleton, Wisconsin

Cycloheximide (Fig. 1) is an antifungal antibiotic produced by *Streptomyces griseus*, an organism which also produces streptomycin (Whiffen, 1948).

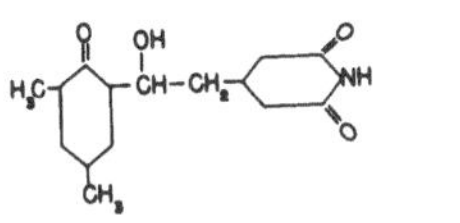

Fig. 1. Structure of cycloheximide.

It is noted especially for its antifungal properties, but it is also toxic to certain higher plants (Ford *et al.*, 1958), animals (Ford and Leach, 1948), protozoa (Loefer and Matney, 1952) and algae (Palmer and Maloney, 1955). Bacteria are generally insensitive to the antibiotic. It has achieved some prominence in plant disease control, but its phytotoxic tendencies have limited its value for this purpose. The agricultural uses of cycloheximide have been thoroughly reviewed by Ford *et al.*, (1958).

SPECIFICITY

Even though cycloheximide is toxic to a wide spectrum of organisms, it exhibits a peculiar specificity toward closely related species. *Saccharomyces pastorianus* , for instance, is completely inhibited by 0.1 ppm of the antibiotic, but under the same conditions other closely related yeasts are unaffected by 1000 ppm (Whiffen, 1948). The rat is much more sensitive to cycloheximide than the mouse (Ford and Leach, 1948). It can be concluded that the systems which are sensitive to cycloheximide are widely distributed among various types of organisms and that a mechanism of resistance to the antibiotic sharply separates closely related species. According to Middlekauff *et al.* (1957), a single pair of genes in *Saccharomyces* controls sensitivity versus resistance to cycloheximide. However, the tolerance of the resistant strains used in their tests does not appear to be nearly as great as that possessed by *S. fragilis* (ATCC 8635) which we have found to be unaffected by 1000 ppm of the antibiotic. The resistance of *S. fragilis* is apparently not based on a capacity of the organism to destroy the antibiotic, because filtrates from cultures of this organism grown for 24 hr in the presence of high concentration of cycloheximide, retain practically as much toxicity to the cycloheximide-sensitive *S. pastorianus* as was present in the medium at the time of inoculation.

Cycloheximide appears to be a fungistatic rather than a fungicidal antibiotic. Cells of *S. pastorianus* exposed for 6 hr to concentrations of the antibiotic much greater than that required to completely inhibit growth, will recover and grow normally upon removal of the antibiotic by washing (Coursen and Sisler, 1960).

181

Apparently it is loosely bound to the cells and no permanent damage results from 6 hr exposures to the toxin. Similar reversal of the toxicity of cycloheximide was obtained by Kerridge (1958) with *S. carlsbergensis*.

MITOTIC EFFECTS

Several cases of mitotic disturbances caused by cycloheximide have been reported. Wilson (1950) and Hawthorne (1952) showed that cycloheximide consistently produces aberrations in mitotic behavior in cells of onion root tips. Berliner and Olive (1953) observed an effect on meiosis in *Gymnosporangium* suggesting that the antibiotic causes a malfunctioning of the spindle, and Whiffen (1951) has noted that it tends to cause a shift from the mitotic to the meiotic state in *Allomyces*. The significance of these observations in relation to the mechanism of toxicity of cycloheximide is not clear. It seems most likely that mitotic disturbances are secondary in nature, arising from more primary disturbances to other phases of cellular activity. The antibiotic is not rated a specific inhibitor of cell division in yeasts by Loveless *et al.*, (1954). Agents such as low doses of alpha, beta, and gamma irradiation, nitrogen mustard, and triethylenemelamine, which stop cell division, but not the growth of yeast, are classified by these authors as specific inhibitors of cell division.

EFFECT OF METALS

Certain divalent metals are reported to affect the toxicity of cycloheximide. Whiffen (1948) noted an increase in the toxicity of the antibiotic to yeast cells when copper sulfate was also present. Copper sulfate likewise increases the inhibitory effect of cycloheximide on fermentation of glucose or of fructose by *S. cereviseae* (Greig *et al.*, 1958). Ferrous sulfate is also reported to increase the effectiveness of the antibiotic in laboratory and field tests (Ford *et al.*, 1958). The basis of these benefical effects of heavy metals remains unexplained. Blumauerova and Starka (1959), on the other hand, demonstrated that heavy metal ions such as Fe, Mn, and Zn are antagonistic to the toxicity of cycloheximide to yeast cells. They suggested that the antibiotic forms chelates with divalent metals and is thus rendered ineffective.

EFFECT ON RESPIRATION AND FERMENTATION

Studies of the effect of cycloheximide on respiration and fermentation in various fungi by Walker and Smith (1952), McCallan *et al.*, (1954), Sisler and Marshall (1957), Greig *et al.*, (1958), and Kerridge (1958), indicate that the antibiotic is only partially inhibitory at toxic concentrations or that it has essentially no effect on respiration or fermentation. In the case of *Fusarium roseum*, a concentration of cycloheximide sufficient to prevent growth, inhibits the oxidation of glucose by approximately 50%, but increasing the concentration of the antibiotic one hundred-fold does not produce any further inhibition of oxygen uptake (Sisler and Marshall, 1957). Its effects are distinct from those of sulfanilamide which inhibits growth, but has no effect on the oxidation of glucose by these cells when a nitrogen source is absent.

Greig *et al.*, (1958) have noted that the antibiotic inhibits the fermentation of glucose by intact yeast cells, but not the fermentation of this sugar by cell-free extracts. They suggested that its action therefore may be at the cell surface. Cycloheximide also inhibits the ability of *Tetrahymena pyriformis* to oxidize certain mono- and disaccharides, organic, and amino acids (Mefferd and Loefer, 1954). It was concluded that the site of action of the antibiotic seems common to many reactions and may be nonspecific in nature.

While these studies of respiration and fermentation provide a useful background for further investigation of the mechanism of action of cycloheximide, they do not give any precise indication of the manner in which it exerts its toxic effects on sensitive cells.

REVERSAL OF TOXICITY

The only substances, of a large number which were tested, that showed appreciable antagonistic activity to the toxicity of cycloheximide were 4-methylcyclohexanone, 5,5-dimethyl-1,3-cyclohexanedione, metamethylcyclohexanone, vitamin A acetate, and vitamin A alcohol (Coursen, 1959). These substances were effective only at low concentrations of the antibiotic. Water extracts of crystalline vitamin A which did not contain detectable amounts of the vitamin were antagonistic. It was postulated that a contaminating cyclic ketone used in the synthesis of vitamin A is the active water-soluble substance in this case (Coursen and Sisler, 1960). The possibility that vitamin A itself is also active was not eliminated in these tests.

The significance of the protection afforded by vitamin A preparations and by certain methylcyclohexanones is not clear, but it is of interest that Cantino and Horenstein (1956) reported an accumulation of *gamma* carotene in cycloheximide-resistant sporangial plants of *Blastocladiella emersonii* when they are treated with the antibiotic. In this connection it might be noted also that the cycloheximide molecule contains a methyl-substituted cyclohexanone ring.

EFFECT ON PROTEIN AND NUCLEIC ACID METABOLISM

The effect of cycloheximide on certain aspects of metabolism in *S. carlsbergensis* was investigated by Kerridge (1957, 1958). In this organism the antibiotic at growth-inhibiting concentrations had little or no effect on respiration or fermentation and it was also possible to show an increase in soluble, 7-min acid label phosphate in the presence of the inhibitor. Thus an interference of cycloheximide with the energy-producing mechanisms in these cells does not appear to be responsible for the inhibition of growth. On the other hand, synthesis of protein and deoxyribonucleic acid is prevented by minimum growth-inhibiting concentrations of the antibiotic. Synthesis of ribonucleic acid continues at a partially inhibited rate for a period in the presence of cycloheximide. The data indicate that cycloheximide interferes with a reaction or reactions involved both in the synthesis of protein and deoxyribonucleic acid. Similar effects of cycloheximide on protein and nucleic acid synthesis in *Aspergillus nidulans* have been observed by Shepherd (1958).

METABOLISM OF C^{14} GLUCOSE

The effect of cycloheximide on the metabolism of C^{14}-labeled glucose by cells of *S. pastorianus* was investigated by Coursen and Sisler (1960). In these tests, untreated cells and cells treated with toxic doses of the antibiotic were fed uniformly labeled glucose for 30-min periods in a medium which would support growth of the cells. The cells were then washed free of the medium and extracted with 80% ethanol to remove metabolic intermediates formed from the radioactive glucose. The intermediates were separated on two-dimensional chromatograms and the radioactivity in specific compounds from treated and untreated cells was compared.

The cells, whether treated or untreated, removed approximately the same quantity of radioactivity from the medium, and the ethanol extracts of the cells in either case contained about the same total quantity of radioactivity. Approximately 35 radioactive compounds were recovered from either treated or control cells. They consisted of amino acids, organic acids, organic phosphorous compounds (both ultraviolet-absorbing and nonabsorbing compounds), and a few unclassified compounds. A comparison of the activities in individual compounds from treated or untreated cells showed that they were essentially the same in many cases. However, the activity of organic phosphorous compounds was generally higher in treated cells. The activity in the amino acids: alanine, aspartic acid, tyrosine, and theonine was also higher in treated than in untreated cells.

It is apparent that a great deal of metabolism proceeds in cells of S. pastorianus in the presence of toxic concentrations of cycloheximide. The excess of certain intermediates in treated cells over that found in the control cells is not unexpected, as these compounds were probably not depleted in the synthesis of protein and nucleic acids as they were in untreated cells.

On the other hand, activity in glutamic acid was lower in treated cells and no detectable radioactivity was ever found in the glutamine from these cells. Glutamine ranked fourth among the compounds recovered from untreated cells in

Table 1. The Distribution of Radioactivity among Amino Acids Extracted from Cells of *Saccharomyces pastorianus* after Exposure for 30 min to C^{14} Glucose in the Presence or Absence of Cycloheximide (3.16 ppm). (From Coursen and Sisler, 1960)

Amino acid[*]	Activity CPM (untreated)	% of total activity recovered	Activity CPM (treated)	% of total activity recovered
Cysteic acid	140	1.0	150	1.1
Aspartic acid	460	3.3	1220	8.5
Glutamic acid	4375	31.0	1135	8.0
Serine	85	0.6	100	0.7
Glycine	95	0.7	140	1.0
Threonine	85	0.6	195	1.4
Alanine	4910	34.8	7370	51.6
Glutamine	650	4.6	0	0.0
Tyrosine	50	0.4	245	1.7
Unidentified	60	0.4	75	0.5
Valine	470	3.3	380	2.7
Leucine-isoleucine	130	0.9	45	0.3

[*]Arginine and histidine were detected by chemical tests, but did not contain detectable amounts of radioactivity in either treated or control samples.

content of radioactive carbon. Judging from the data of these experiments, it appears that glutamine synthetase is strongly inhibited by cycloheximide. The amount of radioactivity in counts/min in various amino acids from treated and untreated cells is shown in Table 1.

It is interesting that the metabolism of glutamic acid and glutamine is affected adversely in treated cells as cycloheximide contains a glutarimide ring and is thus related chemically to glutamic acid and glutamine. A simple hydrolysis of the glutarimide ring of the antibiotic would yield a glutamine derivative which might act as an inhibitory analog of either glutamic acid or glutamine. If such is the case, the selectivity of cycloheximide might well be controlled by the capacity of an organism to hydrolyze the imide ring. There is no evidence, however, that the antibiotic undergoes such a conversion in sensitive organisms.

A failure to synthesize glutamine would undoubtedly be sufficient to cause inhibition of growth. Glutamine is found in considerable quantities in protein and it is used in the synthesis of such vital metabolites as purines (Sonne et al., 1956) histidine (Neidle and Waelsch, 1956) and in the synthesis of glucosamine-6-phosphate (Leloir and Cardini, 1953). However, it does not appear that an inability to synthesize glutamine is the sole reason for the failure of treated cells to grow, because 1000 ppm of glutamine added to the culture medium containing minimum growth-inhibitory amounts of the antibiotic gave no detectable protection from the toxicity of cycloheximide. Glutamic acid and *alpha* ketoglutaric acid were likewise ineffective antagonists.

The synthesis of glutamine from glutamic acid in which a peptide bond is formed between ammonium and glutamic acid has been studied as a model reaction of peptide synthesis. Since the data of Coursen and Sisler (1960) indicate that cycloheximide interferes with glutamine synthesis, it was suggested that the antibiotic may likewise interfere with peptide bond formation in the incorporation of amino acids into protein and that this may be the principal means by which it inhibits cell growth. This suggestion is supported by the results of Kerridge (1958) who showed that glycine is not incorporated into proteins in S. carlsbergensis in the presence of toxic doses of cycloheximide.

Further investigations will be needed to determine whether a relationship exists between the interference of cycloheximide with the synthesis of glutamine, and its interference with protein and nucleic acid synthesis.

REFERENCES

Berliner, M. D., and Olive, L. S. 1953 Meiosis in *Gymnosporangium* and the cytological effects of certain antibiotic substances. Science, 117, 652-653.

Blumauerova, M., and Starka, J. 1959 Reversal of antibiotic action of cycloheximide (actidione) by bivalent metal ions. Nature, 183, 261.

Cantino, E. C., and Horenstein, E. A. 1956 The stimulatory effect of light upon growth and CO_2 fixation in *Blastocladiella*. I. The S. K. I. Cycle. Mycologia, 48, 777-799.

Coursen, B. W. 1959 Effect of the antibiotic, cycloheximide, on the metabolism and growth of *Saccharomyces pastorianus*. Ph.D. Thesis, University of Md.

Coursen, B. W., and Sisler, H. D. 1960 Effect of the antibiotic, cycloheximide, on the metabolism and growth of *Saccharomyces pastorianus*. Amer. J. Botany, In Press.

Ford, J. H., and Leach, B. E. 1948 Actidione, an antibiotic from *Streptomyces griseus*. J. Amer. Chem. Soc., 70, 1223-1225.

Ford, J. H., Klomparens, W., and Hamner, C. L. 1958 Cycloheximide (actidione) and its agricultural uses. Plant Dis. Reptr., 42, 680-695.

Greig, M. A., Walk, R. A., and Gibbons, A. J. 1958 Effect of actidione (cycloheximide) on yeast fermentation. Jour. Bact., 75, 489-491.

Hawthorne, M. E. 1952 The cytological effects of the antibiotic actidione. Publication No. 3077, Dissertation Abstract, Univ. Microfilms, Ann Arbor, Michigan.

Kerridge, D. 1957 The effect of actidione on protein and nucleic acid synthesis in *Saccharomyces mandshuricus*. Jour. Gen. Microbiol., 16, v.

Kerridge, D. 1958 The effect of actidione and other antifungal agents on nucleic acid and protein synthesis in *Saccharomyces carlsbergensis*. Jour. Gen. Microbiol., 19, 497-506.

Leloir, L. F., and Cardini, C. E. 1953 The biosynthesis of glucosamine. Biochem. Biophys. Acta, 12, 15.

Loefer, J. B., and Matney, T. S. 1952 Growth inhibition of free-living protozoa by actidione. Physiol. Zool., 25, 272-276.

Loveless, L. E., Sporel, E., and Weisman, T. H. 1954 A survey of effects of chemicals on division and growth of yeast and *Escherichia coli*. J. Bact., 68, 637-644.

McCallan S. E. A., Miller, L. P., and Weed, R. M. 1954 Comparative effect of fungicides on oxygen uptake and germination of spores. Contrib. Boyce Thompson Inst., 18, 39-68.

Mefferd, R. B., Jr. and Loeffer, J. B. 1954 Inhibition of respiration in *Tetrahymena pyriformis* S. by actidione. Physiol. Zool., 27, 115-118.

Middlekauff, J. E., Hino, S., Yang, S., Lindegren, G., and Lindegren, C. C. 1957 Gene control of resistance vs. sensitivity to actidione in *Saccharomyces*. Genetics, 42, 66-71.

Neidle, A., and Waelsch, H. 1956 Participation of glutamine in the biosynthesis of histidine. Jour. Amer. Chem. Soc., 78, 1767.

Palmer, C., and Maloney, T. E. 1955 Preliminary screening for potential algicides. Ohio J. of Sci., 55, 1-8.

Shepherd, C. J. 1958 Inhibition of protein and nucleic acid synthesis in *Aspergillus nidulans*. J. Gen. Microbiol., 18, IV.

Sisler, H. D., and Marshall, N. L. 1957 Physiological effects of certain fungitoxic compounds on fungus cells. Jour. Washington Acad. Sci., 47, 321-329.

Sonne, J. C., Lin, I., and Buchanan, J. M. 1956 Biosynthesis of purines. IX. Precursors of the nitrogen atoms of the purine ring. Jour. Biol. Chem., 220, 369-378.

Walker, A. T., and Smith, F. G. 1952 Effect of actidione on growth and respiration of *Myrothecium verrucaria*. Proc. Soc. Exptl. Biol. Med., 81, 556-559.

Whiffen, A. J. 1948 The production, assay, and antibiotic activity of actidione, an antibiotic from *Streptomyces griseus*. Jour. Bact., 56, 283-291.

Whiffen, A. J. 1951 The effect of cycloheximide on the sporophyte of *Allomyces arbuscula*. Mycologia, 43, 635-644.

Wilson, G. B. 1950 Cytological effects of some antibiotics. Jour. Heredity, 41, 227-231.

EFFECTS OF ELEMENTAL SULFUR, DITHIOCARBAMATES, AND RELATED FUNGICIDES ON ORGANIC ACID METABOLISM OF FUNGUS SPORES*

Robert G. Owens
Boyce Thompson Institute for Plant Research, Inc., Yonkers, N.Y.

Various aspects of the mechanisms of action, insofar as they are known, of sulfur, dithiocarbamates, and related fungicides have been reviewed by McCallan (1957), Horsfall (1945), and Martin (1959). The present paper reports some comparative biochemical effects of these materials on metabolism of organic acids by way of the citric acid cycle. The purpose of the study was to determine whether all of these fungicides affect the same or different biochemical processes and to relate the observed effects to the structure and chemical properties of the compounds and the enzymes.

The acetate-metabolizing system of *Neurospora sitophila* spores, including the citric acid cycle, was selected for preliminary evaluations. This system is attractive as an experimental indicator because most of the intermediates are acids and can be assayed rapidly and simultaneously by simple techniques. Moreover, the chemistry and group requirements of many of the enzymes are fairly well known, providing a basis for analyzing toxicant interactions in terms of similarities and differences in enzyme and toxicant structure and properties.

MATERIALS AND METHODS

Intact, washed conidia of *N. sitophila* (Mont.) Shear & Dodge were used as the experimental subject. The fungus was cultured and conidia were harvested, washed, and prepared for testing as described previously (Owens *et al.*, 1958).

All data are based on dry weight of the spores determined on 2-ml aliquots taken from freshly prepared aqueous spore suspensions that served as stocks for experimental samples. The 2-ml samples were dried at 100 C for 8 hr and weighed. Dry weights of experimental samples were calculated on the basis of volume relative to the 2-ml aliquots. The spores were always uniformly suspended with a mechanical stirrer when aliquots were taken.

Experimental systems were prepared in 250-ml Erlenmeyer flasks as follows: 400 to 500 mg of spores (dry weight) in 20 ml of distilled water, 200 mg of anhydrous sodium acetate in 20 ml of distilled water, various doses of toxicants as desired (see table and figures) in 2 ml of acetone or water, and distilled water to make a total volume of 60 ml. Controls lacked acetate and/or toxicant, as appropriate.

The amounts of the various toxicants required to kill all spores under this particular set of experimental conditions were determined in preliminary experiments. Subsequently several doses, ranging from that which had no effect on germination up to one that killed all spores, were tested simultaneously to corre-

---
* This work was supported in part by a grant from the Rockefeller Foundation.

late effects on acid metabolism with effects on germination in terms of toxicant concentration.

The samples were incubated overnight on a reciprocating shaker. The spores did not germinate until diluted and supplied with biotin. After incubation, a drop of spore suspension from each sample was transferred to 50 ml of distilled water. A drop of the diluted suspension was then mixed with a drop of Fries solution (Ryan *et al.*, 1943) on a microscope slide and held in a moisture chamber until the controls were fully germinated. Germination counts were made in duplicate as recommended by the Committee on Standardization of Fungicidal Tests of The American Phytopathological Society (1943).

After removal of the drop for determining germination, the spores in the experimental samples were handled in two ways, depending upon whether keto or nonketo acids were to be assayed. For nonketo acids, the spores were extracted five times with about 5 volumes of 6N formic acid each time. The extracts and incubation solution were combined and the water and formic acid were removed in a stream of air under partial vacuum. The residue was taken up in distilled water and passed through a 1×15 cm column of formate-form Dowex 1-X10 resin, 200 to 400 mesh, which retains the acids. The acids were then removed from the column differentially by gradient elution with formic acid (Palmer, 1955), with the aid of a fractionator. The rate of elution for the first 50 fractions was 2 ml/5 min, and 2 ml fractions were collected. The flow rate was then increased to 1 ml/min and 70 3-ml fractions were collected. The water and formic acid were removed in a stream of air. The residual acids were taken up in 1 ml of distilled water and titrated with standardized 0.01N NaOH (Palmer, 1955).

For determination of keto acids, the spore suspensions after incubation were made 2N with respect to HCl, and 100 μg of 2,4-dinitrophenylhydrazine were added for each ml of solution. The samples were stored overnight at 2 C and then filtered to remove the spores. The spore residue was washed with about 100 ml of ethyl acetate, which was then used to extract the aqueous filtrate. After separation of the phases, the water layer was extracted repeatedly with additional portions of ethyl acetate until all of the colored material was in the organic phase. The ethyl acetate was then extracted with several 10-ml portions of 10% aqueous sodium carbonate until no further color appeared in the aqueous phase. The aqueous solution was acidified with HCl and extracted with several portions of ethyl ether until all colored material was in the ether phase. Known aliquots of the ether extract were spotted quantitatively on Whatman No. 1 chromatographic paper along with appropriate standards. The 2,4-dinitrophenylhydrazones were then separated by descending chromatography with the solvent mixture described by Cavallini and Frontali (1954), with addition of 1% ammonia. After chromatography, the spots were cut into small pieces and put into test tubes. The hydrazone derivatives were eluted with 50% aqueous ethanol and quantitated colorimetrically at 380 mμ by reference to appropriate standards.

All toxicants were 99% pure or better except for maneb and Na_2S_x. The maneb was a 70% wettable powder, but results are given in terms of active ingredient. Na_2S_x was a technical grade obtained from Mallinkrodt Chemicals. Elemental sulfur was in colloidal form, prepared according to Miller *et al.*, (1953).

RESULTS

Figure 1 shows a characteristic organic acid elution curve obtained with extracts of untreated spores of *N. sitophila* and of spores incubated overnight with excess sodium acetate. A peak is always found at about tubes 15 to 20. This peak is made up of acid amino acids, but it is of no quantitative significance because some amino acids are destroyed in the extraction and elution procedure. A small peak due to pyrrolidone carboxylic acid is usually found in the vicinity of tube 25. This acid is not considered in the present discussion because it is not involved directly in citric acid cycle interconversions. The major detectable acids in the cycle are succinate, recovered in tubes 30 to 45; malate, found in tubes 45 to 55; and citrate, found in tubes 70 to 85, all ±5, depending upon the total amount of each, present. Fumarate, when detectable, elutes after citrate and is followed by a mixture of organic and inorganic phosphates (curves not shown). The amount of acid in tubes comprising each peak was determined by titration.

The principle used to evaluate effects of the toxicants was that used for many years in the study of biochemical genetics, i.e., elucidation of enzyme deficiencies in terms of accumulation or depletion of biochemical intermediates. In the present investigation, however, enzyme deficiencies were induced with toxicants (chemical inhibitors) rather than by alteration of genes. Chemically induced enzyme deficiencies in the citric acid cycle were particularly obvious when the spores were incubated with sodium acetate because acetate induces substantial increases in citrate, succinate, and malate, as shown in Fig. 1. Thus, failure of any of the

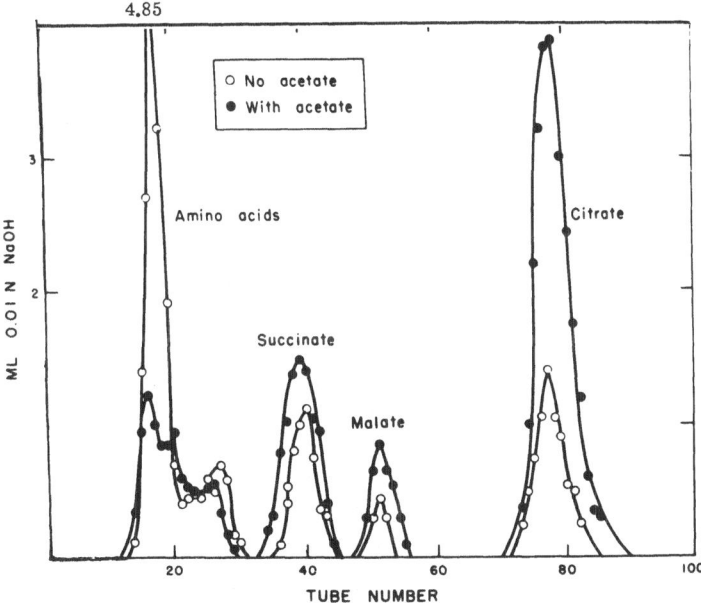

Fig. 1. Typical elution curves for organic acids in extracts of conidia of *Neurospora sitophila* after incubating the spores in the presence and absence of excess sodium acetate (about 400 mg/g spores).

intermediates to increase normally in the presence of excess acetate indicates inhibition of one or more enzymes preceding that intermediate in the pathway. Conversely, abnormally high accumulation of any intermediate reflects inhibition of one or more enzymes following that intermediate in the chain of interconversions. Simultaneous measurement of all detectable components in this particular series of transformations makes it possible to locate the inhibited enzyme within two or three steps at most. In some instances the precise enzyme is indicated.

The data in Table 1 show the effects of the various toxicants at lethal and sublethal doses on organic acid levels. The doses given are the minimum lethal and maximum sublethal doses tested. It will be recognized that values for lethal and

Table 1. Effects of Various Toxicants at Minimum Lethal and Maximum Sublethal Doses Tested on Levels of Certain Organic Acids in Conidia of *Neurospora sitophila*

Toxicant	Dose, mg/g spores	Inhibition of germination, %	Acid, mg/g spores				
			Citric	Succinic	Malic	Pyruvic	α-Ketoglutaric
Group I							
Colloidal sulfur	41	92	1.6	3.8	0.3	0.02	0.01
"	26	0	8.3	3.3	2.3	0.01	0.01
None	0	0	11.7	2.5	1.8	0.01	0.01
Na$_2$S	105	98	2.4	3.8	2.9	0.11	0.15
"	42	0	11.0	3.0	2.1	0.06	0.08
None	0	0	11.3	2.4	1.9	0.08	0.08
Na$_2$S$_x$	75	100	6.2	4.2	1.5	-	-
"	37	0	9.5	5.6	2.1	-	-
None	0	0	12.6	4.4	2.0	-	-
Thiram	90	94	4.8	3.5	2.1	0.05	0.03
"	1.9	0	9.2	2.9	2.8	0.06	0.21
None	0	0	9.4	2.6	1.2	0.05	0.05
Ferbam	27	100	1.4	3.7	4.1	0.13	0.05
"	1.8	0	11.3	2.6	1.2	0.05	0.07
None	0	0	9.6	2.3	1.1	0.05	0.18
Group II							
Ziram	40	100	21.6	2.2	2.1	0.62	1.88
"	2.5	0	9.7	2.7	1.4	0.03	0.12
None	0	0	9.2	2.2	1.8	0.05	0.16
Maneb	78	99	9.3	3.0	2.5	0.09	0.06
"	7	0	6.3	1.8	1.4	0.05	0.08
None	0	0	6.9	4.0	2.4	0.04	0.13
Nabam	60	100	15.8	2.7	3.4	0.04	0.07
"	2.2	0	11.0	1.0	0.9	0.03	0.06
None	0	0	8.4	4.4	1.2	0.01	0.01
8-Quinolinol	17.5	97	21.8	4.2	3.4	-	-
"	1.1	0	13.2	3.2	2.0	-	-
None	0	0	11.6	3.6	3.0	-	-
Group III							
Methyl iso-thiocyanate	2.7	96	6.0	0.7	3.0	0.88	0.17
"	1.4	0	7.9	3.4	1.9	0.14	0.10
None	0	0	6.7	1.5	0.4	0.03	0.02

sublethal doses are only approximate and subject to rather wide variations because of biological variation. Moreover, the spread between concentrations was necessarily rather wide in order to cover the range of doses necessary to kill all spores and still keep the size of the experiments within manageable bounds.

Colloidal sulfur prevented the normal increase in citrate in the presence of acetate, but tended to cause succinate to increase. This indicates blockage of enzymes in the pathway between acetate and citrate and at least partial inhibition of the succinoxidase system. The magnitude of the effect on the succinoxidase system can be expected to be somewhat obscured because of the decrease in citrate, since citrate precedes succinate synthesis in the conventional pathway. Pyruvate and α-ketoglutarate metabolism appeared to be affected little, if any, by sulfur. Although some slight increases in pyruvate were induced by a number of the toxicants, including sulfur, these are not considered significant in most cases because strong inhibition of keto acid metabolism usually results in very large increases, such as produced by ziram and methyl isothiocyanate (Table 1).

Acid patterns almost identical with those induced by sulfur resulted when the spores were poisoned with Na_2S, Na_2S_x, thiram (tetramethylthiuram disulfide), and ferbam (ferric dimethyldithiocarbamate), designated as Group I, Table 1. All of these compounds inhibited at least one enzyme in the pathway from acetate to citrate. This suggests that these compounds have at least one chemical property in common, which is reflected by inactivation of the same enzyme(s). Some theoretical implications of this will be discussed in a subsequent section.

The second group of compounds, consisting of ziram (zinc dimethyldithiocarbamate), nabam (disodium ethylenebisdithiocarbamate), maneb (manganous ethylenebisdithiocarbamate), and 8-quinolinol, induced substantially the same patterns of acid accumulation within the group, but the patterns were distinctly different from those caused by compounds in Group I, Table 1. All compounds in Group II caused accumulation of citrate, whereas those in Group I retarded citrate synthesis. Thus, compounds in Group II apparently had little or no effect on enzymes in the pathway from acetate to citrate, but inhibited instead at least one enzyme in the conversion of citrate to other intermediates in the citric acid cycle. Since neither aconitate nor isocitrate accumulated, the inhibited enzyme apparently was aconitase, which converts citrate to aconitate.

In addition to powerful blockage of aconitase, ziram also induced more than tenfold increases in pyruvate and α-ketoglutarate levels. This suggests substantial inhibition of keto acid carboxylase and oxidases.

Other salts of dithiocarbamic acid and ethylenebisdithiocarbamic acid showed a slight tendency to retard metabolism of pyruvate at the minimum lethal doses. This was more obvious when larger, hyperlethal doses were used. However, since larger doses are in excess of the amounts required to inhibit germination, there seems to be little justification for assuming a primary role for inhibition of keto acid metabolism in toxicity of these particular salts. The effect does suggest, however, that all of the salts may have a chemical property in common, although the data show clearly that compounds in Groups I and II have certain properties that are dissimilar. This follows from the fact that Group I toxicants retard citrate synthesis, whereas Group II compounds do not. Those in Group II, on the other hand, inhibit aconitase whereas those in Group I do not. These differences will be discussed with regard to some possible chemical differences in a subsequent section.

The third category consists at present of only one representative, methyl isothiocyanate. Investigations on other isothiocyanates are in progress, but data on their effects are not yet available. Methyl isothiocyanate produced an acid pattern quite different from any other toxicant tested. Citrate synthesis and metabolism apparently were affected little, if at all. However, malate, pyruvate, α-ketoglutarate, and, at low doses of toxicant, succinate accumulated. Thus, while methyl isothiocyanate inhibited keto acid metabolism as did ziram, it did not affect citrate synthesis as did Group I compounds nor metabolism of citrate as did Group II toxicants. On the other hand, it strongly inhibited malate and succinate metabolism, which were affected slightly, if at all, by other toxicants.

The pattern of methyl isothiocyanate action on succinate levels warrants comment (Fig. 10). As the dosage was decreased, succinate increased. This suggests that the toxicant inhibited succinate metabolism, but that the level of succinate did not increase at the higher doses because of the powerful block in the conversion of α-ketoglutarate to succinate. These data, and those available in the literature on isothiocyanates, suggest that the primary action of methyl isothiocyanate is on dehydrogenases, since a dehydrogenase is operative in normal spores at each point in metabolism where a block is indicated by the data.

CORRELATIONS BETWEEN INHIBITION OF GERMINATION AND METABOLISM

An essential part of all studies on toxicant action are data relating effects on metabolism to effects on life or growth, which in this study were observed in terms of spore viability. Correlations between germinability and relative changes in levels of the various acids at various doses of each toxicant are given in Figs. 2 through 10. Different amounts of each toxicant were required to affect germination, but at lethal doses, effects on organic acid metabolism were similar within the categories given in Table 1. Since toxicants may interact biochemically in similar ways at widely different doses, obviously nothing about mechanism of action can be inferred from relative lethal doses. Mode of action can be elucidated only in terms of effects on metabolism. Moreover, effects on metabolism and on germination must be mutually related to toxicant dose, although quantitative differences in magnitude of effects on the two functions can be expected. These stem from the fact that specific processes may be partially inhibited without observable effects on germination. In a sense, therefore, germination tests are not strictly quantitative because they are usually based on the number of spores germinated or ungerminated at a specific time and not on the absolute rate of germination.

Data in Fig. 2 show that citrate synthesis and germination were inhibited by sulfur to about the same extent over a small range of doses, and the dosage-response curves were similar, showing concomitance of effects on the two functions. The curve for inhibition of citrate synthesis was displaced slightly toward the lower doses, however, suggesting that there was some retardation of citrate synthesis that was not accurately reflected quantitatively by inhibition of germination. Slight to moderate inhibition of citrate synthesis apparently was not lethal.

Succinate also accumulated to some extent in sulfur-poisoned spores, indicating a partial block in the succinoxidase system. This could not be correlated with germination possibly because effects on succinate synthesis were obscured by the block preceding succinate in the system. In any event, the critical locus of sulfur action in acetate metabolism appears to be in the pathway from acetate to citrate, since effects in this area paralleled effects on germination.

Similar correlations were found for Na$_2$S, thiram, and ferbam (Figs. 3, 4, and 5). The data for thiram and ferbam (Figs. 4 and 5) were highly fortuitous in that the values for inhibition of citrate synthesis fall precisely on the curves for inhibition of germination.

Inhibition of germination by ziram, nabam, maneb, and 8-quinolinol (Figs. 6, 7, 8, and 9) corresponded to *increases* in citrate at the various doses of toxicants. In general the characteristics of the dosage-response curves for accumulation of citrate and inhibition of germination covered the same range of toxicant doses and were, therefore, concomitant.

In case of ziram-treated spores the relation between toxicant concentration and citrate accumulation was linear and did not follow the slope of the germination curve. However, it should be noted that ziram also had a powerful inhibitory action

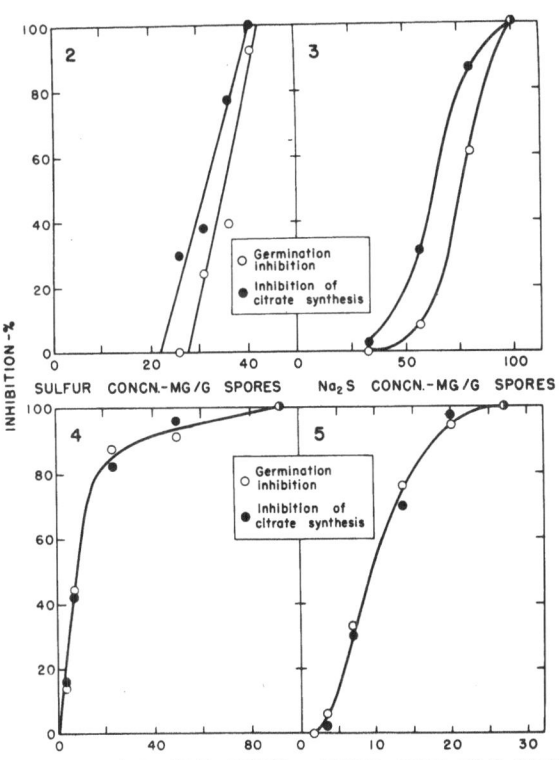

Figs. 2–5. Respectively, effects of colloidal sulfur, Na$_2$S, thiram, and ferbam on germination of conidia *Neurospora sitophila* and on citrate synthesis.

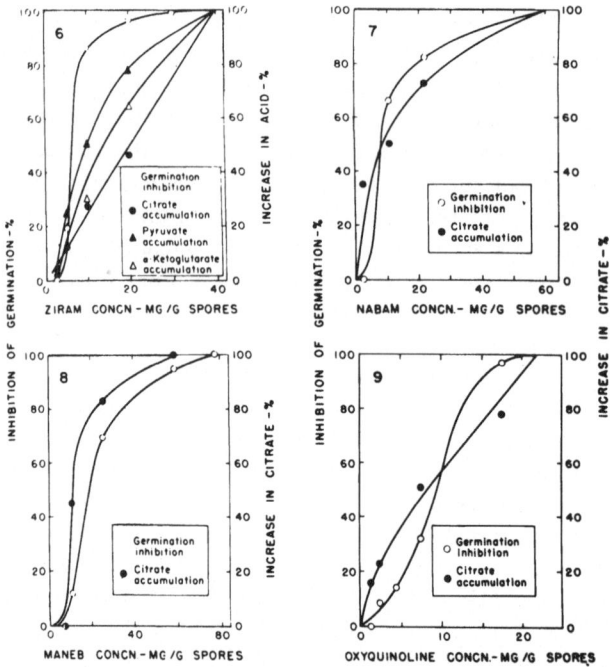

Figs. 6—9. Respectively, effects of ziram, nabam, maneb, and 8-quin-
olinol on germination of conidia of *Neurospora si tophila* and on organic acid metabolism. 8-Quinolinol is called oxyquinoline in Fig. 9.

on keto acid metabolism, so that effects on several systems were manifested
simultaneously. The dosage-response curve for keto acid accumulation was, in
fact, more similar to the dosage-response curve for germination. In such cases
of simultaneous action on different enzymes, it seems reasonable to suspect that
the individual effects might be somewhat additive. If this is true, one might expect
a very steep dosage-response curve for germination accompanying somewhat flat-
ter curves for the individual enzyme processes, as shown in Fig. 6.

Some possible support for this point of view seems to exist in the data for
methyl isothiocyanate (Fig. 10). The slope of the dosage-response curve was found
to be extremely steep, and again, evidence was obtained that several enzymes were
inhibited simultaneously. As with ziram, metabolism of pyruvate and α-ketoglutarate was strongly inhibited and inhibition was concomitant with effects on germination. Unlike ziram, however, methyl isothiocyanate did not inhibit aconitase
but instead apparently inactivated succinic and malic dehydrogenases. As pointed
out in the previous section, accumulation of succinate was inversely proportional
to inhibition of germination. Succinate was detectable only after the dosage of
isothiocyanate was reduced to relatively low levels. This was probably due to the
strong blockage of α-ketoglutarate metabolism at the higher doses, which would
prevent succinate formation. Increase in succinate at the lower doses indicates
that the succinoxidase system was still partially inhibited at doses permitting

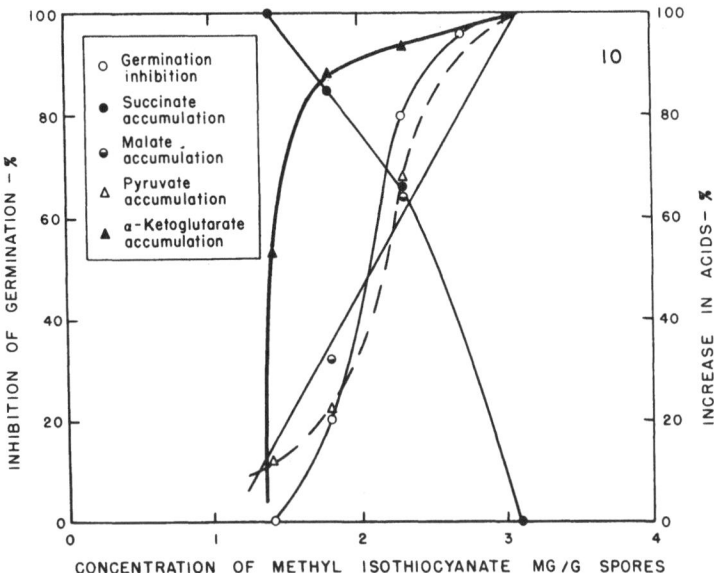

Fig. 10. Effects of methyl isothiocyanate on germination of conidia of
Neurospora si tophi la and on organic acid metabolism.

partial conversion of α-ketoglutarate to succinate. However, this partial inhibition
of succinoxidase did not prevent germination of the spores.

DISCUSSION

Many studies have been made in an effort to elucidate the mode of action of
sulfur and dithiocarbamates (Horsfall, 1945; McCallan, 1957; Martin, 1959), but
comparable data for each toxicant were not previously available because of the
variety of test subjects and experimental conditions employed by the various investigators. Consequently, similarities and differences in interaction patterns in
metabolism were not obvious. The present data, in many cases, confirm and add
to the results obtained by others, but as will be seen in the ensuing discussion,
they do not always confirm conclusions arrived at on the basis of chemical structure, theoretical chemistry, observed decomposition products, or unobserved
theoretical decomposition products.

Elemental sulfur has not been shown to inhibit enzymes *in vitro*, and for many
years its mode of action has been an enigma. As pointed out by McCallan (McCallan,
1957), its toxic properties have been variously attributed to H_2S, SO_2, H_2SO_4, thio-
sulfuric acid, and pentathionic acid. All of these have been disavowed as the toxic
agent ultimately for one reason or another (McCallan, 1957). In view of present
knowledge on inhibition of certain enzymes by sulfur or its derivatives, and the
similarities and differences in its patterns of interaction compared with other
toxicants, it might be profitable at this point to reexamine the known chemical
transformations of this and related toxicants.

It is well established that many organisms and tissues reduce sulfur to hydrogen sulfide (Miller *et al.*, 1953). McCallan and Wilcoxon (1931) considered the reduction to be enzymatic, for they found that there was a temperature optimum at 35 C. This belief has been strengthened by the fact that the sulfur-reducing system is sensitive to sodium arsenite and to 2,3-dichloro-1,4-naphthoquinone (dichlone), which are potent sulfhydryl enzyme inhibitors, and by the competitive interaction of methylene blue in the reducing system (Miller *et al.*, 1953). Moreover, McCallan and Miller (1957) have shown that reduction of sulfur is coupled directly or indirectly with enzymes responsible for CO_2 formation from endogenous substrates.

Establishment of the enzymatic nature of sulfur reduction is an important step forward in formulating some concept of the "active principle" derived from sulfur, for it requires reexamination of the sequence of events in reduction in terms of one-electron-transfer mechanisms characteristic of enzymatic electron transfer (Michaelis, 1951). From room temperature to temperatures somewhat above its melting point, sulfur exists as an 8-membered, puckered ring, although metastable 6-membered rings and linear chains are also known (Walling, 1957, p. 334). Since enzymatic reduction of sulfur occurs by successive steps involving transfer of one electron at a time, reduction must entail ring fission:

$$\begin{matrix} S\!:\!S\!:\!S \\ \ddot{S} \quad \ddot{S} \\ S\!:\!S\!:\!S \end{matrix} \quad \xleftarrow{\ +1e\ } \quad :S\!:\!S\!:\!S\!:\!S\!:\!S\!:\!S\!:\!S\!:\!S \quad \underset{\substack{-1S \\ +2H}}{\overset{+1e}{\xleftarrow{\hspace{1cm}}}} \quad \overset{\bullet}{S}\!:\!S\!:\!S\!:\!S\!:\!S\!:\!S\!:\!\overset{\bullet}{S} \ + \ H_2S \qquad (1)$$

The expected products of ring fission and subsequent dissociation of one sulfur atom are H_2S and a polysulfide free radical. The free radical may react in several ways: a) It may be further reduced to H_2S and progressively shorter polysulfide chains; b) The terminal electrons may pair with ring closure. The resulting ring, however, is still subject to reopening and further reduction, as evidenced by the fact that almost quantitative yields of H_2S are obtained when certain fungus spores are treated with small amounts of sulfur (Miller *et al.*, 1953); c) Two or more of the free radical chains might polymerize to form longer chains. Polysulfide chains up to 60 sulfur atoms have been found to form from S_8 under some conditions (Gee, 1952; Powell and Eyring, 1943); d) Finally, the free radicals may react with various cytoplasmic components including enzymes and coenzymes, for free radicals in general tend avidly to obtain or to share an electron.

Evidence for the existence of sulfur free radicals under certain conditions is now almost irrefutable. It has been found that the viscosity of liquid sulfur rapidly increases as the temperature is increased, reaching a maximum at around 180 C (Gee, 1952; Powell and Eyring, 1943). Above this the viscosity decreases. The viscosity increase has been correlated with increases in chain length, which also reaches a maximum at about 180 C (Gee, 1952; Powell and Eyring, 1943). Polymerization proceeds apparently by free radical additions, for Gardner and Fraenkel (1954; 1956) detected free radical concentrations reaching 6×10^{-3} moles/L at elevated temperatures by electron paramagnetic resonance studies.

Further evidence of the participation of elemental sulfur in reactions of organic free radicals consists in its effectiveness as an inhibitor in vinyl polymerization. Vinyl acetate radicals were found to attack sulfur 470 times as rapidly as they add monomer (Bartlett and Kwart, 1950). Farmer and Shipley (1947) showed that simple olefins react with S_8 to form a mixture of polysulfides. Farmer et al. (1954) reported formation of allylic polysulfides at 0-83 C upon treatment of olefins with SO_2 and H_2S. These reagents apparently interact to produce sulfur in a highly reactive metastable state. All of these investigations and many others point convincingly to the participation of free radicals in reactions of sulfur (Walling, 1957, pp. 334-341).

Although most of the studies cited above were caried out at elevated temperatures enzymatic reduction at physiological temperatures would be expected to produce similar intermediates which can react by the same mechanisms as radicals produced at higher temperatures.

Since free radicals tend to share the unpaired electrons with other molecules bearing unpaired electrons, the electronic state of enzymes and coenzymes during the performance of natural functions is pertinent to a consideration of the chemical nature of enzyme-toxicant interaction. Haber and Willstätter (1931) proposed that a great majority of enzymatic dehydrogenations involve a step in which free radicals exist. Michaelis and Smythe (1938) proposed that all oxidations of organic molecules, although they are bivalent, proceed in two successive univalent steps, the intermediate being a free radical. By electron paramagnetic spectroscopy, free radicals have been detected in several types of biological systems (Calvin and Sogo, 1957; Commoner et al., 1954; Commoner et al., 1957; Ehrenberg and Ludwig, 1958; Sogo et al., 1957; Tollin and Calvin, 1957). Although much still needs to be done to establish the relationship, if any, between sulfur toxicity and its existence as a free radical, these developments suggest a new avenue of approach to possible elucidation of the chemistry underlying the biocidal activity of sulfur.

The concept of free radical interaction becomes increasingly intriguing in view of the similarity in biochemical inhibition patterns obtained with S_8, Na_2S, Na_2S_x, ferbam, and thiram. It can be postulated with considerable justification that each of these toxicants can give rise to free radicals. Under appropriate conditions of pH, oxidation and concentration, solutions of Na_2S, H_2S, and Na_2S_x deposit free elemental S_8. In order for S_8 to form, sulfur atoms from sodium or hydrogen sulfide must polymerize with subsequent ring closure. Polysulfides with fewer than 8 sulfur atoms must also add on a sufficient number of S atoms to permit ring closure. Thus, in biological systems, S_8, Na_2S, H_2S, and Na_2S_x may be considered interconvertible, and it is probable that there exists a common intermediate in the interconversion from one to the other. This intermediate is very likely the reactive free radical discussed above. The interconversion can be illustrated thus:

$$
\begin{array}{c}
\text{Donor} \rightharpoonup \text{H} \qquad \text{Donor} + \text{H}^+ \qquad\qquad \text{Donor} \rightharpoonup \text{H} \qquad \text{Donor} + \text{H}^+ \\[4pt]
S_8 \rightleftharpoons S_8^{\,\cdot} \rightleftharpoons S_7 + H_2S \qquad (2)
\end{array}
$$

A similar series of reductions of S_7, S_6, etc. could eventually convert the whole sulfur molecule to H_2S. The reverse process could give rise to elemental sulfur from H_2S or any of the postulated intermediates.

It is conceivable that the $S_x^{\bullet}$ radicals would react subsequently with proteins and perhaps olefinic components of the cells to form relatively stable cross-linkages of the types found in vulcanized rubber:

$$S_x^{\bullet} \quad Enzyme-R^{\bullet} \longleftrightarrow Enzyme-R-S_x-R-Enzyme \qquad (3)$$

where R is any group on the protein bearing an unpaired electron. The moiety represented as the "enzyme" might also be nonenzymatic protein, coenzymes and/or perhaps olefins.

Although H_2S is considerably less toxic to some fungi than is colloidal sulfur (McCallan, 1957) in terms of concentration, this does not invalidate, but strengthens, the above hypothesis, because conditions of concentration, pH, oxygen tension, and perhaps others, must be conducive to maintenance of an adequate concentration if $S_x^{\bullet}$ to react rapidly with cytoplasmic components. Theoretically, 7 or 8 moles of H_2S would be required to produce the same quantity of free radicals as 1 mole of S_8, assuming quantitative conversion of S_8 to free radicals of the species illustrated above. However, the actual difference probably is not that great, since only a small fraction of the total sulfur applied is in solution and the actual number of molecules from either sulfur or H_2S that react with vital sites is probably a small fraction indeed of the total present. Moreover, toxic free radicals presumably could exist momentarily in the form of $S_1^{\bullet}$, $S_2^{\bullet}$, etc.

Although thiram and ferbam bear little resemblance structurally to S_8, Na_2S, or Na_2S_x, the biochemical data show that they inhibit the same enzymes. This suggests that they have similar properties or give rise to intermediates with similar properties. Thiram was shown by Kern (1955) to produce free radicals capable of initiating vinyl polymerization. Hirshon et al. (1953) have noted that many sulfur compounds such as thiophenol and diphenyl disulfide give deeply colored solutions in sulfuric acid or in the presence of aluminum chloride, which exhibit paramagnetic resonance spectra, thus proving their potentialities for free radical formation. Hoshino et al. (1953) recently showed that methyl mercaptan undergoes an addition reaction with olefins such as vinyl acetate, vinyl chloride, and allyl alcohol upon irradiation with a mercury lamp. The rate in O_2 was much faster than in N_2, indicating the free radical nature of the reaction in which O_2 acts as the catalyst. Many sulfides and disulfides catalyze vulcanization of rubber with S_8, a process now generally believed to involve cross-linkages formed through free radical intermediates (Walling, 1957). These findings strongly suggest a free radical intermediate in some of the reactions of thiram and are in accord with the observed similarities in biochemical effects of thiram and the sulfur compounds, assuming the validity of this mechanism for the sulfur derivatives. Formation of dithiocarbamate free radicals from thiram would involve simply dissociation at the disulfide linkage or partial reduction of this linkage:

$$\begin{array}{ccc} CH_3 & S & \\ \diagdown N-C-S^- & + & \end{array} \begin{array}{cc} CH_3 & S \\ \diagdown N-C-S \end{array} \underset{-1e}{\overset{+1e}{\rightleftharpoons}} \begin{array}{ccc} CH_3 & S & S & CH_3 \\ \diagdown N-C-S-S-C-N\diagup \end{array} \longleftrightarrow 2 \begin{array}{cc} CH_3 & S \\ \diagdown N-C-S^{\bullet} \end{array}$$

$$(4)$$

At present there is no evidence of free radicals from ferbam. However, the biochemical data suggest that ferbam has a chemical property in common with thiram and the sulfur derivatives. Furthermore, this property is not shared by the other salts of dithiocarbamic acid tested at the minimum lethal dose, since the zinc salt and the ethylenebisdithiocarbamates apparently do not inhibit citrate synthesis from acetate. The possibility that ferbam and thiram inactivate enzymes in the pathway from acetate to citrate through ionic interactions is unlikely, because ziram also yields dithiocarbamate anions on dissociation but it does not block citrate synthesis. The possibility that iron from ferbam inactivates the enzymes can be eliminated because of the relatively low toxicity of most iron salts at the dosage used.

On theoretical grounds, a reaction can be written which yields a free radical from ferbam:

$$\left(\begin{array}{c} CH_3 \\ CH_3 \end{array}\!\!> N\text{-}\overset{\overset{S}{\|}}{C}\text{-}S\right)_{\!3} Fe \longleftrightarrow \left(\begin{array}{c} CH_3 \\ CH_3 \end{array}\!\!> N\text{-}\overset{\overset{S}{\|}}{C}\text{-}S\right)_{\!2} Fe \;+\; \begin{array}{c} CH_3 \\ CH_3 \end{array}\!\!> N\text{-}\overset{\overset{S}{\|}}{C}\text{-}S^{\cdot} \tag{5}$$

In this reaction the iron atom is reduced by acquisition of one electron from one of the three dithiocarbamate anions, so that the resulting products are ferrous dimethyldithiocarbamate and a dithiocarbamate free radical. A similar electron transfer can not be written for zinc, manganous or sodium salts because the cations are already in their lowest physiological valence states. The biochemical data show that these latter salts exhibit quite different patterns of action from those toxicants for which free radicals can logically be postulated.

If ziram, nabam, and maneb in fact cannot interact as free radicals with enzymes, as suggested, their interaction with aconitase must be explained in terms of some other chemical property one not exhibited at the minimum lethal dose by Group I compounds. A possible clue exists in the chemical make-up of aconitase. This enzyme has been purified to a point where reliable data on some of its chemical features have been obtained (Ochoa, 1951). The active protein requires Fe^{++} (Dickman and Cloutier, 1950). The Fe^{++} atom, however, is rather loosely bound and can be removed and restored without denaturing the protein moiety (Dickman and Cloutier, 1950). Removal of the iron either by dialysis at the appropriate pH or with sequestering agents results in complete loss of enzymatic activity.

The present data strongly suggest that ziram, maneb, nabam, and 8-quinolinol, a known chelating agent, inactivate aconitase. The mechanics of inactivation very likely involve removal of the Fe^{++} atom, since relatively nondissociated Fe^{++} salts or chelates are known to form with each of these compounds. Moreover, removal of the Fe^{++} atom from the enzyme probably involves cation exchange with establishment of an equilibrium, characteristic of the physical conditions of the environment:

$$Enzyme\text{ - }Fe^{++} \;+\; Me\;(S\text{-}\overset{\overset{S}{\|}}{C}\text{-}N\text{-}R)_x$$
$$\updownarrow$$
$$\underset{(inactive)}{Enzyme\text{ - }Me} \;+\; Fe\;(S\text{-}\overset{\overset{S}{\|}}{C}\text{-}N\text{-}R)_2 \tag{6}$$

The hypothesis involving this kind of cation exchange is strengthened by the fact that ferbam did not inactivate aconitase. This follows because ferbam contains iron as the cation, which, upon reduction, is identical with the atom bound by the protein and which imparts enzymatic activity.

As early as 1944, Zentmyer proposed that certain fungicides, 8-quinolinol in particular, function by precipitating or chelating essential heavy metals in the cell. This hypothesis was also suggested as a possible explanation for the action of salts of dithiocarbamic acid and ethylenebisdithiocarbamic acid (Horsfall, 1945). It was shown later that dithiocarbamates inhibit heavy metal enzymes such as polyphenol oxidase, a copper enzyme, and even catalase in which iron is complexed in a heme group (Owens, 1953; Owens, 1954). Manten, Klöpping, and van der Kerk (1951), on the other hand, found that heavy metals did not antagonize the toxicity of thiram or 8-quinolinol and therefore concluded that heavy metal complexes were not involved. The present biochemical data also suggest that heavy metals are not involved in the toxicity of thiram but do not support their conclusions with regard to 8-quinolinol. Moreover, agreement with regard to thiram is highly fortuitous since their conclusions are based on attempts to reverse inhibition by addition of heavy metals *singly*. This approach cannot be expected to succeed because of the fact that the dithiocarbamates complex or form insoluble salts with practically all heavy metal ions and heavy metal enzymes. Thus to add one metal cannot be expected to restore activity of enzymes specifically requiring other metals.

It is suggested that dithiocarbamate salts will inactivate heavy metal enzymes containing a heavy metal species different from the cation of the salt, provided the dissociation constant of the derivative resulting from interaction is lower than that of the original enzyme-metal complex. This is borne out by the inactivation of aconitase by all except the iron salt and by a previous study (Owens, 1954) in which it was shown that polyphenol oxidase, a copper enzyme, is strongly inhibited by all dithiocarbamates tested except the copper salts.

While certain differences in interaction patterns of dithiocarbamates have been emphasized up to now, all of these compounds were found to affect keto acid metabolism at hyperlethal doses (unpublished data). This suggests that they have a third potentially toxic property in addition to the possible free radical interaction and heavy metal inactivation already mentioned. The degree of accumulation of pyruvate and α-ketoglutarate varies considerably among the different derivatives and in most cases does not appear to be a primary factor in toxicity. Nevertheless, it illustrates that all dithiocarbomates have at least one chemical property in common, as might be expected, regardless of the cation associated with the anion. Studies still in progress suggest that inhibition of keto acid metabolism at high dosages results from interaction of the dithiocarbamate anion with cocarboxylase. Further details of the interaction must await results of current studies.

Inhibition of pyruvate metabolism by sodium diethyldithiocarbamate was reported previously by Sijpesteyn and van der Kerk (1954b; 1956) and inhibition of α-ketoglutarate metabolism by ziram was observed by Sisler and Marshall (1957). The latter investigators, however, found no increase in citrate or pyruvate in the presence of ziram, such as reported in the present paper (Table 1). Their failure to do so is probably due to the fact that they used glucose instead of acetate as a substrate. Since ziram strongly inhibits enzymes requiring Fe^{++}, aldolase, an Fe^{++} enzyme (Meyerhof, 1951), was probably inactivated in their experiments. This would block conversion of glucose to pyruvate or citric acid cycle intermediates. With acetate as a substrate, as in the present experiments, no heavy metal enzyme is encountered in the pathway before aconitase. The increase in α-ketoglutarate obtained by Sisler and Marshall (1957) probably resulted from blockage of metabolism of α-ketoglutarate derived from endogenous glutamic acid and glutamine by deamination.

Sijpesteyn and van der Kerk (1956) and Goksøyr (1955) suggested that coenzyme A is involved in the action of sodium diethyldithiocarbamate. This is not consistent with the fact that citrate synthesis from acetate, which involves coenzyme A, was not inhibited by ziram, maneb, or nabam at the minimum lethal doses in the present experiments. Whether coenzyme A or some other component of the acetate-oxalacetate condensing system is inactivated by sulfur, Na_2S, Na_2S_x, thiram, and ferbam has not been established.

Methyl isothiocyanate was distinctly the most toxic of the materials tested. It was included to test the validity of the theory that isothiocyanate is the ultimate toxicant when fungi are treated with bisdithiocarbamates (Sijpesteyn and van der Kerk, 1954a). This theory holds that ethylenebisdithiocarbamates are converted to ethylene iso- or diisothiocyanates which are supposed to exist in solution in equilibrium with ethylenethiuram monosulfide. The latter compound has been isolated and its chemical structure determined by Ludwig et al., (Ludwig et al., 1955). That these products form in small quantities under appropriate conditions seems well founded. However, their contribution to or responsibility for the toxicity of bisdithiocarbamates is open to question. No conclusive evidence of their taking part in the same cytoplasmic interactions as the parent bisdithiocarbamates has been advanced. The evidence on which the theory rests consists of observed antagonism of the action of isothiocyanates and bisdithiocarbamates by the same thiols. Antagonism of toxicant action in general shows only that the antagonist and the toxicant can react to form a nontoxic or less toxic product. In some cases, the interaction results in decomposition of the toxicant (Weed et al., 1953). Thiols, particularly, are known to detoxify a number of different fungicides, including captan and dichlone, but considerable difference has been demonstrated in a number of aspects of their modes of action (Owens and Novotny, 1958; Owens and Novotny, 1959). Thus, there seem to be few experimental data to justify the conclusion that toxicants which are antagonized by the same reagent necessarily affect the same metabolic site at the minimum lethal dose.

The evidence that nabam is biologically active by virtue of conversion to diisothiocyanate needs confirmation based on metabolic data before it can be fully accepted. Sijpesteyn and van der Kerk (1954a) compared the effects of nabam, ethylene diisothiocyanate, and ethylenethiuram monosulfide on growth of Botrytis allii, Penicillium italicum, Aspergillus niger, and Rhizopus nigricans. They found that nabam and diisothiocyanate were equally toxic to R. nigricans, since equivalent amounts, 10 ppm, of each toxicant were required to inhibit growth. This means, of course, that nabam would have to be converted quantitatively to diisothiocyanate, if the latter were the toxic agent. If nabam was converted quantitatively to diisothiocyanate in the medium in which *R. nigricans* was tested, it is logical to expect quantitative conversion in the same medium in which *B. allii, P. italicum,* and *A. niger* were grown, and the two compounds should have been equally toxic. However, the data show that 20 to 40 times more nabam than diisothiocyanate was required to inhibit growth of these fungi. Moreover, the data show that *R. nigricans* was inhibited by 5 ppm of ethylenethiuram monosulfide, whereas 10 ppm of diiso-thiocyanate was required. Thus, even if the ethylenethiuram monosulfide were quantitatively converted to diisothiocyanate, the resulting concentration of diiso-thiocyanate would be only about 50% of that required to inhibit growth completely. On the basis of these data, the authors concluded that "ethylenethiuram mono-sulfide fits entirely in the pattern found for nabam and the corresponding ethylene diisothiocyanate." In view of the differences pointed out above, however, it is suggested that the data cannot be used in support of the hypothesis unless it can be shown that differences in absorption by the various organisms account for the differences in toxicity. The data show that ethylenethiuram monosulfide was toxic at concentrations one-half of those of nabam for each fungus. Thus, nabam could possibly act through conversion to the monosulfide. However, even in this case, the data are merely suggestive, and the hypothesis will have to be confirmed by data showing inactivation of the same enzymes by the two toxicants.

The present biochemical data obtained with methyl isothiocyanate (Table 1) are inconsistent with the idea that ethylenebisdithiocarbamates act by virtue of conversion to isothiocyanates. Whereas nabam and maneb inhibited aconitase, there was no indication that methyl isothiocyanate affected this enzyme. The iso-thiocyanate inhibited malic and succinic dehydrogenases, whereas nabam and maneb did not. The only enzymes affected in common in acid metabolism were pyruvate and α-ketoglutarate metabolizing systems, which in some cases are inhibited by dithiocarbamates not included in the isothiocyanate theory.

During the past 20 years many references have appeared in the literature establishing the potency of isothiocyanates as inhibitors of various dehydrogenases and certain other sulfhydryl enzymes (see Chemical Abstracts). The present data (Table 1) are in accord with these, since a dehydrogenase is known to function at each point in acetate and pyruvate metabolism and in the citric acid cycle where a block is indicated by the data. There seems to be little in the present data or in the literature to indicate the same mechanism of action for isothiocyanates and ethylenebisdithiocarbamates. However, at this point the isothiocyanate theory is still worthy of consideration, for it must be recognized that some peculiarities in the fungicidal action of nabam exist. It has been known for a number of years that foliage treated with nabam is protected from diseases for much longer periods than would be expected for a water-soluble salt. Moreover, McCallan *et al*. (1954) have shown that nabam is nontoxic to fungus spores under anaerobic conditions. These facts suggest that on foliage nabam might undergo a change whereby it is converted to a product that is less water-soluble. Ludwig *et al*. (1955) have shown that ethylenethiuram monosulfide is formed from nabam in dilute solutions in the presence of oxygen, especially if metal catalysts such as manganese are present. This material is more toxic than nabam to certain fungi and is less water-soluble than the parent nabam. It might, therefore, account for the practical efficiency of nabam on foliage. They also showed that small amounts of isothiocyanate form from the monosulfide which still leaves the possibility that the mode of action of nabam on foliage might be different from that in solution, as in the present experiments. Further investigations on this point are being carried out.

SUMMARY

The present studies show that sulfur, Na_2S, Na_2S_x, thiram, and ferbam, or some derivative from them, have certain similar properties that are reflected in their ability to inactivate at least one enzyme in the metabolic pathway from acetate to citrate. It is postulated that this property is related to formation of free radicals which are expected to act alike and at the same loci in metabolism.

Ziram, nabam, and maneb, which cannot logically be expected to form free radicals, do not inhibit the enzyme affected by the above group of toxicants, but instead they inactivate aconitase, an Fe^{++} enzyme. It is postulated that these toxicants inactivate the enzyme by removing the essential Fe^{++} atom, probably by exchange with the nonfunctional cation of the dithiocarbamate salt.

Isothiocyanate apparently inactivates dehydrogenases and perhaps other sulfhydryl enzymes wherever they occur in metabolism. The dissimilarities between biochemical effects of methyl isothiocyanate and ethylenebisdithiocarbamates indicate that the latter do not act as isothiocyanates under the conditions of the present experiments.

The effects elucidated with the present model system are not assumed to constitute the over-all mode of action of any of the toxicants tested. They are presented to illustrate similarities and differences in the chemical properties of the toxicants and they serve as a test of existing theories about the mechanism of action of sulfur and dithiocarbamates. They also serve to point out that these toxicants may react with cytoplasmic components, including enzymes and coenzymes, by several different mechanisms, some perhaps as free radicals and as ions, others as ions and metal complexing agents, and perhaps simply as nucleophilic displacement reagents. Although the demonstrated effects on the various processes and enzymes in this small area of metabolism probably do not represent the over-all picture of toxicity, the fact that the effects are concomitant with inhibition of germination of the spores indicates that they are involved to a considerable degree in the over-all picture. In all probability, other enzymes in other areas of metabolism with similar properties and functional groups are also affected concomitantly by these materials, perhaps by more than one of the chemical interactions postulated above. The large number of chemically similar enzymes in living organisms and the variety of reactions possible for the dithiocarbamates precludes specificity of action and the possibility of resistance developing by a simple metabolic bypass.

REFERENCES

American Phytopathological Society. 1943 Committee on Standardization of Fungicidal Tests. The slide-germination method of evaluating protectant fungicides. Phytopathology, 33, 627-632.

Bartlett, D., and Kwart, H. 1950 Dilatometric studies of the behavior of some inhibitors and retarders in the polymerization of liquid vinyl acetate. I. J. Am. Chem. Soc., 72, 1051-1059.

Calvin, M., and Sogo, P. B. 1957 Primary quantum conversion process in photosynthesis: electron spin resonance. Science, 125, 499-500.

Cavallini, D., and Frontali, N. 1954 Quantitative determination of keto acids by paper partition chromatography. Biochem. et Biophys. Acta, 13, 439-445.

Commoner, B., Townsend, J., and Pake, G. E. 1954 Free radicals in biological materials. Nature, 174, 689-691.

Commoner, B., Heise, J. J., Lippincott, B. B., Norberg, R. E., Passonneau, J. V., and Townsend, J. 1957 Biological activity of free radicals. Science, 126, 57-63.

Dickman, R., and Cloutier, A. A. 1950 Activation and stabilization of aconitase by ferrous ions. Arch. Biochem., 25, 229-231.

Ehrenberg, A., and Ludwig, G. D. 1958 Free radical formation in reaction between old yellow enzyme and reduced triphosphopyridine nucleotide. Science, 127, 1177-1178.

Farmer, E. H., and Shipley, F. W. 1947 The reaction of sulphur and sulphur compounds with olefinic substances. Part I. The reaction of sulphur with mono-olefins and with $\Delta^{1:5}$-diolefins. J. Chem. Soc., 1947, 1519-1532.

Farmer, E. H., Ford, J. F., and Lyons, J. A. 1954 The interaction of sulphur and sulphur compounds with olefinic substances. VII. Low-temperature sulphuration of trialkylethylenes with hydrogen sulphide-sulphur dioxide, and with a sulphur-zinc dithiocarbamate system. J. Appl. Chem. (London) 4, 554-561.

Gardner, D. M., and Fraenkel, G. K. 1954 Paramagnetic resonance in liquid sulfur. J. Am. Chem. Soc., 76, 5891-5892.

Gardner, D. M., and Fraenkel, G. K. 1956 Paramagnetic resonance of liquid sulfur: determination of molecular properties. J. Am. Chem. Soc., 78, 3279-3288.

Gee, G. 1952 The molecular complexity of sulphur in the liquid and vapour. Trans. Faraday Soc., 48, 515-526.

Goksøyr, J. 1955 The effect of some dithiocarbamyl compounds on the metabolism of fungi. Physiol. Plantarum, 8, 719-835.

Haber, F., und Willstätter, R. 1931 Unpaarigkeit und Radikalketten im Reaktionsmechanismus organischer und enzymatischer Vorgänge. Ber. Deut. Chem. Ges., 64B, 2844-2856.

Hirshon, J. M., Gardner, D. M., and Fraenkel, G. K. 1953 Evidence for the existence of radicals in the presence of Lewis acids. J. Am. Chem. Soc., 75, 4115.

Horsfall, J. G. 1945 Fungicides and their action. 239 pp. Chronica Botanica Co., Waltham, Mass.

Hoshino, T., Yamagishi, K., and Ichikawa, Y. 1953 On the photochemical addition reaction of methylmercaptane with vinylacetate, vinylchloride, and allylalcohol. Proc. Japan. Acad., 29, 55.

Kern, R. J. 1955 Some sulfur compounds as polymerization sensitizers. J. Am. Chem. Soc., 77, 1382-1383.

Ludwig, R. A., Thorn, G. D., and Unwin, C. H. 1955 Studies on the mechanism of fungicidal action of metallic ethylenebisdithiocarbamates. Can. J. Botany, 33, 42-59.

McCallan, S. E. A. 1957 Mechanisms of toxicity with special reference to fungicides. Proc. Second Internat. Plant Prot. Conf. 1956. Fernhurst, England, pp. 77-95.

McCallan, S. E. A., and Miller, L. P. 1957 Equimolar formation of carbon dioxide and hydrogen sulfide when fungus tissue reduces sulfur. Contrib. Boyce Thompson Inst., 18, 497-506.

McCallan, S. E. A., Miller, L. P., and Weed, R. M. 1954 Comparative effect of fungicides on oxygen uptake and germination of spores. Contrib. Boyce Thompson Inst., 18, 39-68.

McCallan, S. E. A., and Wilcoxon, F. 1931 The fungicidal action of sulphur. II. The production of hydrogen sulphide by sulphured leaves and spores and its toxicity to spores. Contrib. Boyce Thompson Inst., 3, 13-38.

Manten, A., Klöpping, H. L., and van der Kerk, G. J. M. 1951 Investigations on organic fungicides. III. The influence of essential trace metals upon the fungitoxicity of tetramethylthiuram disulphide and 8-hydroxyquinoline. Antonie van Leeuwenhoek, 17, 58-68.

Martin, H. 1959 The Scientific Principles of Crop Protection. 359 pp. Edward Arnold (Publishers) Ltd., London. 4th ed.

Meyerhof. O. 1951 Aldolase and isomerase. *In*: Sumner, J. B., and Myrbäck, K., editors. The Enzymes. Vol. II, pt. 1, pp. 162-182. Academic Press, Inc., New York.

Michaelis, L. 1951 Theory of oxidation-reduction. *In*: Sumner, J. B., and Myrbäck, K., editors. The Enzymes. Vol. II, pt. 1, pp. 1-54. Academic Press, Inc., New York.

Michaelis, L., and Smythe, C. V. 1938 Biological oxidations and reductions. Ann. Rev. Biochem., 7, 1-36. 1938.

Miller, L. P., McCallan, S. E. A., and Weed, R. M. 1953 Quantitative studies on the role of hydrogen sulfide formation in the toxic action of sulfur to fungus spores. Contrib. Boyce Thompson Inst., 17, 151-171.

Ochoa, S. 1951 Fumarase and aconitase. *In*: Sumner, J. B., and Myrbäck, K., editors. The Enzymes. Vol. 1, pt. 2, pp. 1217-1236. Academic Press, Inc., New York.

Owens, R. G. 1953 Studies on the nature of fungicidal action. I. Inhibition of sulfhydryl-, amino-, iron-, and copper-dependent enzymes *in vitro* by fungicides and related compounds. Contrib. Boyce Thompson Inst., 17, 221-242.

Owens, R. G. 1954 Studies on the nature of fungicidal action. III. Effects of fungicides on polyphenol oxidase *in vitro*. Contrib. Boyce Thompson Inst., 17, 473-487.

Owens, R. G., Novotny, H. M., and Michels, M. 1958 Composition of conidia of *Neurospora sitophila* . Contrib. Boyce Thompson Inst., 19, 355-374.

Owens, R. G., and Novotny, H. M. 1958 Mechanism of action of the fungicide dichlone (2, 3-dichloro-1,4-naphthoquinone). Contrib. Boyce Thompson Inst., 19, 463-482.

Owens, R. G., and Novotny, H. M. 1959 Mechanism of action of the fungicide captan [N-(trichloromethylthio)-4-cyclohexene-1,2-dicarboximide]. Contrib. Boyce Thompson Inst., 20, 171-190.

Palmer, J.K. 1955 Chemical investigations of the tobacco plant. X. Determination of organic acids by ion exchange chromatography. Conn. Agr. Expt. Sta. Bull. 589. 31 pp.

Powell, R. E., and Eyring, H. 1943 The properties of liquid sulfur. J. Am. Chem. Soc., 65, 648-654.

Ryan, F. J., Beadle, G. W., and Tatum, E. L. 1943 The tube method of measuring growth rate of *Neurospora* . Am. J. Botany, 30, 784-799.

Sijpesteyn. A. K., and van der Kerk, G. J. M. 1954a Investigations on organic fungicides. VIII. The biochemical mode of action of bisdithiocarbamates and diisothiocyanates. Biochim. et Biophys. Acta, 13, 545-552.

Sijpesteyn, A. K., and van der Kerk, G. J. M. 1954b Investigations on organic fungicides. IX. The antagonistic action of certain imidazole derivatives and of α-keto acids on the fungitoxicity of dimethyldithiocarbamates. Biochim. et. Biophys. Acta, 15, 69-77.

Sijpesteijn [sic], A.K., and van der Kerk, G.J.M. 1956 Investigations on organic fungicides. X. Pyruvic acid accumulation and its relation to the phenomenon of inversion growth as affected by sodium dimethyldithiocarbamate. Biochim. et Biophys. Acta, 19, 280-288.

Sisler, H. D., and Marshall, N. L. 1957 Physiological effects of certain fungitoxic compounds on fungus cells. J. Wash. Acad. Sci., 47, 321-329.

Sogo, P. B., Pon, N. G., and Calvin, M. 1957 Photo spin resonance in chlorophyll-containing plant material. Proc. Nat. Acad. Sci. U.S., 43, 387-393.

Tollin, G., and Calvin, M. 1957 The luminescence of chlorophyll-containing plant material. Proc. Nat. Acad. Sci. U.S., 43, 895-908.

Walling, C. 1957 Free Radicals in Solution. 631 pp. John Wiley & Sons, Inc., New York.

Weed, R. M., McCallan, S. E. A., and Miller, L. P. 1953 Factors associated with the fungitoxicity of ferbam and nabam. Contrib. Boyce Thompson Inst., 17, 299-315.

Zentmyer, G. A. 1944 Inhibition of metal catalysis as a fungistatic mechanism. Science, 100, 294-95.

VARIATION IN MICROORGANISMS

Dr. Joseph G. Becker, presiding

Panelists: Dr. Robert Fuerst
Dr. Carl C. Lindegren
Dr. Max R. Zelle
Dr. Waclaw Szybalski

INTRODUCTION — WHAT IS MICROBIAL VARIATION?

Joseph G. Becker

Microbiological Research & Development, Applied Research Laboratory
Fine Chemicals Division, Nopco Chemical Company, Harrison, New Jersey

The diversity of form and function among microorganisms is unparalleled among other groups of organisms. Microorganisms exhibit an enormous capacity to evolve new potentialities. This is associated with a: 1) relatively undifferentiated soma in which all or most nuclei are potentially germinative rather than somatic; 2) short generation time; 3) manifold means of variation; 4) existence of a haploid vegetative phase allowing expression of recessive genes.

Variation refers to change in an organism relative to its parent or former state. From the viewpoint of the industrial microbiologist, only the *expressed* change in appearance or function—the phenotype—is of interest. From the viewpoint of the organism. even latent changes in the gene pattern, the genotype may have evolutionary significance, ultimately perhaps, for the microbiologist as well.

Visual changes at the cellular level in turn reflect changes in the genes or cytoplasm. Thus, differences in phenotype between different organisms grown under identical conditions would, in general, be accounted for by a difference in genes present. However, the absence of expressed differences does not preclude differences in genotype.

Variation is environmental or hereditary, depending on whether the change results from interplay of gene and environment, or from the introduction of new genes (by intra-change or inter-change). Since heredity is the transfer of genes, hereditary (gene) variation is transmissible to the progeny. Environmental variation by definition does not alter the genes, and is therefore considered nontransmissible.

Structurally and functionally, a microbial cell is composed of genes (or nucleus containing genes) and a surrounding cytoplasm. The genes have been described as sites on a bipartite deoxyribonucleic acid (DNA) chain. As cellular determinants, the genes apparently leave their imprint on a ribonucleic acid (RNA) chain, which passing to the cytoplasm, acts as a template for the polymerization of activated amino acids into enzyme molecules. The enzymes, thereby synthesized under gene control, mediate the activities of the cell.

In considering environmental variation, the environment may be pictured as controlling the *level* of expression within the range permitted by the gene, by reacting directly or indirectly with the gene in a reversible physicochemical reaction. Since the immediate environment of the gene is the cytoplasm, environmental stimuli originating outside the cell may be modified here. A change in pH level of the medium, for example, could alter solubility of metabolites, membrane permeability, enzyme activity through dissociation of enzyme-substrate complexes, equilibria, hydrogen bonding, etc. These effects in turn would trigger other reactions or reaction sequences. The effected changes cease when equilibrium is restored in the reaction between gene and environmental reactant.

Environmental stimuli include response to specific medium constituents and their concentration, pH, temperature, moisture, light, aeration, etc. Time may be looked upon as an environmental factor, in that it allows accumulation of metabolites, and wear and tear, resulting perhaps in maturation. Differentiation of cells of presumably identical heritage could reflect local threshold concentrations of metabolites resulting from differences in time of exposure or distance from stimuli.

Those engaged in fermentation have noted that an organism's past environmental history may markedly affect the future course of development of the culture--in the absence of any demonstrable change in its heredity. There is no transfer of acquired characters, since the new traits are seldom permanent. A plausible hypothesis for which evidence would be welcome is presented here. Namely, carryover of mitochondria, microsomes, and enzyme molecules with high turnover could have significant activity, though diluted extensively in concentration by growth and cell division. It is the transfer of such substances, rather than the gene pattern to which it bears semblance, that is operative in these instances. In all cases, however, the perpetuation of this pattern depends ultimately on replication and carryover of the genes.

When the environmental stimulus seems to have a directed effect in the elaboration of an enzyme not normally produced, the latter is said to be an adaptive or induced enzyme. It is conceivable that the so-called constitutive enzymes are induced by endogenously formed metabolites. The sum total of responses or changes the organism undergoes in approaching a new equilibrium in a new environment is adaptation. Environmental variation is only one facet of adaptation, since the range of ability to adapt, which is gene-controlled, is also subject to variation.

Unlike environmental variation, in which the reaction of the environmental stimulus with the gene or an intermediary is a reversible one, hereditary variation may be hypothesized to be an irreversible reaction directly involving the gene. Since the self-replicating gene is affected, and is also transferable, this type of variation tends to be perpetuated.

Hereditary variation can occur in microorganisms by numerous means. In mutation, a gene may be modified, inactivated or deleted. As part of a linkage group or chomosome, it may be abnormally replicated (polyploidy), translocated to another chromosome, or inverted on its own. Its position with respect to other genes affects its activity. When yield of a fermentation process is improved by mutation, it is just as likely that genes controlling competing reactions or suppressor genes were inactivated, as it is that the genes controlling the desired reaction were favorably modified. The vast majority of mutations would be expected to be undesirable, in that the organism has over a period of time adapted to its environment. Sudden changes in the genes would throw the organism out of equilibrium. Only a minute fraction of the random mutations that occur could be expected to be desirable from either the standpoint of the organism or the investigator. The cause of spontaneous mutation is not known. Certainly wear and tear, and cumulative cosmic radiation among others, cannot be ignored. Inherent differences in mutability between genes may reflect their relative exposure and ability to withstand such stress. This in turn is a function of the location of the gene (its chemical environment) and its chemical nature.

Hereditary variation may occur in the asexually reproductive stage of microorganisms. The cytoplasm of two cells of different strains (presumably established by mutation) may fuse. If the nuclei remain distinct in the common cytoplasm, the newly synthesized organism, called a heterokaryon, may be at an advantage over either homokaryotic strain, if the allelic genes of the respective nuclei are complementary. A heterokaryon may also be formed by mutation of one or more nuclei in a homokaryotic multinucleate cell. Hereditary variation is frequently encountered when the different nuclei of a heterokaryon are separated into distinct cells during spore formation, each spore then representing the strain of the respective nucleus it bears.

Hereditary variation of an unusual "parasexual" type may occur by mitotic rather than meiotic crossing-over and mitotic reduction in diploids produced by fusion of heterokaryotic nuclei.

Hereditary variation of a sexual nature can occur by recombination of genes from haploid nuclei to form a diploid heterozygote. Recombinant types, somewhat limited by linkage, are partially restored by crossing-over. If the organism possesses a haploid stage, variation may occur upon segregation.

In transformation, DNA extracted from one strain and exposed to another becomes incorporated in the genetic apparatus of the receptor. It is interesting to speculate whether this type of hereditary variation could occur in nature by autolysis of one strain in the presence of a suitable receptor.

In transduction, phage carries genetic information, apparently single genes, from the cell of one strain to another. The transferred genes become part of the genotype of the receptor cell.

The above instances of hereditary variation concern the nuclear genes. In certain organisms containing the normal complement of nuclear genes, other particulate matter in the cytoplasm has been found to control certain characters, analogous to nuclear genes. These particles are also mutable and self-replicating.

Therefore, transfer of these so-called cytogenes or plasmagenes during cell division can be considered cytoplasmic inheritance. No orderly transfer of cytogenes occurs as with the nuclear genes, and the latter still exert considerable control over the presence and activity of the cytogenes. The basic concepts of microbial genetics remain unchanged even though the scope of microbial variation keeps enlarging. As the search for more specialized properties of microorganisms is intensified, a greater disparity among microorganisms is likely to be revealed.

VARIATIONS IN *NEUROSPORA CRASSA*

Robert Fuerst

Department of Biology, Microbiological Research, Texas Woman's University
Denton, Texas

In the year 1896, E. B. Wilson wrote, "Inheritance is the recurrence in successive generations, of like forms of metabolism." Yet, because of differences in metabolism, there are probably no two individuals of any species completely identical. Although individuals may have exactly the same genetic complement, *if* that is ever actually possible, as soon as cells divide, the environment determines what limitation of pH, substrate, temperature, competitions, etc., exist for the enzyme systems to act upon.

Variation was so well defined in the introduction to this symposium (Becker, 1960), that perhaps the following discussion may seem redundant, yet it seems important to the context of this presentation.

Unlikeness between parents and offspring may be the result of mutations. These mutations, if transmitted to the next generation, are permanent changes. Recombinations—the association of genes, new and different from parent type combinations—are one of the commonest sources of *hereditary* variability. One can thus speak of hereditary or genotypic variation resulting in unlikeness between parent and offspring.

The second type of variation is *environmental* variation and persists only as long as conditions for it prevail. It is indeed selective but *not* so in the Lamarckian sense. Environmental and hereditary variation occur in nature side by side, so that the differences observed between individuals are usually partly environmental and partly hereditary in origin.

This presentation is primarily concerned with environmental variations. The organism selected for this study was *Neurospora crassa*, the pink bread mold. The first known historic account of *Neurospora* was published by Léveille in 1843 in France, who named the organism *Oidium aurantiacum*. Shear and Dodge (1927) gave an excellent account of this and other early work on *Neurospora*, and in addition described some of its natural history.

It is indeed a pleasure to mention the name of a great scientist who appears next on this program, Dr. C. C. Lindegren. Dr. Lindegren started his work on *Neurospora* in the laboratory of T. H. Morgan in 1932. C. C. Lindegren and G. Lindegren (1941) were first to recognize the importance of *Neurospora* as a genetic tool and published papers on x-ray- and ultraviolet-induced mutations in *Neurospora*. The result of these publications was the later extensive program of genetic investigations carried on by Beadle and Tatum (1941) which resulted in the establishment of "biochemical genetics" as a field of study and in the so-called "one gene, one enzyme theory" (Beadle, 1946).

*This work was supported by the National Institutes of Health, Research Grant CY-3853 (C2).

211

Two mating types exist in *Neurospora*. They are accurately called mating types and not sexes, because each one is actually a hermaphrodite having the characteristics of both sexes. Conidia of each mating type mutually fertilize the protoperithecia of the other type. Mycelial fragments may also fertilize protoperithecia of the opposite mating type. In each of these crosses, the sexual spore is called an ascospore, being one of eight ascospores derived from a single zygote in one ascus, which develops inside a perithecium where many asci are located. In every ascus containing eight ascospores, there are at least two different genetic constituents represented. In asexual reproduction, a "single strain" reproduces by means of conidia, or mycelial fragments, but never forms a ripe perithecium with ascospores.

Slide cultures of *Neurospora* were studied extensively, and the growth recorded by means of phase-contrast cinematography (Hsu and Fuerst, 1958).

The photographs were taken from slide cultures which were prepared in the following way: A chamber of paraffin was built up on a glass slide. A cover slip was painted with complete *Neurospora* agar medium, and dusted with *Neurospora* conidia. or ascospores were dropped on it. The chamber received a droplet of water to keep it moist and the cover slip was inverted over it and sealed airtight. Time of setting was always noted. The slide was then placed under a time-lapse machine equipped with a phase-contrast microscope and a Kodak 16-mm movie camera. Again the exact time was noted. The frames were projected by means of a camera lucida on graph paper, and each movie frame was examined individually.

In most preparations, some normal fragments of mycelium could be seen. Occasionally a big fat spore showed up, a giant in a culture of mostly small conidia. Other conidia seen on the same slide were all of the normal-size variety. Mycelial fragment strands stretched across the field.

Often, conidial spores of a different variety show multiple germination tubes, which is not the normal condition. Usually a conidium sends out a single germination · tube, which, perhaps later on, may fuse with other such germination tubes, making the entire culture in essence, a single individual with protoplasmic flow going through the entire culture. This indicates clearly the multinuclear condition of *Neurospora*. This condition, together with protoplasmic flow, necessitates the picking of individual ascospores, or the use of the spore-shooting technique, in order to pick mutants after the culture has been exposed to a mutating agent. The chances of picking a mutant from mycelial fragments are obviously almost zero, since all normal nuclei or other mutants would mask the perhaps existing and isolated mutant. Thus in order to get mutants, one has to proceed through the single nucleus ascospore stage. In other words, one has to pick spores. Occasionally, workers attempt to establish so-called microconidial strains. To think that only one nucleus would persist per conidium is indeed wishful thinking. Protoplasmic flow proceeds even through the conidia, and with it, redistribution of the nuclei in the culture occurs.

Inside the mycelium, there are septa between the individual cells, with holes in them. Bubbles and cellular inclusions together with nuclei move through these septa from cell to cell. Time-lapse studies have shown that this flow occurs in both directions. (Hsu and Fuerst, 1958). It is also possible to follow a single bubble throughout a great portion of the culture and perhaps see it return to its starting place. It is really fascinating to see a big fat bubble squeeze through one of these small holes.

Conidia form at the tips of mycelial hyphae. Later on, the pink conidia form the entire surface growth of the culture. Often, a zone of change in the agar is observed. This may be brought about by the heat of the growing cells or may be the result of a celluloselike enzyme which has been reported in *Neurospora*.

During germination of an ascospore, both ends may germinate at the same time. This is not always the case. When a number of germinated ascospores were counted, about twice as many germinated at one end than did at two ends.

After ascospore germination, the growth of individual hyphal tips can be observed. Often the hyphal tip grows into the microscopic field at the bottom of the frame, to anastamose later on with the mycelia on the field. It is most interesting to watch how mycelia will grow together in this way. Fusion of mycelial tips occurs in a matter of minutes.

Although any analogy to the slime molds is purely coincidental, the question arises. "What attracts one hyphal tip to another one, so that it practically grows out of its way to fuse in anastamosis with another tip?"

A heterocaryon is a culture containing more than one type of nucleus but being of the same mating type. Obviously in *Neurospora*, a culture containing the nuclei of a certain vitamin-requiring mutant will grow without this vitamin, if wild type nuclei are also present that produce the enzymes needed for the synthesis of this growth factor or vitamin.

Vegetative cultures of *Neurospora* have to be considered to be *one* organism. The metabolic system of the organism is the product of all of the different chromosomal determinants in all of the different nuclei of this culture.

Several different strains of *Neurospora* were examined with respect to spore size. This included Em5256A, Em5297a, which are both wild types, and T34A, an adenine-less purple pigment mutant. Conidiospores in these strains varied from 5μ to 16μ in size for five-day-old cultures. The largest spores recorded for Em5297a were 11μ, while T34A reached only 9μ.

Of course the age of the culture makes quite a difference. The rate of growth does vary between individual spores but is close to about 1μ every 5 min. The size of the conidia is also in direct correlation to the time required to germinate. Spores 16μ in size germinated in about 1 hr, while those 6 in size would require 3.5 hr. The three different strains exhibited the same trend in size-germination relationship. It is not clear if the difference in spore size between the Em5256A wild type and T34A, the adenine-less mutant, is due to actual spore-size differences or to the selection of the test sample. It was necessary, after all, to select a field containing only a few spores, for photography.

Ascospores were heat-activated in order to germinate and then did send out germination tubes. These tubes were measured in square micra. Between 1.5 hr and 3.5 hr, all spores that would germinate after heat-activation did germinate.

The rate of growth of the germination tubes was almost constant for all spores. Germination at both ends would be delayed in one of the tubes after it had started.

About $100\mu^2$ of growth would occur in 45 min. The second tube, however, would rest for about 3 hr and then proceed at close to the same rate as the first tube.

The inheritance of nutritional mutants of *Neurospora* has been studied most extensively, although morphological mutants have also received their share of attention. These mutants differ from each other not only in their growth requirements, but also in some phase of their metabolic pattern, and as the author has shown, in their pattern of free intracellular components, in the case of free intracellular amino acids. (Fuerst and Wagner, 1957).

Free intracellular amino acids were demonstrated to be present as a definite cell fraction of *N. crassa*. Some 35 ninhydrin-positive substances were encountered amoung 28 different strains analyzed; however, none of the strains had all of these substances present in their extracts. In some mutants internal free amino acids were found to be associated in their expressions with mutant characteristics. In other mutants, certain amino acids were consistently absent. Other apparent differences disappeared in backcrosses to wild type, and new amino acids showed up. Table 1 shows the amino acid patterns of some of the strains tested.

When certain strains were backcrossed to wild type to the F_5 generation, results showed that a characteristic free intracellular amino acid pattern is asso-

Table 1. Free Amino Acid Content of Various *Neurospora* Strains

Strain[b]	Alanine	Threonine	Glutamine	Serine	Glycine	Lysine	Glutamic acid	Aspartic acid	Tyrosine	γ-Aminobutyric acid	Glutathione	β-Alanine	α-Aminobutyric acid	Arginine	Methionine-valine (Meth. stnd.)	Leucine-phenylalanine (Leu. stnd.)	Alanine for standard					
																	Basic amino acids	Ethanolamine?	Unknown E	Unknown 22 (taurine?)	Unknown D	Unknown F	Unknown G
Em5256A	127	49	43	203	4	75	94	14	36	32	55	-	-	3	35	86	75	166	-	--	24	-	-
Em5297a	164	34	73	131	9	47	103	12	16	5	93	16	18	-	56	89	37	5	-	2	17	-	-
36104a	92	200	120	634	38	85	100	37	-	-	206	-	-	5	29	21	52	11	30	--	40	2	-
T8a	138	196	125	39	17	109	139	37	-	9	-	29	-	-	160	105	127	32	-	16	--	-	-
T77a	143	68	0	266	24	191	118	67	116	28	-	11	-	93	91	141	80	29	-	8	-	-	-
T145A	338	68	334	563	23	57	151	26	34	-	-	72	35	-	72	105	83	24	-	-	48	-	-
T34A	138	62	117	113	10	39	121	36	31	7	-	-	-	-	88	91	43	-	3	-	-	-	10
T72A	211	57	167	159	16	52	139	38	16	5	33	-	18	11	40	63	24	-	-	-	32	-	-
70007A	143	117	157	144	7	90	118	24	-	52	-	-	-	-	27	56	80	16	2	-	-	-	-
15300a	122	34	99	131	0	11	88	10	5	-	96	-	-	-	29	56	24	2	3	-	-	-	-
T111	211	61	68	371	45	113	100	12	116	41	415	-	-	19	96	192	99	40	-	10	18	-	-
T150	100	72	54	296	40	67	54	62	120	32	728	-	18	40	160	515	57	8	-	-	-	-	-
T7A	175	57	261	300	47	65	121	24	19	28	-	59	18	35	40	54	52	17	-	-	-	-	-
T102	83	99	52	159	8	71	103	38	-	18	179	-	-	-	75	91	80	30	-	-	-	-	13

[a] Readings given in micrograms of amino acid/100 mg dry mycelial weight.

[b] All strains except Em strains (wild types) are mutants.

ciated with the genotype of the mutant that controls the mutational deficiency. Up into the F_5 generation, differences in the amino acid pool stayed consistent and were not lost on backcrosses to wild type. This indicates that the amino acid pattern must be related to the particular phenotype brought about by a single gene mutation.

From a cross between a methionine-less mutant and an adenine-less purple pigment mutant, all possible F_1 segregants were obtained. Several strains of each recombination of segregants were tested by chromatography for both qualitative and quantitative differences in their free intracellular amino acid pattern. Differences appeared to be peculiar to each of these types of F_1 progenies. Some minor differences existed between strains that had the same mating type and growth factor requirement, but in the general pattern, this experiment proved again the conclusion that was drawn from the F_5 experiment with backcrosses to wild type. Results from one experiment of this type are shown in Table 2. When a strain is grown on different media, the composition of the medium determines to some extent the relative composition of the free intracellular fraction.

Several compounds have the ability to colonize or restrict *Neurospora* growth. Experiments conducted by Tatum *et al.* (1949) revealed that some surface-active anionic compounds, such as Tergitol-7 or Sodium Desoxycholate, caused colonial growth in *Neurospora* and *Syncephalastrum*. No direct relation of surface activity to the colonial paramorphogenic activity was observed in all of the cases tested.

The author has tested a number of other colonizing agents on *Neurospora* grown on minimal medium (2% agar). Normal growth of mycelium produces a certain amount of branching. Colonial mutants show heavy, thick growth with considerable curling of the hyphae and thus colonial growth. There is also another reason for colonization. This type of colonization is brought about by 1% sorbose in the medium. The organism grows; then suddenly the pressure becomes too great and the growing tip bursts. The mycelial content discharges into the medium and gelatin occurs immediately. Many chemical substances and conditions were investigated; among them, colonial growth was induced by: gammexane, sodium desoxycholate, L-sorbose, dupanol, p-hydroxybenzophenone azine, 6,8-diamino-2,3-dimethyl quinoxaline, and pH 3.0.

Recently, Fuerst and Higgins (1960) investigated several thiophenes and found some that would also colonize *Neurospora*. 2-(2-Ethylhexyl)thiophene would cause the organism to grow in perfect circles that felt smooth and silky to the touch, while 2-(2-methylpropyl)thiophene, and 2-heptylthiophene would feel coarse to the touch. Obviously, the reasons for colonizing are different for these different compounds. Casein hydrolysate relieved the colonizing effect of these latter two compounds.

When *Neurospora* was exposed to high sucrose concentrations of 10% or 20%, curling of the mycelia would occur, except that one or two hyphae per colony would always grow out in normal fashion.

In all cases observed, the curling of the mycelia was in clockwise direction.

During the last four years, considerable time was spent in the author's laboratory, testing some of the known and some new antitumor antibiotics against

Neurospora . Most of these inhibited *Neurospora* growth remarkably. Two methods were perfected for testing. One is a flask method, the other a plate-well method. The latter method was used to adapt *N. crassa* Em 5256A a wild type and drug-sensitive strain to inhibitory doses of different inhibitors.

The procedure used consists of the following: One hundred milliliters of sterile minimal *Neurospora* (3% agar) medium was poured into 150×20 mm petri dishes. Not more than five evenly spaced holes were punched into the solidified agar, straight to the bottom of the petri dish. The punching was done with a cork borer No. 3 (8 mm diameter). The borer was sterilized by dipping it in ethanol and flaming it with an alcohol burner. After removal of the agar plugs, 15 μl of a heavy microconidial suspension of *N. crassa* was pipetted into each well from a 50-μl micropipette, followed by a range of, for instance, 10 μl to 70 μl of each solution to be tested. The *Neurospora* suspension was prepared by pipetting 10 ml of sterile distilled water onto a slant of *Neurospora* in order to wash the conidia from the slant.

After incubating the plates for 18 hr at 25 C, the zones of growth around the wells were measured in millimeters with the aid of calipers. The plates were

Table 2. Free Amino Acids and Other Ninhydrin-Positive Substances in Some
F_1 Segregants from a Cross between 36104a $\times$ T34A

All optical-density values reported correspond to 40 μl of solution as based on 100 mg dry mycelial weight material/ml of extract.

Phenotype	Wild type			Methionine-less			Purple adenine-less			Double mutants		
Weight per flask, mg	109	120	126	131	107	106	12	12	11	13	11	13
Alanine[a]	0.47	0.65	0.73	0.97	0.82	0.89	0.62	0.64	0.76	0.77	0.61	0.99
Threonine	0	0	0	0.50	0.57	0.60	0	0	0	0.12	0.08	0.10
Glutamine	0.21	0.23	0.23	0	0.36	0.30	0.36	0.30	0.46	0.47	0.26	0.38
Methionine-valine	0.03	0.05	0.05	0.15	0.12	0.23	0.16	0.23	0.31	0.20	0.26	0.57
Phenylalanine-leucine-iso-leucine	0.09	0.07	0.04	0.16	0.19	0.22	0	0	0	0.29	0.06	0.18
α-Aminobutyric acid	0	0	0	0.02	0.13	0	0	0	0	0	0	0
γ-Aminobutyric acid	0.11	0.22	0.10	0.24	0	0.53	0	0	0	0.10	0	0
Serine	0.06	0.05	0.08	0.11	0.16	0.15	0.10	0.08	0.15	0.14	0.12	0.11
Glycine	0	0	0	0.24	0.14	0.43	0	0	0	0	0	0
Glutamic acid	0.24	0.31	0.28	0.35	0.55	0.40	0.42	0.37	0.37	0.53	0.32	0.54
Aspartic acid	0.13	0.16	0.14	0.27	0.34	0.36	0.12	0.14	0.14	0.14	0.26	0.31
Glutathione	0	0	0	0	0	0.16	0	0	0	0.32	0.18	0.13
Unknown F	0	0	0	0	0	0	0.02	0.02	0.13	0.11	0.06	0.06
Unknown 20	0	0	0	0	0	0	0	0	0	0.11	0.06	0.07
Taurine	0	0	0	0	0	0	0	0	0	0.12	0.09	0
β-Alanine	0	0	0	0	0.17	0	0	0	0	0	0	0
Tyrosine	0	0	0	0	0.05	0	0	0	0	0	0	0
Unknown B	0	0	0	0.23	0	0.29	0	0	0	0	0	0

[a]All the values for the free intracellular amino acids are listed as optical densities.

stored in a cold room (10 C) for about 5 hr in order to facilitate the convenience of the experiment with regard to the time element involved. In this way plates can be inoculated from other plates every 24 hr.

The shortest and the longest individual *Neurospora* mycelial strands were then specifically selected and were transferred to individual wells in another plate. The transfer was accomplished with the aid of a sterile, very sharp, steel surgical needle. An increased amount of the solution to be tested was then pipetted into the well with a 50-μl micropipette and the plate was again incubated for 18 hr at 25 C. The experiment was carried on in this way until *N. crassa* had become adapted to very large inhibitory doses of the compound. Throughout the experiment, occasional *Neurospora* mycelial strands were transferred to *Neurospora* minimal (2% agar) medium slants containing the full inhibitory dose of the compound. The slants were allowed to conidiate at 25 C and were then stored in the refrigerator.

Table 3. Attempts to Adapt *Neurospora crassa* Em 5256 to Inhibitory Doses of Inhibitors by Using the Plate-Well Method

Compounds adapted to higher inhibitory concentrations	Initial				number of consec. transfers	Final		Tube number	Tube inoculated from	
	inhib. conc., mg	mycel. growth, mm	trans. conc., mg	mycel. growth mm		tested conc., mg	mycel. growth, mm		tested conc., mg	mycel. growth, mm
8-azaguanine	0.35	0	0.30	3	1*	0.30	3			
					2*	0.61	9	1	0.61	9
					3*	0.46	3			
					4*	0.80	11	2	1.25	5
					5*	1.52	13	3	1.52	10
					6*	2.13	11			
					7*	2.28	7	4°	2.28	7
					8*	4.26	5	4	4.26	5
					9*	5.48	8	4'	5.47	8
4-aminopyrazolo (3,4-d)pyrimidine	0.24	5	0.10	11	1	0.24	8			
					2	0.14	4			
					3	0.21	6			
					4	0.51	3			
					5	0.68	4			
					6	1.01	1			
					7	4.60	32			
					8	4.06	9			
					9	5.48	2			
					10	2.03	4			
					11	4.06	1			
					12	4.06	10			
					13	4.06	5	A	4.06	5
					14	4.46	1			
					15	3.65	1			
					16	4.06	5			
					17	4.46	0			

*Selected for longest strand; all unmarked transfers were selected for the shortest available strand.

Table 3 shows some results obtained with 8-azaguanine and with 4-aminopyrazolo (3,4-d)pyrimidine, (4-APP). Control growth in this experiment was 22 mm. It can be seen from these data that the mycelial growth was as good on higher concentrations of these compounds as previously on low quantities. The organism became adapted to more than 15 times the originally tolerated concentration of these compounds.

Table 4. Attempts to Adapt *Neurospora crassa* Em 5256A to Inhibitory Doses of Inhibitors by Using the Plate-Well Method

Compounds adapted to higher inhibitory concentrations	Initial				number of consec. transfers	Final		Tube number	Tube inoculated from	
	inhib. conc., mg	mycel. growth, mm	trans. conc., mg	mycel. growth, mm		tested conc., mg	mycel. growth, mm		tested conc., mg	mycel. growth, mm
Aminopterin	3.21	11	4.58	14	1	3.21	11			
					4	9.17	11	1	9.17	11
					1*	4.58	14			
					3*	10.66	15			
					4*	10.31	15	1'	10.31	15
2,6-diamino-purine sulfate	1.17	3	1.41	18	1*	0.43	14			
					4*	12.30	14	1	12.30	14
					6*	17.51	25	2	17.51	25
					1	0.43	14			
					4	12.30	14	1'	6.40	5
					6	15.74	11	2'	15.74	11
Methotrexate	0.15	0	0.14	2	1	0.12	2			
					4	2.40	13	3	2.40	13
					1*	0.12	2	1	.12	2
					5*	4.50	18	4	3.90	17
4-aminopyrazolo (3,4-d)pyrimidine	0.24	5	0.10	11	1	0.24	8			
					6	1.01	1	2	1.01	5
					9	5.48	2			
8-azaguanine	0.35	0	0.30	3	1*	0.30	3			
					9*	5.48	8	4'	5.47	8
4-(2-thienyl) butanoic acid	0.18	0	0.15	6	Eight attempts to increase the concentration and to get growth failed.					

*Selected for longest strand; all unmarked transfers were selected for the shortest available strand.

Table 4 shows some other inhibitors to which *Neurospora* has also been successfully adapted. Only a limited amount of all data collected is presented here. Not shown in the table are other compounds like the 4-(2-thienyl)butanoic acid (which is listed) that were tested, but which would not grow after the first transfer and thus repeatedly defied all attempts to grow on higher than inhibitory doses.

Some of these compounds were 4-APP derivatives, thiophenes, and aminopterin, actinobolin, and DON which is 6 diazo-5-oxy-nor-L-leucine.

Several days after the adapted strains were placed as slant cultures in the refrigerator, tests were set up in flasks and later on in plates in order to determine whether these strains are actually permanently drug-resistant. Results on both liquid culture and plates were disappointing. Most of the cultures had reverted again to drug-sensitive. Some exceptions are shown in Table 5.

Table 5. Tests for Resistance of Drug-Resistant Strains
of *Neurospora* on Agar Plates

*	Number of adapted strain	Chemical compound	Amount of chemical compound, mg	Growth of adapted strain, mm	Growth of control strain, mm
I.	4° (Resistant to 2.3 mg)	Azaguanine	0	29	23
			0.5	20	11
			0.8	14	8
			1.0	12	5
	1' (Resistant to 21.6 mg)	4-APP	0	41	30
			0.5	35	18
			0.8	30	17
			1.0	36	3
II.	4° (Resistant to 2.3 mg)	Azaguanine	0	6	23
			0.5	18	6
			0.8	0	5
			1.0	14	4
	1' (Resistant to 21.6 mg)	4-APP	0	25	18
			0.5	0	6
			0.8	0	4
			1.0	0	3

* I.- Strand-inoculated plates.
  II.- Drop-inoculated plates.

The azaguanine strain, for example, grown on 1 mg of azaguanine gave 14 mm growth as compared to 4 mg for the control. 4-APP strain 1' grew 36 mm compared to 3 mm for the control in one of the experiments, but in another series (II), it failed to grow on higher concentrations of 4-APP.

It has been possible to adapt *Neurospora* to very high doses of usually inhibitory substances. Some enzyme systems may have been altered in this way, apparently most of them temporarily only, some perhaps longer. The selected short hyphal strands were adapted to the drugs. Later, after growing and dividing, other nuclei and enzymes expressed themselves again, so most strains selected and now under different environmental conditions failed to retain their drug resistance. A few, however, did retain it. How long, is, of course, another question.

How about crossing these strains with others How about cross-resistance to other drugs Indeed, this problem of variations in *Neurospora* has just begun.

REFERENCES

Beadle, G.W., and Tatum, E.L. 1941 Genetic control of biochemical reactions in *Neurospora*. Proc. Natl. Acad. Sci. 27, 499-506.

Beadle, G.W. 1946 Genes and the chemistry of the organism. Am. Scientist, 34, 31-53.

Becker, J.G. 1960 Introduction. What is microbial variation?

Fuerst, R., and Wagner, R.P. 1957 An analysis of the free intracellular amino acids of certain strains of *Neurospora*. Archives of Biochemistry & Biophysics 70, 311-326.

Hsu, T. C., and Fuerst, R. 1958 The Biology of *Neurospora crassa* (motion picture with sound). Released by the M. D. Anderson Hospital, Houston, Texas.

Lindegren, C. C., and Lindegren, G. 1941 X-ray- and ultra-violet-induced mutations in *Neurospora*. I. X-ray mutations. J. Heredity, 32, 405-412.

Shear, C. L., and Dodge, B. O. 1927 Life histories and heterothallism of the red bread-mold fungi of the *Monilia sitophila* group. J. Agric. Res., 34, 1019-1042.

Tatum, E. L., Barratt, R. W., and Cutter, Jr., V. M. 1949 Chemical induction of colonial paramorphs in *Neurospora* and *Syncephalastrum*.Science, 109, 509-511.

Wagner. R. P.. Haddox, C. H., Fuerst, R., and Stone, W. S. 1950 The effect of irradiated medium. cyanide and peroxide. on the mutation rate in *Neurospora*.Genetics, 35, 237-249.

PHYSIOLOGICAL AND CULTURAL VARIATION OF GENETIC ORIGIN IN YEASTS

(ABSTRACT)

Carl G. Lindegren

Biological Research Laboratory, Southern Illinois University, Carbondale, Illinois

In the yeast *Saccharomyces*, various granules are present. Those in the cytoplasm are assorted in a random manner; those in the nucleus are more regularly assorted. In budding, unequal passage of granules to daughter cells tends to produce variants.

The vegetative stages sporulate; then the resulting spores germinate to reform the vegetative phase. Spores of the a and alpha mating types may fuse and produce a new hybrid. Two vegetative cells may also fuse. A haploid nucleus from each cell enters the bud and fusion of the two nuclei occurs.

Cells heterozygous for two genes may produce several new genotypes because of ploidy.

Another type of variation involves respiratory capacity. Respiratory-sufficient organisms are grown in glucose agar, which is then overlaid with agar containing one per cent triphenyl tetrazolium chloride. The colonies are all white initially, but later, those that are respiratory-sufficient turn red, while the respiratory-deficient mutants remain white.

The respiratory potentialities are inherited both through the nucleus (genes), and factors in the cytoplasm. The same gene seems to be involved in all the respiratory mutants.

A zymophage attacking these yeasts is the first phage to be reported in the higher fungi.

MODIFICATION OF RADIATION EFFECTS IN BACTERIA

M. R. Zelle

Division of Biology and Medicine, U.S. Atomic Energy Commission
Washington 25, D.C.

Radiobiological studies in bacteria began in 1877 when Downes and Blunt published their observations of the killing of bacteria by sunlight. A voluminous literature has been accumulated since that date, and in a short discussion it is impossible to accomplish more than to mention briefly some aspects of the subject which, hopefully, are of interest.

There are several reasons for the widespread interest in radiobiological effects in bacteria and other microorganisms. They are of obvious practical importance in such processes as food preservation and pasteurization, sterilization of vaccines, and many other industrial or commercial applications. Despite the literally thousands of papers which have been published since 1877, even today we know very little about the exact kinds of damage responsible for the inactivation of bacteria and very little about the mechanisms whereby this damage is produced. In addition, microorganisms serve as model systems for the study of cellular radiobiology and such studies have contributed greatly to our understanding of radiobiological effects in cells of higher organisms. For example, mammalian tissue-culture techniques have borrowed much from bacterial genetics and the utilization of tissue-culture techniques has only begun to be exploited in studies of mammalian radiation biology, genetics, carcinogenesis, and other problems. Bacteria and other microorganisms have been utilized as screening agents for chemical radiation-protectant compounds, and detailed and quantitatively precise studies of radiobiological phenomena in bacteria are possible which are presently illuminating radiobiological processes in cells in general.

The wide range of cellular properties which may be affected by radiation in bacteria complicates any consideration of radiobiological effects. Conversely, however, these cellular processes may be studied with radiation as an experimental tool. These effects include metabolic processes such as respiration, nucleic acid synthesis, protein synthesis, adaptive enzyme synthesis, specific amino acid uptake, and virus synthesis, to name only a few. All such studies are obviously important and may shed much light on normal cell function as well as on the radiobiological processes. However, the greatest attention has been focused on two effects: loss of ability to divide, i.e., killing or inactivation; and viable genetic changes or mutation. Since the problem is so broad, this discussion will be confined largely to these latter two radiation effects.

The problem is further complicated by the wide variety of radiations including very low and very high quantal energies which can produce cellular effects in bacteria. Low-energy radiations are of interest because of their selective absorption by different molecules of biological importance. Interesting and important effects have been observed following infrared, visible, near ultraviolet, and ultraviolet irradiation (UV). Perhaps the greatest interest has centered about the ultraviolet effects (2100-3000 A). Action spectra techniques very early (Gates, 1928) focused attention on the nucleus and deoxyribonucleic acid as the chromophore and probable site of lethal radiation damage in bacteria. Even earlier, Henri (1914) had concluded from crude action spectrum studies utilizing filters that the lethal effects of UV were due to the absorption of the energy in the proteins of the nucleus and reasoned that sublethal doses might lead to heritable modifications. In experiments expressly designed for this purpose, Henri observed both stable and unstable genetic changes in *Bacillus anthracis* following exposure to UV radiation.

A wide variety of high-energy radiations has been studied ranging from the softest x-rays through high-energy gamma rays and including neutrons, electrons, protons, deuterons, alpha particles, and, recently, heavy nuclei accelerated to high energies. Perhaps the most important variable in considering the high-energy radiations is the ionization density or rate of linear energy transfer along the track of the particle.

There are many points of similarity between the effects of high and low energy radiations, or between UV and x-rays with which the vast majority of such studies with microorganisms have been concerned, but there are also many differences. For example, both UV and x-rays can inactivate cells and produce viable genetic changes, but UV is somewhat more effective in producing viable genetic changes than x-rays at a given level of survival. Both x-ray and UV effects can be modified by a variety of experimental conditions. In some cases, both UV and x-rays are affected, in others only one or the other. For example, ultraviolet effects are generally partially reversed or prevented by exposure to so-called photoreactivating light (Kelner, 1949) but no appreciable photoreactivation has been reported for any x-ray-produced effect (Jagger, 1958). The oxygen concentration during irradiation greatly affects the efficiency of x-rays in killing bacteria (Hollaender *et al.*, 1951). In certain bacterial strains, the postirradiation plating conditions have a pronounced effect on the survival following both x-rays and ultraviolet (Stapleton, 1955a; Gillies and Alper, 1959; Doudney and Haas, 1958). Especially interesting in this connection is the inhibition of protein synthesis by various means, most commonly by the incorporation of chloramphenicol into the plating medium. In this connection the inhibition of protein synthesis and other alterations of postirradiation metabolism greatly influence the frequency of mutations following ultraviolet radiation (Haas and Doudney, 1958; Witkin, 1956, 1958; Lieb, 1959), but it is not clear that similar effects on mutation frequency are obtained following high-energy radiations. In any case, since various strains of bacteria differ in their response to postirradiation plating conditions (Alper and Gillies, 1958) and since different specific mutations in the same strain of organism, *E. coli* B/r, vary in their mutagenic response as influenced by postradiation metabolism and protein synthesis (Zelle *et al*. 1958; Witkin, 1958) too ready generalization from a particular study utilizing one radiation and one strain or mutation system to all radiations and mutation in general may serve to becloud rather than clarify the total picture.

DIRECT VERSUS INDIRECT EFFECT

Many radiobiological phenomena have been interpreted in terms of the target theory of which Lea (1947) was the strongest proponent. We have already referred to the action spectrum technique in which clues as to the biologically important molecules affected by ultraviolet radiation could be obtained by noting similarities between the absorption spectra of various compounds and the action spectrum or relative efficiencies of different wavelengths in producing the effect. In both the action spectra and target theory techniques, the assumption was made that the biological effect was produced in the molecules absorbing radiation. Since, however, it is known that a major proportion of ultraviolet effects are photoreversible (Kelner, 1949) and that oxygen greatly increases the efficacy of x-rays (see Gray, 1959, for review) the assumption of a direct effect in the molecule absorbing the energy is not valid under all conditions since a major portion of both ultraviolet and x-ray effects may be indirect. The discovery of the influence of oxygen and of photoreactivation has been a powerful stimulus to radiobiological research and attempted analyses of these phenomena have contributed greatly toward the understanding of radiobiological mechanisms.

It is clear, therefore, that indirect mechanisms exist for both ultraviolet- and x-ray-produced effects. It has commonly been assumed that the residual effects, i.e., the proportion of x-ray effect not influenced by oxygen concentration and the proportion of ultraviolet effect which is not photoreversible, occur by direct mechanisms. Although under certain experimental conditions, such as the absence of oxygen and dehydration, it is certain that direct effects of radiation are produced in enzymes *in vitro*, it is not clear that an appreciable proportion of the total cellular radiobiological effect of radiation is produced by similar direct effects. Of interest in this connection is the observation of an oxygen effect in crystalline trypsin in which the inactivation is generally presumed to be direct (Alexander, 1957) and of protection by glutathione against the presumed direct effects of x-radiation on tobacco mosaic virus (Ginoza and Norman, 1957). The difficulty of complete removal of water and the possible importance of bound water (Wood and Taylor, 1957), must be remembered. Until recently, all studies of chemical protection against radiation in bacteria have utilized oxygenated systems. Kohn and Gunter (unpublished) have observed a chemical protecting effect for bacteria irradiated under anaerobic conditions. This is further evidence that the presumed direct effects not influenced by oxygen concentration may actually involve other, indirect mechanisms.

MODIFICATION OF RADIATION EFFECTS IN BACTERIA
CULTURAL AND PHYSIOLOGICAL CONDITIONS

Hollaender *et al*. (1951) observed that cultivation of *E. coli* B/r under anaerobic conditions significantly increased the resistance to x-rays and furthermore that the inactivation curve became sigmoidal rather than exponential as observed for aerobically grown cells. The effect of anaerobic cultivation was additive to the effect of oxygen concentration at the time of radiation. Although these results have been confirmed by others (Howard-Flanders and Alper, 1957; Sargeant, 1958) the underlying causes are still obscure. Whereas Sargeant (1958) reports that the deoxyribonucleic acid content per cell is greater, suggesting that the cells are multinuclear or actually multicellular, Adler and Stapleton (unpublished data) found no evidence of a greater DNA content per cell in $E.$ $coli$ B/r. It is not clear that the increased resistance and change in inactivation kinetics is a reflection of a change in genetic make-up. Rather it is conceivable that the increased resistance is due to a fermentation product which acts as an intracellular chemical protecting agent in much the same manner as reported by Kilburn $et\ al.$ (1958) for the resistant Sarcina in which a mercaptoalkyl-amine is formed and which may be responsible for the extremely high resistance of this organism.

Stapleton (1955b) and Brownell (1955) have shown that the stage in the culture growth cycle greatly influences the resistance of the organism and the inactivation kinetics observed. Both authors note that the change in sensitivity parallels changes in number of nuclear bodies, although it is not clear in $E.\ coli$ (see Zelle, 1955) that inactivation by radiation in $E.coli$ is due to genetic or nuclear damage.

The enzymatic constitution of $E.\ coli$ is known to influence the inactivation by x-rays. Thus Stapleton (1955a) has shown that $E.\ coli$ cells grown on nutrient broth exhibit a lower survival following x-radiation if plated on a minimal medium as compared to nutrient agar. Cells grown in a minimal medium before irradiation do not exhibit this difference in survival on minimal medium, presumably because they already possess the enzyme systems required for growth on the minimal medium. Adler (1958) has studied a heminless mutant of $E.\ coli$ in which all porphyrins are absent. Such cells are more sensitive to x-rays and exhibit a post-radiation sensitization to hydrogen peroxide produced in the extracellular medium.

EXPERIMENTAL MODIFICATION OF RADIATION SENSITIVITY

It has long been known that the water content profoundly influences the radio-sensitivity of a variety of biologically important molecules and organisms (Lea, 1947). Thus the observations on the inactivation of plant viruses in aqueous and dried states (Lea, 1947) and the protective effect of proteinaceous material in the aqueous systems are a classical example of indirect effects of ionizing radiations. The difference in water content may be partially responsible for the greater radiation-resistance of spores as compared to vegetative cells. Stapleton and Hollaender (1952) have shown that the sensitivity of $Aspergillus$ spores decreases with lower water content and that the mutagenic response as measured by morphological mutations is also decreased, but to a lesser degree. Interrelated with the effect of water content is the quantitatively smaller but still significant oxygen effect in dried spore systems observed by Powers $et\ al.$ (1960). The influence of water has been assumed to be due to the greater yield of free radicals in aqueous systems a yield which may be further increased by the presence of oxygen. In the same interpretation, the protective influence of proteins and other compounds may be by scavenging for free radicals or by forming compounds with oxygen to reduce the free oxygen concentration.

Although a number of investigators have observed changes in radiosensitivity with a change in phase state, some of the more interesting and critical data are those of Powers $et\ al.$ (1960) and Wood (1959). Powers $et\ al.$ have shown a linear increase in radiation sensitivity with temperature in the range from about 125 K to 325 K in dried *Bacillus megaterium* spores. No discontinuity was observed at 273 K. Wood has observed decreased radiosensitivity in the solid state and has shown an influence of rate of freezing upon the sensitivity of yeast cells.

The influence of the gaseous environment during radiation has already been mentioned in connection with the oxygen effect. The oxygen effect is quite general for a wide range of radiobiological effects in a number of different organisms. Recently, Howard-Flanders and Alper (1957) have shown that the maximal oxygen effect is obtained at considerably lower concentrations of oxygen than previously believed, as a result of improved equilibration techniques between the micro-organisms and their environment. Gray (1959) has recently compiled a list of dose-reduction factors and oxygen concentrations for maximal effect for a variety of radiation effects in a number of species, and the constancy of the observed values leads him to conclude that they are independent of the biological source and related to a fundamental process along the particle track. Gray has reviewed much of the recent evidence dealing with the oxygen effect. Reference has been made above to one interpretation of the oxygen effect by enhancing the yield of free radicals in aqueous systems under radiation. Howard-Flanders and Alper (1957) are proponents of an alternative hypothesis that oxygen is effective by reacting with ionized target molecules. In their interpretation a target molecule is ionized with the resultant formation of a carbon radical somewhere within it. Irreparable damage will result if this carbon radical reacts with a second radical which may be molecular oxygen, nitric oxide, or a radical produced by a densely ionizing radiation nearby. The nitric oxide observation is very important since it shows no additivity to oxygen and is equally effective in their system. Hutchinson (1957) has supplied evidence that oxygen probably does not produce its effect through enhancing the formation of HO_2 radicals. He observed a value of about 30 A for the mean action radius of various free radicals possibly involved in enzyme inactivation in yeast cells. This value is considerably smaller than had been previously estimated and is much smaller than the 700 A average distance between oxygen molecules at 10μ molar oxygen concentration where the oxygen effect is still fully operative. Hutchinson points out that the probability that an H radical will collide with an oxygen molecule during the migration time estimated to be 5×10^{-9} sec is negligibly small.

Although Howard-Flanders (1957) has found that nitric oxide is equally effective as oxygen in the vegetative bacterial and yeast cells studied, Powers *et al*, (1960) have not observed a similar equivalence in their studies of the oxygen effect in dried *B. megaterium* spores. On the contrary, they observed a protective effect of nitric oxide which appears to be quite similar to the decrease in sensitivity to heat applied after radiation and in the absence of oxygen. These investigators have partitioned the oxygen effect into two components, an immediate effect which increases sensitivity by a factor of 1.25 and a latent or postirradiation oxygen effect with a factor of 2.1. The product of these two factors is 2.6 which compares favorably with the factors of about 3 found in aqueous metabolizing systems.

Although there are many perplexing aspects of the oxygen and nitric oxide effects, continued analysis along the lines of Howard-Flanders and Alper (1957),

Wood (1959), and Powers *et al.* (1960), promise to contribute greatly to our understanding of the immediate postirradiation processes.

A wide variety of types of compounds have been utilized in studies of chemical protection against radiation. Stapleton (unpublished) has classified these into three categories: 1. monofunctional which act either by removing oxygen or by permitting cells to deplete themselves of oxygen (oxidizable substrates); 2. multifunctional which remove oxygen and compete for radiochemical products of H_2O ($NA_2S_2O_4$ and many SH compounds); 3. those that specifically protect sensitive sites or promote recovery. Space does not permit any attempt to examine in detail the results of chemical protection studies. Briefly, the same compound varies sometimes quite widely in effect in different strains of the same species or in different species or genera. The results of comparative studies of the protection against both lethal and mutagenic effects of radiation are confusing, the mutagenic protection afforded being quite variable. Many more studies of this type are needed and could be valuable in elucidating the similarities and differences in the processes leading to radiation-induced lethality and viable genetic change.

RECOVERY FROM RADIATION-PRODUCED DAMAGE

Although the term "recovery" has been used widely, there is still no clear-cut evidence, so far as the writer is aware, for a real recovery from any lethal or mutational radiobiological damage once it is fully accomplished. Rather, the term "prevention" may be more appropriate and a variety of postradiation conditions are known to influence the physiological processes which ultimately lead to the radiobiological damage. Stapleton *et al.* (1953) have shown that postradiation temperature of incubation has a profound influence on survival of *E. coli* and that different strains exhibit different optimal temperatures. A number of other postradiation conditions such as starvation, deprivation of certain nutrilites, and incorporation of certain metabolic inhibitors have been shown to influence the survival following a given radiation exposure. Such recovery phenomena have been observed in yeasts, bacteria, protozoa, and other microorganisms. Reference has already been made to the apparent recovery from ultraviolet radiation effects by exposure to photoreactivating light of longer wavelengths. Similarly, the influence of enzymatic constitution on postradiation survival has been discussed earlier (Stapleton, 1955a). Gillies and Alper (1959) and Doudney and Haas (1958) have shown that protein synthesis affects survival following both ultraviolet and x-ray radiation. Alper and Gillies (1958) found that survival of certain strains of bacteria are affected by "Difco" versus "Oxoid" media. These observations are reminiscent of the early observations of Roberts and Aldous (1949) on *E. coli* B where postradiation incubation in aqueous, nonnutritive media and other postirradiation treatments greatly influence the ultimate survival.

Some of the more interesting and important studies of this type have been those of Witkin (1951, 1958), Lieb (1959), and Haas and Doudney (1958) who have studied the influence of postirradiation metabolism and protein, DNA, and RNA synthesis on the ultimate rate of radiation-induced mutations in *E. coli*. Since radiation is known to inhibit DNA synthesis (Kelner, 1953), since DNA, RNA, and protein synthesis are interrelated in a fashion not clearly understood, and since

DNA is known to constitute the genetic material, it is not surprising that these synthetic processes influence the ultimately observed rate of mutation and that the interrelationships are so complex that there are certain points of apparent disagreement in the data of the different investigators. There is general agreement that inhibition of protein synthesis immediately following radiation reduces the ultimate yield of mutants. Haas and Doudney (1958) have postulated that three different processes related to RNA, DNA, and protein synthesis, are involved in the ultimate stabilization and fixation of potential mutations. Witkin (1956, 1958) has observed that there is a postirradiation sensitive period during which protein synthesis can affect the mutation rate and that treatments which prolong the first postirradiation division time have a corresponding effect on the length of the sensitive period. She concludes that the process leading to the termination of the sensitive period and to the irreversible fixation of the mutant frequency is associated with the synthesis of nucleic acids. It is important to note as Witkin (1958) points out, that not all mutation systems in *E. coli* B/r are similarly influenced by protein synthesis following radiation and that the greater than linear increase of mutation rate with dose sometimes observed in *E. coli* may be due to an increase in number of potential mutants induced along with an increase in the proportion of potential mutations actually expressed due to an increase in the length of the sensitive period by the greater prolongation of the lag phase with increasing ultraviolet dose. Studies of the interrelationships of RNA, DNA, and protein synthesis as affecting mutation rates are very important, not only on contributing to the understanding of radiation genetic effects, but also because of their implications on the question of duplication of genetic material and the interrelationships between nucleic acid and protein synthesis in general.

CONCLUSION

It is difficult to summarize this brief and sketchy survey of the current status of the modification of radiation effects in bacteria. There are few if any, sweeping generalizations or solid conclusions to be reached at this time.

Although the oxygen effect has been known and studied intensively for more than ten years, the exact nature of the mechanisms is still obscure and current concepts are quite different from those of just a few years ago. Thus HO_2 radicals are now considered of little importance and there is increasing evidence that there may be more than one mechanism whereby oxygen influences radiobiological effect. Although quite a little is known concerning indirect effects of radiation, very little is known of direct effects or even if direct effects constitute a significant proportion of the total effect of radiation. One of the big oversights has been the lack of data concerning protection against and modification of the presumed direct effects of radiation. Although in yeast and *Neurospora*, for example, there is good evidence that the lethal effects of radiation are due primarily to nuclear damage, in bacteria that exact nature of the lethal effects is obscure and it is uncertain if nuclear damage is involved. The intermediate steps between the absorption of the radiation energy and the final biological effect are almost completely unknown as yet. Different strains and different species react differently to various treatments. Lethal and mutagenic damage have some similarities and some differences. There seems to be almost total lack of generality. However, the situation is far from hopeless and, in fact, is encouraging since with the recognition that such wide differences exist, the utilization of more critical experimental controls and genetically specified strains in comparative studies of lethal and mutational damage promises to contribute greatly to our ultimate understanding of this complex problem.

REFERENCES

Adler, H. I. 1958 An after-effect of ionizing radiation on a mutant of *Escherichia coli*. Radiation Research, 9, 451-458.

Alexander, P. 1957 Effect of oxygen on inactivation of trypsin by the direct action of electrons and alpha-particles. Radiation Research, 6, 653-660.

Alper, T., and Gillies, N. E. 1958 Dependence of the observed oxygen effect on the post-irradiation treatment of microorganisms. Nature 181, 961-963.

Brownell, A. S. 1955 The effect of physiological and morphological changes on radiation sensitivity of *Escherichia coli*. Univ. of California Radiation Laboratory Technical Reports, U.C.R.L. 3055.

Doudney, C. O., and Haas, F. L. 1958 Modification of ultraviolet mutation frequency and survival in bacteria by postirradiation treatment. Proc. Natl. Acad. Sci. U.S., 44, 390-401.

Downes, A., and Blunt, T. P. 1877 Researches on the effect of light upon bacteria and other organisms. Proc. Roy. Soc. London, 26, 488-500.

Gates, F. L. 1928 On nuclear derivatives and the lethal action of ultraviolet light. Science, 68, 479-480.

Gillies, N. E., and Alper, Tikvah 1959 Reduction in the lethal effects of radiations on *Escherichia coli* B by treatment with chloramphenicol. Nature, 183, 237-238.

Ginoza. W., and Norman, A. 1957 Radiosensitive molecular weight of tobacco mosaic virus nucleic acid. Nature, 179, 520-521.

Gray, L. H. 1959 Cellular radiobiology. Radiation Research, Supplement 1, 73-101.

Haas. F. L. and Doudney, C. O. 1958 Interrrelations of nucleic acid and protein synthesis in radiation-induced mutation induction in bacteria. Proc. of the Second United Nations International Conference on the Peaceful Uses of Atomic Energy, 22, 336-341.

Henri. V. 1914 Etude de l'action métabiotique des rayons ultraviolettes. Productions de formes de mutation de la bactéridie charbonneuse. Compt. rend., 158, 1032-1035.

Hollaender, A., Stapleton, G., and Martin, F. L. 1951 X-ray sensitivity of *E. coli* as modified by oxygen tension. Nature, 167, 103-104.

Howard-Flanders, P. 1957 Effect of nitric oxide on the radiosensitivity of bacteria. Nature, 180, 1191-1192.

Howard-Flanders, P., and Alper, T. 1957 The sensitivity of microorganisms to irradiation under controlled gas conditions. Radiation Research, 7, 518-540.

Hutchinson, F. P. 1957 The distance that a radical formed by ionizing radiation can diffuse in a yeast cell. Radiation Research, 7, 473-483.

Jagger, J. 1958 Photoreactivation. Bacteriol. Reviews, 22, 99-142.

Kelner, A. 1949 Effect of visible light on the recovery of *Streptomyces griseus* conidia from ultraviolet irradiation injury. Proc. Natl. Acad. Sci. U.S., 35, 73-79.

Kelner, A. 1953 Growth, respiration, and nucleic acid synthesis in ultraviolet-irradiated and in photoreactivated *Escherichia coli*. J. Bacteriol., 65, 252-262.

Kilburn, R. E., Bellamy, W. D., and Terni, S. A. 1958 Studies on a radiation-resistant, pigmented Sarcina sp. Radiation Research, 9, 207-215.

Lea, D. E. 1947 Actions of Radiations on Living Cells. The Macmillan Company, New York. (Also Cambridge University Press, London, 1946).

Lieb, M. 1959 Factors influencing the postirradiation appearance of mutants in *Escherichia coli*. Genetics, 44, 523.

Powers, E. L., Webb, R. B., Ehret, C. F. 1960 Storage, transfer and utilization of energy from x-rays in dry bacterial spores. Bioenergetics Symposium, Brookhaven National Laboratory, Oct. 12-16, 1959. Proceedings to be published in supplement to Radiation Research.

Roberts. R. B.. and Aldous. E. 1949 Recovery from ultraviolet irradiation in *Escherichia coli*. J. Bacteriol., 57, 363-375.

Sargeant. T. 1958 Effects of pH and anoxia on growth and x-ray sensitivity of *E. coli*. Univ. of California Radiation Laboratory Technical Reports, U.C.R.L. 8513.

Stapleton, G. E. 1955a The influence of pretreatments and post-treatments on bacterial inactivation by ionizing radiations. Annals N. Y. Acad. Sci.. 59, 604-618.

Stapleton, G. E. 1955b Variation in the sensitivity of *Escherichia coli* during the growth cycle. J. Bacteriol., 70, 357-362.

Stapleton, G. E., Billen, D., and Hollaender, A. 1953 Recovery of x-irradiated bacteria at suboptimal incubation temperatures. J. Cellular Comp. Physiol., 41, 345-357.

Stapleton, G. E., and Hollaender, A. 1952 Mechanism of lethal and mutagenic action of ionizing radiations on *Aspergillus terreus*. II. Use of modifying agents and conditions. J. Cellular Comp. Physiol., 39, Suppl. 1, 101-113.

Witkin, E. M. 1956 Time, temperature and protein synthesis: A study of ultraviolet-induced mutation in bacteria. Cold Spring Harbor Symposia on Quantitative Biology, 21, 123-138.

Witkin, E. M. 1958 Post-irradiation metabolism and the timing of ultraviolet-induced mutations in bacteria. Proc. of the 10th International Congress of Genetics, 1, 280-299.

Wood, T. H. 1959 Inhibition of cell division. Radiation Research, Supplement 1, 332-346.

Wood, T. H., and Taylor, A. L. 1957 Dependence of x-ray sensitivity of yeast on phase state and anoxia. Radiation Research, 6, 611-625.

Zelle, M. R. 1955 Radiation induced mutations and their implications on the mechanisms of radiation effects on bacteria. Bacteriol. Reviews, 19, 32-44.

Zelle, M. R., Ogg, J. E., and Hollaender, A. 1958 Photoreactivation of induced mutation and inactivation of *Escherichia coli* exposed to various wave lengths of monochromatic ultraviolet light. J. Bacteriol., 75, 190-198.

THE MECHANISM OF CHEMICAL MUTAGENESIS WITH SPECIAL REFERENCE TO TRIETHYLENE MELAMINE ACTION

Waclaw Szybalski[*]

Institute of Microbiology, Rutgers, the State University, New Brunswick, New Jersey

The interaction between a bacterial cell and a chemical mutagen, leading to the production of a mutant cell, has many properties characteristic of a complex chemical process. We have attempted to evaluate the mechanism of this reaction sequence in a particular case, namely, triethylene melamine (TEM) induced mutations from streptomycin (SM) dependence to SM independence in *Escherichia coli* strain Sd4.

MUTAGEN AND MUTATIONAL SYSTEM

The rationale behind the selection of the mutagen and the mutational system can be summarized as follows: Under the experimental conditions employed in mutagenicity tests, TEM is a relatively stable compound endowed with high mutagenicity in association with low bactericidal activity (Iyer and Szybalski, 1958a). The system of mutation from SM dependence to SM independence in *E. coli* Sd4 (Bertani, 1951) also has many operational advantages, including simplicity of manipulation. SM-dependent cells are grown in SM-containing broth, and appropriate dilutions are plated on SM-supplemented nutrient agar (as a measure of total cell number) and on SM-free agar (as a measure of the SM-independent mutant cell number). The numerical ratio of mutants to total cells *in this particular case* is roughly proportional to the mutation rate, by virtue of the fact that most of the nondependent mutants, being SM-sensitive, do not survive in the SM-containing broth. Moreover, the parental dependent cells, when transferred to the SM-free medium, can undergo a predetermined number of divisions (depending on the SM content of the broth) until growth ceases. This period of residual growth permits the full phenotypic expression of SM independence and initiation of mutant colony formation. The loss of the SM requirement seems to be associated with a large series of suppressor mutations (Iyer and Szybalski, 1958a; Demerec *et al.*, 1957-1958), an advantageous feature when screening for nonspecific mutagens (or for their metabolic intermediates) is of primary interest. Thus, this system has many characteristics in common with mutations from prototrophy to various auxotrophies, but is amenable to simple quantitative evaluation.

KINETIC STUDIES

Kinetic studies on the mutagenic action of TEM were initiated in cooperation with Dr. V. N. Iyer (Iyer and Szybalski, 1958a). The experiments were designed in such a manner as to dissociate the chemical reactions between the constituents

[*]Present address: McArdle Memorial Laboratory, University of Wisconsin, Madison 6, Wis.

and the mutagen from bacterial growth and multiplication. The cells were treated with TEM while in the "resting" state so that the only components of the reaction mixture were the cells, the mutagen, and distilled water. The progress of the mutagenic process was followed by periodically taking samples, freeing them of mutagen, plating on nutrient agar, and counting the number of mutant colonies. It was soon found (Fig. 1) that the frequency of mutants increased during mutagen treatment at a rate dependent on temperature, but the maximum value (over one-thousandfold increase in the number of mutants) was *independent* of temperature. Moreover, after a short initial TEM treatment period (30-60 min), the course of the reaction at 2 C was not altered (Fig. 1) by removal of the mutagen by centrifugation, washing, and resuspension of the cells in TEM-free distilled water (Iyer and Szybalski, 1958a).

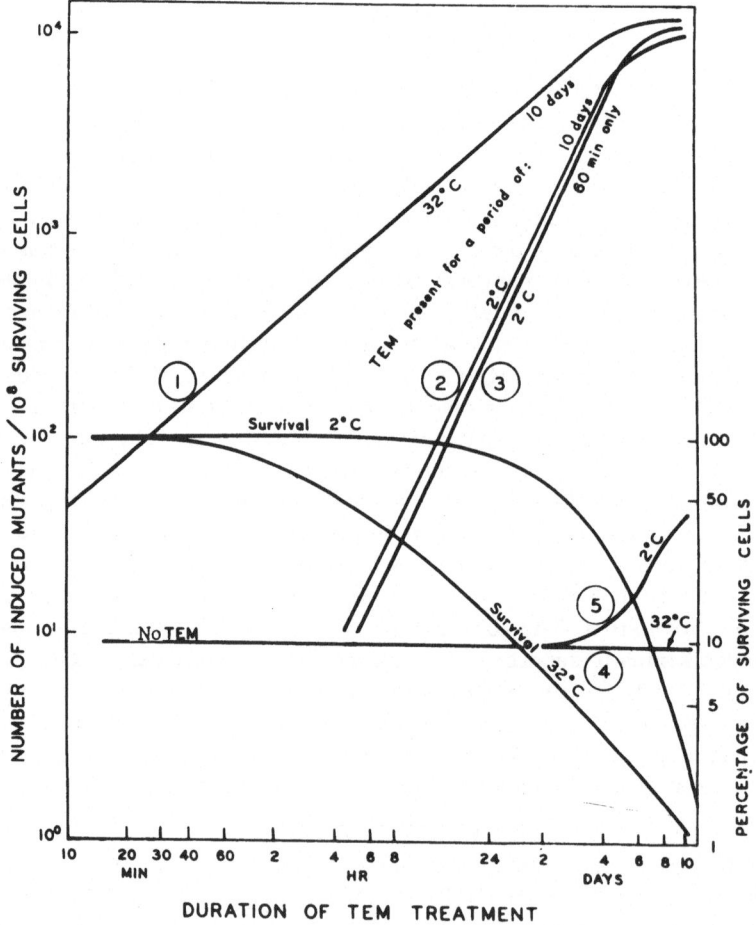

Fig. 1. Effect of continuous and intermittent exposure to TEM (10 μg/ml) on the number of induced mutants (Sd4 reversions) and on the percentage of survivors, at 32 C and at 2 C. TEM was present throughout experiments 1 and 2, but only during the first hour in experiment 3. Survival curves correspond to experiments 1 (32 C) and 2 (2 C). TEM was not present in experiments 4 and 5.

The following conclusions may be drawn from the above results: (1) Enough mutagen is semireversibly bound by the cells during the short preliminary contact to produce a maximal number of mutants. This absorption or adsorption process will be designated *reaction* I. (2) Mutagenesis is not accomplished if the cells are immediately transferred to conditions under which synthesis of macromolecular cell components (DNA?) resumes instantly. These conditions are fulfilled on nutrient agar, where even in the absence of SM, the Sd4 cells can undergo a few residual divisions. (3) For mutagenesis to be accomplished, a reaction must take place between the absorbed mutagen and some component of the *resting* cells, *prior* to plating on nutrient agar. The rate of this reaction, but not the final "equilibrium," is temperature-dependent. Further studies were aimed at elucidation of the nature of the latter reaction, which we shall designate as the *mutagenic reaction* or *reaction* II.

Before proceeding any further, two properties of the TEM-Sd4 system which relate to its temperature-dependence should be briefly described. While the frequency of mutants remained constant upon storage of the cells for prolonged periods at 32 C in mutagen-free distilled water, a five- to tenfold increase in the number of SM-independent revertants was observed at 0-4 C under otherwise identical conditions (Fig. 1). This low-temperature "mutagenesis" was observed also by Ryan (1955) and Wainwright (1956). It could be interpreted as the effect of endogenous mutagen, which is inactivated at 32 C (but not at 0-4 C) before it can react with some other cell component to produce a mutagenic intermediate active during subsequent cell multiplication.

This interpretation is in agreement with the effects of cell-absorbed TEM upon subsequent storage of cells at 2 or 32 C in TEM-free water. As mentioned above, the maximum frequency of mutants (over 10^4 mutants per 10^8 surviving cells) was reached under these conditions at 2 C (Fig. 1). At 37 C, in the absence of extracellular TEM, the frequency of mutants at first increased, with kinetics characteristic of the faster 32 C reaction, but subsequently leveled off at the fiftyfold lower equilibrium of approximately 2×10^2 mutants per 10^8 survivors (Iyer and Szybalski, 1958a, Fig. 5). The latter results could indicate the loss, breakdown, or inactivation of cell-bound TEM at elevated temperature (32 C).

Studies on the nature of the cell component involved in the reaction with cell-bound TEM were initiated in cooperation with Mr. S. Mashima. Two broad classes of cell components were evaluated as possible participants in the mutagenic reaction: structural polymers (including DNA, RNA, and proteins) and their low-molecular intermediates.

EFFECT OF THE RELATIVE NUCLEIC ACID
AND PROTEIN CONTENT OF THE CELL

It is tempting to consider the possibility that TEM reacts directly with the DNA component of the cells and by modifying its structure causes duplication errors and mutations during subsequent DNA replication. If this were true, one would expect that cells containing more DNA, i.e., a larger number of functional genetic complements, at the time of TEM treatment would not only be more prone to undergo spontaneous mutation, but would also contribute toward the achievement of a proportionally higher plateau of TEM-induced mutants.

In experiments designed by Dr. A. W. Kozinski and performed by Mr. S. Mashima, the DNA content of the cells was regulated either by harvesting the cells at different growth phases (Fig. 2), or by growing them for various period of time in the presence of chloramphenicol (Fig. 3). Although the average DNA content per cell varied over a wide range ($1.2 - 7 \ \mu g/10^8$ cells), the final frequency of mutants after a 7-day exposure to TEM at 0-2 C (Figs. 2 and 3) was relatively constant.

It seems safe to conclude that direct reaction between TEM and DNA plays only an insignificant role, if any, and cannot explain the increase in the frequency of mutants observed during storage of the cells in TEM solution. The small variation in the final frequency of mutants and in the rate of the reaction, if significant, can be more easily understood in terms of variation in the content of other cell components, under the experimental conditions used to achieve the variation in DNA content.

The above experiments and their modifications did not indicate any significant or consistent correlation between the progress of reaction II and the cell content of other biopolymers, including RNA and protein.

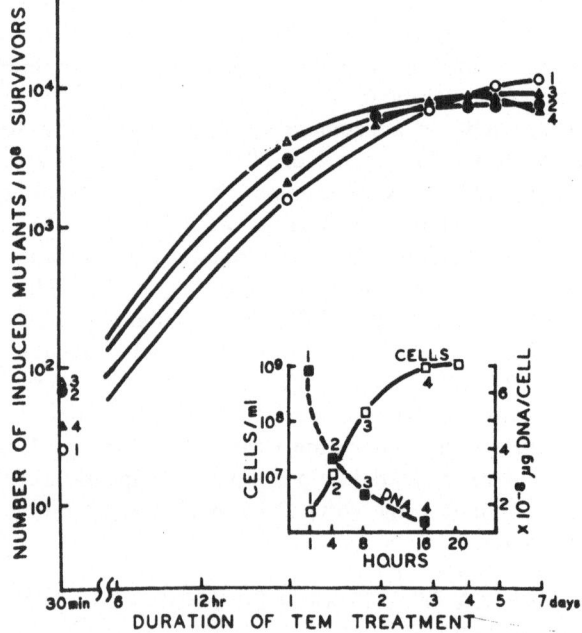

Fig. 2. Effect of the DNA content per cell on the number of TEM-induced mutants (Sd4 reversions) at 2 C. Small insert diagram (lower right corner) indicates the average DNA content (■) of Sd4 cells harvested at different stages of the growth cycle (□) (nutrient broth; 20 μg/SM/ml; 37C). Curves 1 (O), 2 (●), 3 (△), and 4 (▲) repre- the increase in the number of mutants obtained by TEM (10 μg/ml) treatment of the cultures, harvested at the first, fourth, eighth, and sixteenth hr of the growth cycle, respectively.

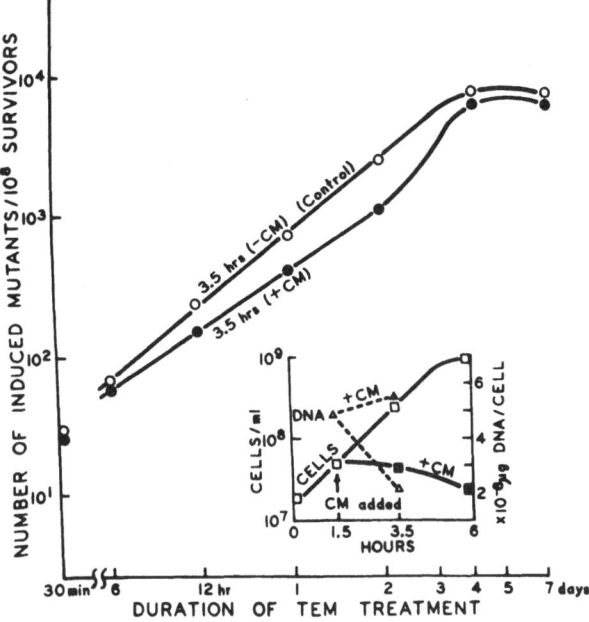

Fig. 3. Effect of the chloramphenicol (CM)-modified DNA content per cell on the number of TEM-induced mutants (Sd4 reversions) at 2 C. Inserted diagram (lower right corner) depicts cell proliferation (viable cell count) in the absence (□) and in the presence (■) of CM. CM (20 µg/ml) was added at the time indicated by the arrow. The DNA content of the cells was determined at the moment of CM addition (1.5 hr) and 2 hr later for the CM-inhibited culture (▲) and for the noninhibited control (△). At the same time (3.5 hr), the cells were harvested, suspended in distilled water, and treated with 10 µg TEM per ml. The curves represent the increase in the number of induced mutants obtained by treatment of 3.5-hr control (○) and 3.5-hr CM-inhibited (●) cells.

EFFECT OF LOW-MOLECULAR CELL COMPONENTS

Since the mutagenic reaction II did not seem to involve directly the structural components of the cells, our attention was directed toward their low-molecular precursors, and especially toward DNA precursors. It was attempted to measure the effect of the intracellular concentration of low-molecular components, at the time of exposure to TEM, on the production of mutants. Osmotic shock, as described by Bolton et al. (1955-1956), was the means for removal of the low-molecular cell components ("cold 0.5 N trichloroacetic acid soluble pool") without killing the cells. According to Bolton et al. (1955)1956; p. 127, Table 16), even phosphorylated intermediates (P^{32}-labeled) are removed by this method with 85% efficiency.

The procedure involved suspending the cells in 0.3M NaCl, collecting them on a Millipore filter, and washing with either (1) distilled water ("shocking-out") or (2) a solution of various precursors ("shocking-in"). Over 90% of the viable cells could be recovered by briefly shaking the fragmented Millipore filter in

5 ml of distilled water. The effect of TEM on cells subjected to such treatment is illustrated in Fig. 4. When compared with the "nonshocked" control (curve 1), there was a delay and a slight decrease in the production of mutants by cells subjected to osmotic shock before TEM treatment (curve 2). "Shocking-in" the yeast extract (curve 4) or thymidine solution (curve 3) produced an increase in the final yield of mutants. "Shocking-out" the cells, previously treated with TEM for a 48-hr period caused a transient drop in the frequency of mutants (curve 1a), while simple transfer of the TEM-treated cells to TEM-free water had no effect on the final yield of mutants (c.f. Fig. 1).

These preliminary experiments suggest the possibility that low-molecular components in general, and thymidine in particular, may play a role in the muta-

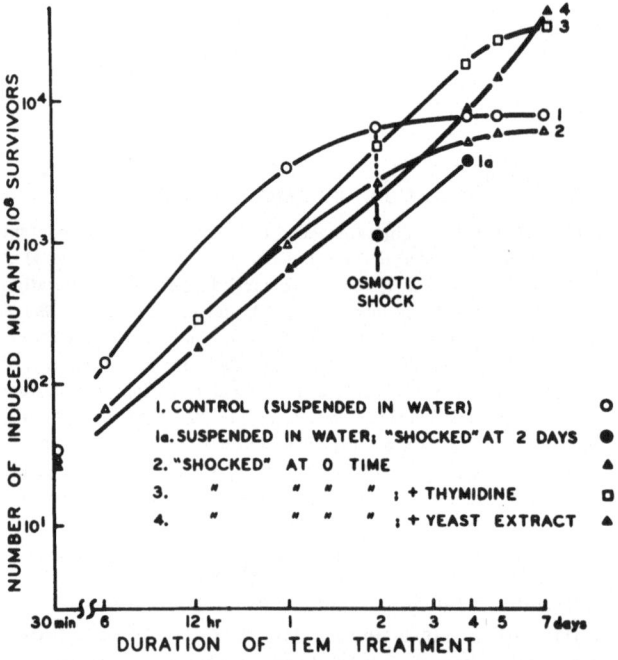

Fig. 4. Effect of variation in the intracellular pool of low-molecular intermediates on the number of TEM-induced mutants (Sd4 reversions) at 2C. Cells were grown overnight in nutrient broth + 20 μg SM per milliliter (37C), washed and suspended in distilled water (2C). Curve 1: 10 μg TEM per milliliter added, stored at 2C. Curve 1a: same treatment until second day when NaCl was added (0.3M), collected on a Millipore filter, washed with distilled water, resuspended in distilled water, and stored at 2C. Curve 2: supplemented with NaCl to obtain 0.3M NaCl, collected on a Millipore filter, washed twice with distilled water, resuspended in distilled water + 10 μg TEM per milliliter, and stored at 2C. Curve 3: procedure similar to curve 2, but the first wash on Millipore filter with 0.1mM thymidine. Curve 4: procedure similar to 2, but the first wash on Millipore filter with 0.5% yeast extract.

genic reaction II. A working hypothesis was formulated, according to which TEM reacts with one or more of the intracellular deoxynucleotides (reaction II), transforming them into fraudulent base analogs, which are incorporated into DNA, causing replication errors leading to cell death or to mutation.

CHEMICAL AND BIOLOGICAL EVALUATION OF THE PRODUCTS
OF REACTION II

To test the above hypothesis, experiments were devised to determine whether TEM per milliliter and 10 to 250 g per milliliter of any one of the following compounds: adenine, guanine, cytosine, thymine, or uracil, their nucleosides or de-reaction, if any, retain their normal or modified mutagenic activity (Lorkiewicz and Szybalski, 1960).

The reaction was performed at 1-4 C in an aqueous solution containing 10μ g TEM per milliliter and 10 to 250μ g per milliliter of any one of the following compounds: adenine, guanine, cytosine, thymine, or uracil, their nucleosides or deoxynucleosides. The pH of the solution was adjusted to neutrality. The progress of the reaction was followed by sampling the mixture (and also solutions of the individual components) at 0 time and after various periods up to 6 days, and (i) analyzing the products chromatographically and spectrophotometrically, as well as (ii) assaying their mutagenic activity. The latter was accomplished by exposing washed Sd4 cells to the reaction mixture and to the individual components for 2 hr at 32 C, followed by assay of survivors and mutants.

The preliminary results of these experiments, which are being conducted by Dr. Z. Lorkiewicz, can be summarized as follows: (1) Chromatographic and spectrophotometric analyses indicated that TEM did not enter into any chemical interaction with natural purines or their derivatives. In agreement with these observations, TEM alone or mixed with adenine, guanine, their ribosides or deoxyribosides, retained its unaltered mutagenic activity throughout the entire reaction period (6 days, 0 to 4 C). (2) Under the same conditions, TEM reacted extensively with pyrimidines and with their nucleosides and deoxynucleosides. The extent of this reaction was visualized by ascending chromatographic analysis (Whatman No. 1 paper; solvent composed of 50 parts methanol, 25 parts ethanol, 6 parts concentrated HCl, and 19 parts water). TEM alone produced an ultra-violet light(UV)-absorbing spot with R_f value 0.000 (probably products of a reaction between TEM and the paper components) and a weakly UV-absorbing streak extending from R_f 0.80 to 1.00. The approximate R_f values for deoxycytidine and thymidine were 0.64 and 0.71 respectively. Chromatography of the reaction mixtures at time 0 gave spots identical with the spots produced by the pure components. After 1 to 6 days, however, chromatographic analysis of the TEM-pyrimidine mixtures revealed the disappearance of the "reactive" TEM spot characterized by an R_f value of 0.00, and the concomitant appearance of a new spot with R_f 0.75. The position of the spots characteristic for the pyrimidines was not significantly altered, but subsequent examination of the eluates revealed profound alteration of their UV-absorption spectra. In general, a few small peaks were superimposed over the normally smooth UV-absorption curve, with the 260 to 280 mμ peaks increasing by a factor of 1.2 to 1.8, when examined in a Cary

Model 14M recording spectrophotometer at pH 1, 2, 7, and 9 and compared with the eluted spots of the unreacted pyrimidine derivatives. (3) When compared with the effect of TEM alone, the mutagenic activity of the products resulting from the cytosine-, cytidine-, or deoxycytidine-TEM reactions was practically abolished, while the mutagenicity of the deoxyuridine-TEM mixture remained unchanged and that of the thymidine-TEM mixture increased by a factor of approximately two.

CONCLUSIONS

Several sequential reaction steps must be postulated to explain the mutagenic effect of TEM. These may include: (1) semireversible absorption of TEM by resting cells (reaction I); (2) chemical reaction between the mutagen and pyrimidine deoxynucleotides, with the production of fraudulent DNA precursors (reaction II); (3) incorporation of these unnatural analogs into DNA during synthesis of the latter; (4) occurrence of permanent but nonlethal errors (mutations) in the base sequence during subsequent replication of the modified DNA. A diagrammatic representation of these four mutagenic steps, in idealized, highly simplified form, may be found in Fig. 5.

At low temperature (0 to 4 C), the absorption of TEM (reaction I) is accomplished within less than 1 hr. The subsequent removal of extracellular TEM hardly affects the progress of reaction II at 0-4 C.

Although direct alkylation of DNA by TEM was first suspected to constitute reaction II, experiments did not corroborate this notion. On second thought (or with some hindsight), direct DNA-TEM interaction was considered rather unlikely, since the reactive sites of the pyrimidine bases are not readily accessible inside the double-stranded, protein-coated DNA molecule.

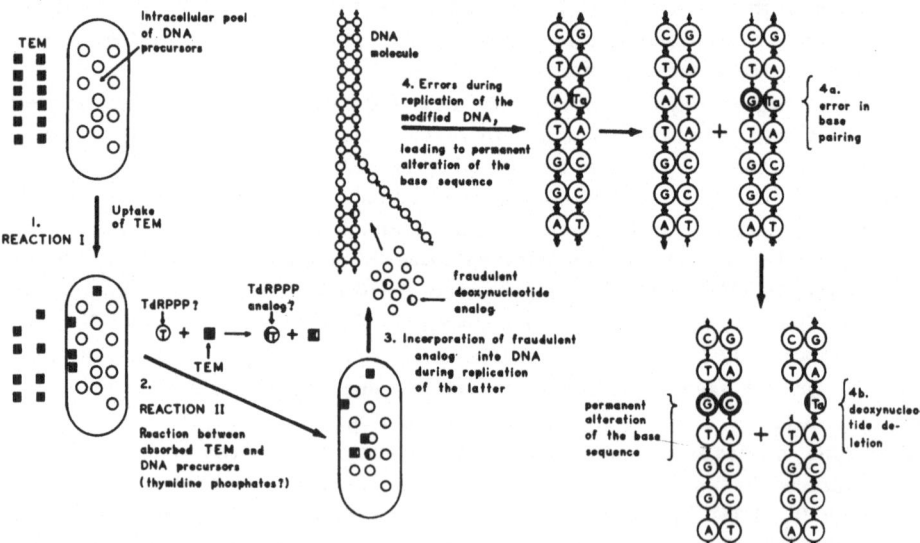

Fig. 5. Diagrammatic representation of the hypothetical mechanism of TEM mutagenesis in idealized, highly simplified form. For discussion see text. (A = deoxyadenylic acid; C = deoxycytydilic acid; G = deoxyguanylic acid; T = thymidilic acid; Ta = TEM-altered thymidilic acid; TdRPPP = thymidine triphosphate; DNA = deoxyribonucleic acid; TEM = triethylene melamine).

The evidence for the participation of thymidine phosphates in reaction II is only circumstantial. It was shown, on the one hand, that TEM reacts *in vitro* with thymidine, and the products are mutagenic to a degree equal to or greater than TEM alone. Thymidine phosphates, however, and not thymidine, are the normal constituents of the cellular pool of DNA precursors. The products of the TEM-thymidilate interaction, on the other hand, cannot be readily tested with intact cells, which are not permeable to phosphorylated intermediates. It must be postulated that the fraudulent chemical configuration of the thymidine-TEM interaction products would still permit them to undergo kinase-catalyzed phosphorylations and thus make them eligible for incorporation into the new DNA strands. Isotopically-labeled TEM and DNA intermediates will be utilized to elucidate further the nature of reaction II, and to correlate the mutagenic effects with the pool size and the incorporation of the putative, TEM-produced thymidine analogs into the pool and into DNA.

Dr. Z. Lorkiewicz is also studying the incorporation of other fraudulent thymidine analogs, including 5-bromo- and 5-iodo-deoxyuridine, into bacterial DNA in the hope of elucidating steps 3 and 4 of the mutagenic process. The hypotheses delineating the probable mode of action of various analogs on the level of steps 3 and 4 have been summarized by Freese (1959).

The similar evaluation of other mutagens might reveal diverse mechanisms of mutagenicity. According to Reiner and Zamenhof (1957), purines, especially guanine, rather than pyrimidines react *in vitro* with various alkylating agents, including nitrogen mustard, dimethyl sulfate, and diethyl sulfate. In their studies, however, and in those of Press and Butler (1952) and of Wheeler *et al.*, (1955a,b), the mutagenic effects of the products of chemical interaction were not evaluated. The concentrations of alkylating agents used in the above studies were many times higher than that found to be effectively mutagenic for *E. coli* Sd4 (Iyer and Szybalski, 1958b). Direct interaction between DNA and mutagenic agents (i.e., according to a scheme in which DNA is the reaction-II substrate and step III is bypassed) is a highly appealing interpretation of the mutagenic effects of ethylmethane sulfonate and diethyl sulfate on extracellular phages (Loveless, 1959), or of HNO_2 on viruses (Mundry and Gierer, 1958; Vielmetter and Wieder, 1959; Tessman, 1959), and on bacteria (Kaudewitz, 1959). On the other hand, reactions I and II seem to be bypassed in the mutagenic process observed with phages grown in the presence of purine and pyrimidine analogs (Freese, 1959). Radiation-initiated mutations, another type of mutagenic response, can have a more complex mechanism. One might postulate chemical modification caused by radiation "hits" affecting (1) the preformed DNA directly, (2) its precursors, or (3) other cell and medium constituents. Reactive chemicals produced in the latter case can act in a manner similar to that hypothesized for TEM.

It would be of utmost theoretical and practical importance to isolate the hypothetical products of mutagen-intracellular precursor interaction. In this manner, it might become possible, especially in the case of highly reactive alkylating agents, to determine the biological activities (carcinogenicity, mutagenicity, antineoplastic effects, toxic effects, among others) of a variety of mutagen- (alkylating agent) modified substances, some of which might possess properties of practical importance and lower toxicity than the nonspecifically reactive parental compounds, e.g., nitrogen mustards, epoxides, and ethylenimino derivatives. The importance of this approach in the designing of antineoplastic drugs, based on previous experience with the alkylating agents, is obvious.

ACKNOWLEDGMENTS

It is a pleasure to acknowledge the collaboration of Dr. V. N. Iyer, Dr. I. D. Steinman, Mr. S. Mashima, Dr. A. W. Kozinski, and Dr. Z. Lorkiewicz in various aspects of this work. The skillful technical assistance of Miss Clara E. Biro throughout the whole study is appreciated.

Parts of this study were aided by grants from the National Science Foundation (Grant No. G-7021) and from the National Institutes of Health, the U. S. Public Health Service (Grants No. CY-3240 and CY-3492).

SUMMARY

A mechanism which involves several steps must be postulated to explain the mutagenic effect of triethylene melamine (TEM). These include (1) semireversible uptake of TEM, (2) chemical reaction between TEM and pyrimidine deoxyribotides with the production of modified deoxynucleotides, (3) incorporation of the TEM-produced, fraudulent analog into DNA during synthesis of the latter, (4) occurrence of self-perpetuating errors in the sequence of natural bases during subsequent replication of the modified DNA strand. The experimental results which support this interpretation of the mutagenic process are presented and discussed.

REFERENCES

Bertani, G. 1951 A method for detection of mutations, using streptomycin dependence in *Escherichia coli*. Genetics, 36, 598-611.

Bolton, E. T., Britten, R. J., Cowie, D. B., Creaser, E. H., and Roberts, R. B. 1955-1956 Biophysics. Carnegie Institution of Washington Year Book, 55, 110-148.

Demerec, M., Lahr, E. L., Miyake, T., Goldman, I., Balbinder, E., Banic, S., Hashimoto, K., Glanville, E. V., and Gross, J. D. 1957-1958 Bacterial Genetics. Carnegie Institution of Washington Year Book, 57, 390-406.

Freese, E. 1959 The difference between spontaneous and base-analogue induced mutations of phage T4. Proc. Natl. Acad. Sci. U.S., 45, 622-633.

Iyer, V. N., and Szybalski, W. 1958a The mechanism of chemical mutagenesis. I. Kinetic studies on the action of triethylene melamine (TEM) and azaserine. Proc. Natl. Acad. Sci. U.S. 44, 446-456.

Iyer, V. N., and Szybalski, W. 1958b Two simple methods for the detection of chemical mutagens. Appl. Microbiol., 6, 23-29.

Kaudewitz, F. 1959 Production of bacterial mutants with nitrous acid. Nature, 183, 1829-1830.

Lorkiewicz, Z., and Szybalski, W. 1960 Reaction between a mutagen and DNA components. Abstracts, Fourth Annual Meet. Biophys. Soc., Feb. 24-26, 1960, Philadelphia, Pa., p. 18. Mechanism of TEM-initiated mutagenic process. Microbial Genetics Bull., No. 17, 10.

Loveless, A. 1959 The influence of radiomimetic substances on deoxyribonucleic acid synthesis and function studied in *Escherichia coli*/phage systems. III. Mutations of T2 bacteriophage as a consequence of alkylation *in vitro* : the uniqueness of ethylation. Proc. Royal Soc., B, 150, 497-508.

Mundry, K. W., and Gierer, A. 1958 Die Erzeugung von Mutationen des Tabakmosaik-virus durch chemische Behandlung seiner Nucleinsäure *in vitro*. Z. Vererbungslehre 89, 614-630.

Press, E. M., and Butler, J. A. V. 1952 The action of ionizing radiations and of radio-mimetic substances on deoxyribonucleic acid. Part IV. The products of the action of di-(2-chloroethyl) methylamine. J. Chem. Soc., 626-631.

Reiner, B., and Zamenhof, S. 1957 Studies on the chemically reactive groups of deoxyribonucleic acids. J. Biol. Chem., 228, 475-486.

Ryan, F. J. 1955 Spontaneous mutation in nondividing bacteria. Genetics, 40, 726-738.

Tessman, I. 1959 Mutagenesis in phages ΦX174 and T4 and properties of the genetic material. Virology 9, 375-385.

Vielmetter, W., and Wieder, C. M. 1959 Mutagene and inactivierende Wirkung salpetriger Säure auf freie Partikel des Phagen T2. Z. Naturforsch. 14b, 312-317.

Wainwright, L. K. 1956 Spontaneous mutation in stored spores of a *Streptomyces* sp. J. Gen. Microbiol., 14, 533-544.

Wheeler, G. P., Morrow, J. S., and Skipper, H. E. 1955a Studies with mustards. I. Ion-exchange chromatography of reaction mixtures of purines and pyrimidines with nitrogen mustard. Arch. Biochem. Biophys., 57, 124-132.

Wheeler, G. P., Morrow, J. S., and Skipper, H. E. 1955b Studies with mustards. II. Paper chromatography and radioautography of reaction mixtures of various compounds with S[35]-sulfur mustard. Arch. Biochem. Biophys., 57, 133-139.

CONTRIBUTED PAPERS

Miss Mary O'Hara, presiding

Panelists: C. L. San Clemente
J. J. Solomon
D. W. McWade
A. R. Drury
Dr. Mary E. Pinkerton
M. E. Simpson
P. B. Marsh
J. M. Sharpley

COAGULASE ACTIVITY OF BOVINE CLINICAL ISOLATES AND PHAGE PROPAGATING STRAINS OF STAPHYLOCOCCUS

(ABSTRACT)

C. L. San Clemente, J. J. Solomon, D. W. McWade, and A. R. Drury

Michigan State University, East Lansing, Michigan

During the past year we have been studying strains of staphylococci isolated from bovine udders, and made quantitative determinations of coagulase activity of these and of two sets of phage-propagating strains. Phage typing was accomplished by the techniques of Blair and Carr (J. Inf. Dis. 93: 1-13, 1953) using 24 phages of the Blair-International series as well as six bovine-adapted phages prepared by Seto and Wilson (Am. J. Vet. Res. 19:241-246, 1958). The tube dilution technique of Tager (Yale J. Biol. and Med. 20:487-501, 1948) was used for the estimation of free coagulase. Staphylococci from 166 milk samples collected from 93 cows on 23 farms were phage-typed with the two sets of phages. From a total of 562 lytic reactions we found:

A. International-Blair Series (58%)
   Miscellaneous (10%
   Group I (10%)   Group II (2%)
   Group III (35%)   Group IV (1%)
B. Seto-Wilson Series (42%)
   S1 (9%)   S2 (9%)   S3 (5%)
   S4 (10%)   S5 (4%)   S6 (5%)

The data showed evidence of colonization of the same strain within a given herd and occasionally cross-infection from man to animal.

Quantitative free coagulase production in the propagating strains varied widely from titers almost zero in the case of P.S. 29, 52A/79, 3C, 42B, and 73, to titers above 1/4096 in P.S. 6, 70, and S3. Wide variation in coagulase production also appeared in the clinical isolates. A number of strains, typeable with neither series of phage, usually displayed low free coagulase production. This phenomenon will require a redefinition of the term "coagulase positive," since the relative amount of free coagulase (Duthie, E.S.J., Gen. Microbiol., 10, 427, 1954) is ignored.

TREATMENT OF SPONTANEOUS RINGWORM IN MICE WITH GRISEOFULVIN*

Mary E. Pinkerton

Department of Dermatology, University of Texas, Medical Branch, Galveston, Texas

During attempts to produce protracted experimental dermatomycosis in mice, spontaneous ringworm was encountered fairly frequently in older C-57 black mice. Since the final objective was an *in vivo* tool for testing fungicides, it was deemed of interest to observe the effects of the new internal medicament for superficial fungus infections—i.e., griseofulvin, on the natural murine infection.

Spontaneously infected C-57 mice of both sexes first exhibit signs of intense pruritus and then rapidly develop an extensive skin involvement. Diffuse alopecia and scabs are noted most frequently on the back of the neck. The whole back may become involved, and the eyelids as well, and opaque plaques frequently develop on the cornea. The facial fur is also of ragged appearance. The general health of the mouse does not seem to be radically affected, as he maintains weight and activity, and only rarely succumbs even with extensive skin involvement. These mice exhibit a nervous behavior, eating and drinking almost constantly. Occasionally, their stools become mucoid. Microscopic examination of feces shows sometimes many trichomonads and, almost always, pinworms (*Syphacia obvelata* and *Aspicularis tetroptera*).

There is no tendency toward recovery and these mice often live for months, becoming progressively more grotesque in appearance. A concomitant feature is the presence of enormous numbers of ectoparasitic mites—*Myobia musculi* in the region of the head, and *Myocoptes musculinus* on the back and abdomen. However, some apparently normal mice are also heavily infested with mites, but do not have ringworm. On the other hand, all of the mice with cutaneous lesions show similar strains of *Trichophyton mentagrophytes* on culture, which have not been found by us on normal mice. Microscopic examinations of hairs and skin in KOH preparations are rather unrewarding. Occasional plaques of spores are the only findings. However, on culture, the fungi appear to grow from the hair. Although we have felt that the fungus might be associated with the mites, cultures from mites and their eggs have been negative, as have been those from the sawdust bedding and feces.

All experimental mice were kept in $8 \times 6\frac{1}{2}$ in glass jars with wire covers. Standard Purina Laboratory Chow,† fed in little metal holders, was the only ration. The infection also developed in stock mice in large cages with wire bottoms. It has not been observed in CF1 and C3H mice of similar age and care.

The identical organism has been isolated in all cases. On Sabouraud's agar, the surface is of a chalky-white to tan color, of powdery-granular consistency, with irregular feathery edges. The reverse shows reddish-brown streaks extend-

---

*This research supported by United States Public Health Service Grant No. E-1385.

† Ralston Purina Co., St. Louis, Mo.

ing from the center, peripherally. Similar strains have been isolated by us from acute human ringworm of known animal origin, and from animals other than the mouse.

Although griseofulvin was discovered some years ago (Oxford *et al.*, 1939), and utilized topically for ringworm without success, its effectiveness via the oral route has only recently been recognized (Blank and Roth, 1959). The drug is now produced from *Penicillium griseofulvum* and several other species of *Penicillium*, and is known to have the formula $C_{17}H_{17}O_6Cl$. It is colorless, neutral and thermostabile in solution. *In vitro* tests against dermatophytes have shown inhibition of growth for 72 hr at levels of 0.2-$0.4\,\mu g/ml$ (Blank and Roth, 1959). It causes a "curling effect" on fungi with chitin walls, but has no effect on yeasts, actinomycetes, bacteria, and oomycetes. *In vivo*, it has a tendency to localize in the skin, where it exerts a fungistatic effect. In large doses it interferes with mitosis at the metaphase stage (Paget and Walpole, 1958). The effective dosage for humans has been established at about 1 g/day (14 mg/kg) given at intervals in 4 tablets of 250 mg each. Treatment with 5 g/day has also been well tolerated. It has been our observation that the drug is relatively ineffective in tinea pedis due to *T. mentagrophytes* . Gentles (1958) was able to cure experimental *T. mentagrophytes* infection in guinea pigs with doses of 15 mg/kg/day.

EXPERIMENTAL

Two spontaneously infected C-57 mice were treated twice daily with griseofulvin,* a total oral dose of 30 mg/kg/day for a period of two months, with cultures becoming negative and hair regrowth occurring. However, one mouse remained in an emaciated condition and died soon thereafter. Death was attributed to infection of a recently prolapsed rectum. The fact that cultures from skin and feces showed heavy growth of a *Mucor* species just before death is interesting. This has also been observed in cancer patients (Hutter, 1959). Four other C-57 mice were treated with 15 mg/kg/day for five weeks, with little clinical effect and with cultures remaining positive.

Because of a suggestion that pinworms might be associated with the spontaneous ringworm infection (Beneke, 1958), two mice were treated with Antepar [†] (piperazine citrate) at a dose of 20 mg/mouse/day for a week. This had no clinical effect on the skin, and cultures remained positive, although the stools were temporarily devoid of pinworms and eggs.

SUMMARY AND CONCLUSIONS

1. Spontaneous ringworm infection in C-57 mice over a year old, suggests that these animals may be a useful tool for *in vivo* testing of fungicides. A natural infection is preferable to an experimental one, especially since it is progressive with no tendency to heal spontaneously.
2. In our experiments, a dosage of 30 mg/kg/day was necessary to cure spontaneous *T. mentagrophytes* infection in C-57 mice.
3. It would be interesting to investigate the metabolic conditions that favor this spontaneous infection. There is, perhaps, a similarity to human adult

---

[*] Obtained from Dr. Harvey Blank for experimental use.

[†] Registered trade name, Burroughs Wellcome Co., Tuckahoe, N.Y.

generalized ringworm due to *T. rubrum*, *T. mentagrophytes*, or *T. tonsurans*, in which we have frequently noted associated internal malignancies (especially lymphomas) or diabetes.

4. The possibility of an intermediate host such as a mite or nematode bears further investigation.

5. Griseofulvin is only the first internal medication for superficial fungus infection. More potent and effective drugs should be sought.

REFERENCES

Beneke, E. S. 1958 *Trichophyton mentagrophytes* of an epidemic proportion in laboratory mice. Paper presented at Mycological Society of America Section of the AIBS, August 1958.

Blank, H., and Roth, F. J. 1959 The treatment of dermatomycoses with orally administered Griseofulvin. A.M.A. Arch. Dermat., 79, 259-266.

Gentles, J. C. 1958 Experimental ringworm in guinea pigs: oral treatment with Griseofulvin. Nature, London, 182, 476-477.

Hutter, R. P. 1959 Phycomycetous infection (Mucormycosis) in cancer patients: a complication of therapy. Cancer, 12, 330-350.

Oxford, A. E., Raistrick, H., and Simonart, P. 1939 Studies in the biochemistry of microorganisms. LX. Griseofulvin, $C_{17}H_{17}O_6Cl$, a metabolic product of *Penicillium griseofulvum* Dierckx. Biochem. J., 33, 240-248.

Paget, G. E., and Walpole, A. L. 1958 Some cytological effects of Griseofulvin. Nature, London, 182, 1320-1321.

THE DECOMPOSITION OF THE CELLULOSE OF COTTON FIBERS
BY FUNGI OF THE GENUS ASPERGILLUS[*]

M. E. Simpson and P. B. Marsh
Crops Research Division, Agricultural Research Service
U.S. Department of Agriculture, Beltsville, Maryland

The work on which we report describes experiments to determine which members of the genus *Aspergillus* can cause cellulosic degradation in a cotton fiber. Prior to these studies, it was known that many fungi, including certain of the *Aspergilli*, could cause degradation. The evidence consisted of strength losses in cotton fabric when a pure culture of the fungus was incubated on the fabric. However, one of the first observations in the present work suggested that these earlier tests might have failed to detect cellulose-decomposing ability in some of the tested fungi. These earlier tests, which were carried out with the fungus incubated on the cotton fabric in the presence of mineral salts, showed that *A. niger* van Tieghem was a non-cellulose-decomposer. This was indicated by the fact that there was no strength loss of the fabric. However, when a small amount of glucose was added to the test medium, the same fungus caused quite a substantial strength loss. It appeared then, that in some way, glucose was responsible for the ability of *A. niger* to decompose cellulose. It seemed possible, therefore, that the addition of glucose and/or other changes of technique might enable other isolates of the *Aspergilli*, also known as non-cellulose-decomposers, to degrade the cellulose of cotton fiber. Further, a more sensitive measure of cellulosic degradation, other than strength loss, might reveal a cellulose-decomposing potentiality in certain fungi where it had not been detected previously. There was some reason to believe that an alkali-swelling technique might be useful as a sensitive test to detect cellulose decomposition. Such a technique was employed in these experiments.

In the present work, the various fungi were generally incubated on cotton fabric in culture bottles containing agar, mineral salts, and some activator, usually glucose. After the incubation period, the fabric was removed from the cultures and tested to determine any changes in strength and in the alkali-centrifuge values. The alkali-centrifuge test involves placing the fiber into an 18% sodium hydroxide solution (4.5N) and determining the amount of alkali solution absorbed. This test is based on the fact that when a cotton fiber is placed into an alkaline solution, the degree of its swelling is limited by a restrictive effect of the outer wall of the fiber and that whenever the outer wall is damaged, a greater degree of swelling occurs. Prior evidence indicated that the swelling measured in this test was actually an index of cellulose decomposition. Specifically, it had been found that cell-free filtrates from certain cellulolytic fungal cultures were always active in causing alkali-centrifuge value increases in cotton fiber when the organism had been grown on a cellulose-containing medium. The filtrates were always inactive when growth had taken place on a noncellulosic medium. In other words, the pro-

[*] Study previously published: Plant Disease Reporter, Vol. 43, No. 9, September 15, 1959.

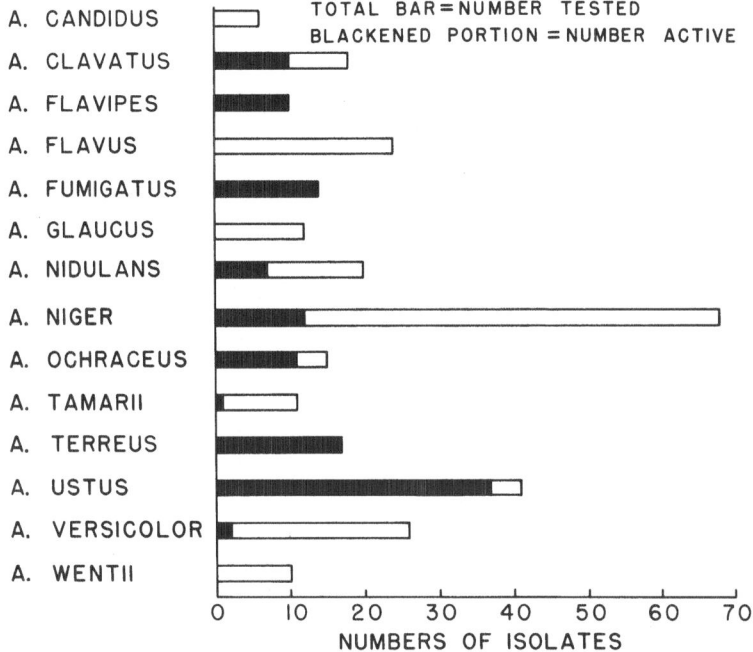

Fig 1. Cellulose-decomposing activity of Aspergilli species.

duction of an enzyme capable of causing an alkali-swelling response had been shown to be adaptive with respect to the presence of cellulose in the medium.

Figure 1 summarizes previous results from the literature relating to the occurrence of cellulose-decomposing activity among the *Aspergilli*. The total length of the bar represents the total number of isolates tested. The shaded portion represents the number of isolates having activity. All these results are from experiments in which no glucose or other activator had been added and in which strength loss of fabric was the criterion for activity. The fabric used in these tests was a gray duck. Similar tests were recorded for bleached sheeting. Those isolates which were active against the bleached sheeting were also active against the gray duck. You will notice that the genus *Aspergillus* is divided into 14 groups based on the classification by Thom and Raper. Each of these groups contains a limited number of species that are closely related taxonomically. In the *A. niger* group, as indicated here, tests on more than 60 isolates are to be found in the literature. All the active isolates were among the so-called purple-brown forms. None of the truly black forms, as typified by *A. niger* van Tieghem, was active. No active isolates were reported in the *A. candidus, flavus, glaucus,* and *wentii* groups. The few active isolates in the **A.** *versicolor* and *tamarii* groups were all in species other than the dominant species of the group.

Now let us examine some of our own results on cellulose-decomposing activity of a few truly black *Aspergilli* (Fig. 2). The strength loss is shown in the lower curve and the alkali-centrifuge value in the upper one. Addition of a small amount of glucose brought about large increases in both values. You will notice that the

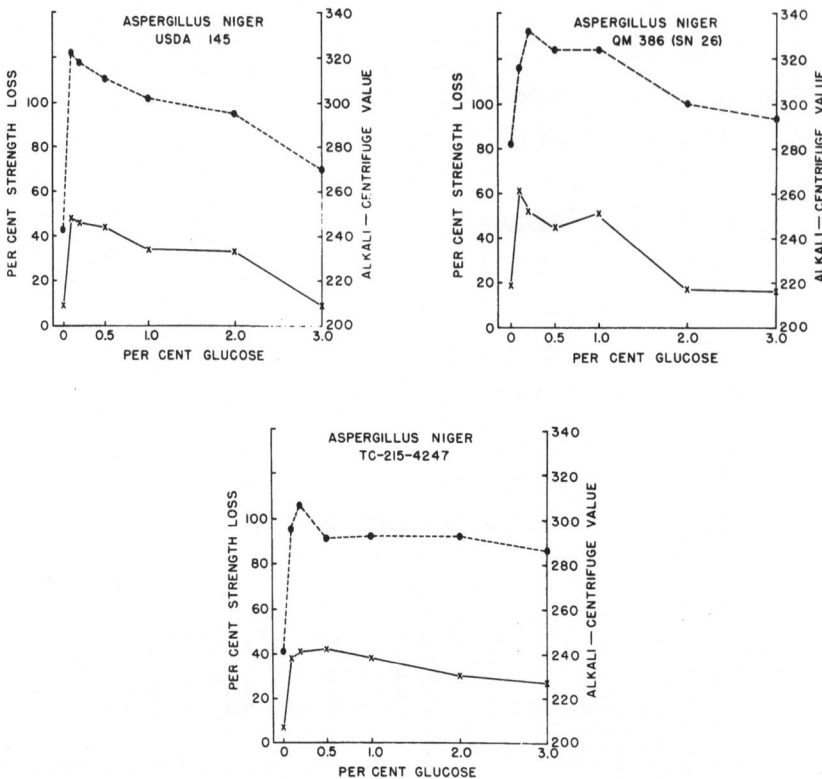

Fig. 2. Effect of glucose content of the test medium on strength loss (—) and alkali-centrifuge value (---) of cotton duck, measured after incubation in pure culture with three of the truly black Aspergilli. TC-215-4247 and SN 26 are the standard mildew-resistance test organisms previously believed to be unable to weaken cotton fabric. USDA 145 was isolated from a field sample of cotton.

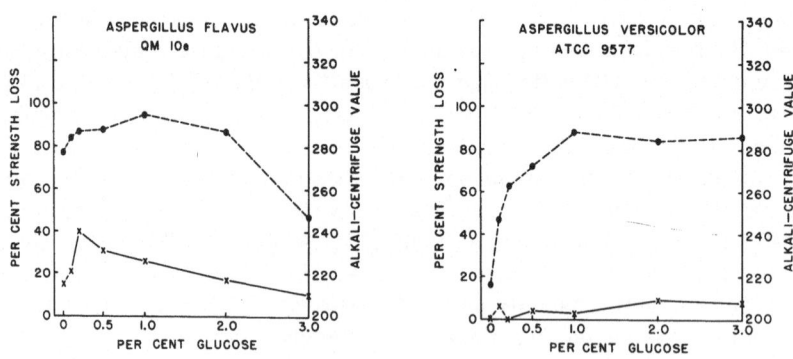

Fig. 3. Effect of glucose content of the test medium on strength loss (—) and alkali-centrifuge value (---) of cotton duck, measured after incubation with A. flavus QM 10e and A. versicolor ATCC 9577.

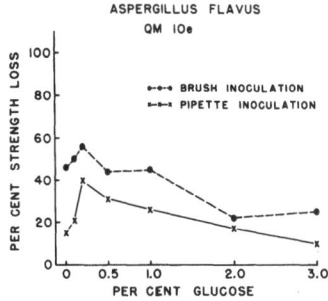

Fig. 4. Effect of glucose content of the test medium on strength loss of cotton
duck as influenced by manner of inoculation

strength loss and alkali-centrifuge value curves show a definite tendency to parallel each other; both properties are measures of cellulose decomposition.

Through the use of the technique of employing glucose as an activator for cellulose decomposition, fabric strength losses were observed for the first time not
only among the truly black *Aspergilli* but also in the *A. flavus* group (Fig. 3). You
will notice that the curves for strength loss (lower curve) and the alkali-centrifuge
value (upper curve) are similar to those of the black *Aspergilli*.

In the *A. versicolor, candidus,* and *wentii* groups the observed activity was not
high enough to be reflected in strength loss of the fabric but was clearly significant in the alkali-centrifuge value increases, indicating the sensitivity of the
alkali-centrifuge test. This result indicates that these organisms are able to
degrade the cellulose of cotton fiber to some extent, and that perhaps under more
favorable conditions they might be able to cause strength loss of fabric.

Strength losses were also increased substantially in the *A. flavus* and *niger*
groups by another device, one whose mechanism is not yet understood. In all the
earlier tests inoculation of the fabric samples was accomplished by pipetting a
spore suspension onto the fabric strip. On the other hand, when spores were
brushed onto the fabric with a dry camel's hair brush, the strength losses were
increased. It can be seen from the results obtained (Fig. 4) that the brush-inocu-
lated fabric had a greater strength loss (upper curve) than the pipette-inoculated
fabric. The rest of the test conditions remained the same.

In conclusion, the data indicate that the ability to cause strength loss in cotton
fabric is present in at least some isolates in 11 of the 14 groups of the genus
Aspergillus . The elevated alkali-centrifuge values in two of the remaining groups,
A. candidus, and *wentii* , indicate that these groups have the potentiality for causing
strength loss in fabrics. The truly black *Aspergilli* of the *A. niger* group, previously
widely known as "non-cellulose-decomposers," have now been shown to be able
to cause strength loss in cotton fabric when glucose is present in an appropriate
concentration as an activator. Glucose served to stimulate activity also in the
A. flavus group. Stimulation of cellulolytic activity was further enhanced by a
second method, namely by brush inoculation of the test fabric with dry spores.

In closing, it may be desirable to place these results on the cellulose-decomposing activity of the *Aspergilli* into a little broader setting. In the earlier work,
many genera of fungi were tested for cellulose-decomposing ability in the presence
of mineral salts. In some of the imperfect genera such as *Chaetomium, Fusarium,
Humicola, Alternaria,* and *Trichoderma* all isolates examined have proven active.
Many of the *Penicillia* have been tested, with approximately half of them showing
activity. On the other hand, in the genera of the order Mucorales—*Rhizopus, Mucor,
Phycomyces*, etc., most efforts upto the present time have failed to reveal evidence
of cellulose-decomposing ability. The present study suggests that some isolates
in the earlier work may have been incorrectly classified as non-cellulose-decomposers. By some simple modification of the testing technique in future experiments
some of these isolates might also prove to be cellulose-decomposers.

A TECHNIQUE FOR THE DIRECT ESTIMATION
OF BACTERIA IN OIL FIELD WATERS

J. M. Sharpley

Buckman Laboratories, Inc., Memphis, Tennessee

The culture of many bacteria found in oil field waters on conventional media is characteristically difficult. Some of the microorganisms require the presence of hydrocarbon gases for adequate growth; many are autotrophic or semiautotrophic with poorly understood growth requirements; others require very low oxidation-reduction potentials for adequate growth; and still others refuse to reproduce in artificial cultures for reasons as yet unknown to us. While the unusual growth habits of these bacteria lead to many very interesting research problems, the difficulties encountered in culturing them in the field have been quite annoying. Moreover, such difficulties have precluded rapid estimations of populations for the purpose of indicating the degree of control of a total microbial population.

We have been interested for some time in the potentials of the membrane filter for the counting of total populations of microorganisms. This has been particularly true in the petroleum industry, since water is characteristically filtered through membrane filters to obtain other information, and the filters are available for microscopic examination. For several years we have periodically attempted to devise techniques for the direct estimation of bacteria on membrane filters but have always been unsuccessful, either because of the loss of bacteria from the surface of the filter or the lack of adequate staining to provide good contrast between the bacteria and the filter. We recently developed a technique that overcomes these objections. We believe it enables us to make more accurate estimates of total population of bacteria in water samples.

Cultures

The stock cultures used in this study were obtained from our laboratory stock culture collection. The sources of the various types of natural waters are identified where discussed. In all cases, the waters were filtered in the field, and there is no inherent population change from shipping or handling of the samples.

Media

The compositions of the various media used during this work are given in the following. Medium 1 and Medium 4 are commercial dehydrated media.[*] The remainder were prepared as indicated.

Medium 1

Peptone	5 g
Beef extract	3 g
Agar	15 g
Distilled water to 1000 ml	
pH 7.1	

[*] Baltimore Biological Laboratory, Baltimore, Maryland.

Medium 2

Sodium lactate	4.0 ml
Yeast extract	1.0 g
Ascorbic acid	0.1 g
Magnesium sulfate	0.2 g
Dipotassium phosphate	0.01 g
Ferrous ammonium sulfate	0.1 g
Sodium chloride	10.0 g
Agar	15.0 g

Distilled water to 1000 ml
pH 7.5

Medium 3

Dextrose	1.0 g
Peptone	1.0 g
Magnesium sulfate	0.2 g
Calcium chloride	0.05 g
Ferric chloride	0.001 g
Agar	12.5 g

Tap water to 1000 ml
pH 7.0

Medium 4

Tryptone	5.0 g
Beef extract	3.0 g
Dextrose	1.0 g
Agar	15.0 g

Distilled water to 1000 ml
pH 7.0

Medium 5

Dextrose	1.0 g
Peptone	1.0 g
Magnesium sulfate	0.2 g
Calcium chloride	0.05 g
Ferric chloride	0.001 g
Agar	12.5 g

Natural water to be examined to 1000 ml
pH adjusted to that of sample water

Medium 6

Medium 5 plus carbon dioxide atmosphere

Medium 7

Sodium lactate	4.0 ml
Yeast extract	1.0 g
Ascorbic acid	0.1 g
Agar	15.0 g

Natural water to be examined to 1000 ml
pH adjusted to that of sample water

Apparatus

The membrane filters and stainless steel membrane filter holder were obtained from commercial sources.* The membrane filter apparatus for pressure filtration was made locally using blueprints obtained from the research laboratories of a major oil company.**

---

*Millipore Filter Corporation, Watertown, Massachusetts. Type PH filters.
**Shell Research and Development Laboratories, Houston, Texas.

When using the direct counting technique in the field, we have found a hand vacuum pump to be indispensable. Sufficient vacuum is easily obtained for rapid filtration. The elimination of breakable equipment in the field by the use of molded polyethylene filtration flasks, graduates, and beakers used in conjunction with a stainless steel membrane filter holder has proved to be very useful. This equipment has also proved to be useful in dealing with many of the highly corrosive waters.

Techniques

The general technique that we have used is to employ one of the classic fixatives on the bacteria prior to their staining with an appropriate dye. We tried several fixing fluids, including alcohol, osmic acid, formalin, formalin—acetic acid—alcohol (FAA), formalin—propionic acid—alcohol (FPA), Bouin's fluid, formalin—acetic acid—picric acid, and picric acid. Picric acid appears to give as good fixation as any of the other materials tried. It is stable in solution, and is easily prepared. A 1 per cent water solution of picric acid was used for all of the work reported in this paper.

We have used several different stains during the course of our work, including 0.2 per cent aqueous crystal violet, alcian blue, lactofuchsin, and acid fuchsin. The best results were obtained with acid fuchsin, and all of the work reported here was done using 0.3 per cent acid fuchsin in water. The use of picric acid as a fixative as well as acid fuchsin as a stain has also been recently reported by Ecker and Lockhart (1959). The acid fuchsin should be freshly prepared for optimum results and should always be prefiltered through a membrane filter before using.

Depending upon the approximate number of bacteria expected in the sample, we have used two methods in the preparation of the membrane filters for examination. Where very large populations are expected, as in a laboratory culture, for example, we have found it best to pipette an aliquot of the material directly into the fixing solution and then filter through a membrane filter. Where the expected counts are lower or unknown, as in a natural water sample, we usually separate prepared filters by filtering 1000 ml, 100 ml, and 10 ml of water. The filter is then chosen for counting that has 25 to 100 bacteria per microscope field present. When the larger quantities of water are filtered, the bacteria are fixed on the surface of the filter using picric acid rather than mixing them with a volume of the fixative. When the bacteria are fixed by adding them to the fixative solution, we have not found the fixing time to be critical. When fixing previously filtered bacteria on the surface of a filter, one should allow a minimum of 5 min for fixation.

Whichever filtration technique is used, we have found that one to two minutes staining with acid fuchsin is sufficient to give good contrast. After staining, the filter is dried, mounted on a slide with immersion oil to render it transparent, and examined.

The presence of particulate matter and oil on the surface of membrane filters used for filtering natural oil field waters proved to be troublesome when we first used these techniques in the field, since such material partially obscures the bacteria. We have found that in the case of iron salts, such as iron oxide or iron sulfide, prewashing the membrane filter with 10 per cent hydrochloric acid after filtration and before fixation removes much of the obscuring material. In the case of oil deposited on the surface of the membrane filter, a prewash of petroleum ether has given satisfactory results. Both the acid wash and the solvent wash somewhat decreased stain absorption by the bacteria, but this does not seem to be sufficient to interfere with counting.

It is possible to prepare permanent or semipermanent mounts of discs cut from membrane filters stained in the manner described for future reference or records. Porter (1959) has recently outlined the preparation of permanent mounts of membrane filters on microscope slides. He suggests placing enough immersion oil into a small petri dish to cover the bottom of the dish. After staining and drying, a disc smaller than the cover glass is cut from the membrane filter and is floated on the immersion oil. The disc then is carefully removed from the surface of the oil with forceps and the excess immersion oil is removed from the underside by drawing it slowly across the edge of the petri dish. Then the disc is rolled onto the center of a clean microscope slide of appropriate size, care being taken to avoid entrapping air bubbles. A No. 1 cover slip is placed over the disc and sealed around the edges with cellulose tape or a mounting medium. This procedure results in semipermanent mounted slides.

It is necessary to calibrate a given microscope for counting bacteria on the surface of a filter. In the present study a modified Whipple-Hausser ocular grid was used in the microscope eyepiece. The area of the object slide covered by the ocular grid was calculated in square millimeters from measurements of the grid with a stage micrometer at the magnification used. The area of the total effective filtering surface of the membrane filter was then measured in square millimeters. The total filtering area divided by the area of the ocular grid provides a factor for calculating the total number of ocular grids superimposed on the object slide. Thus it is only necessary to find the average number of bacteria per ocular grid, multiply by the area factor and divide by the number of milliliters of water filtered to find the number of bacteria/ml.

Results

Prior to the use of this technique in the field, its accuracy was compared with conventional plating techniques in the laboratory. Using cultures of *Aerobacter aerogenes* , as shown by the data in Table 1, a good correlation was obtained between the numbers of bacteria/ml when counted by conventional plating techniques and when counted using membrane filters. In order to obtain the degree of corre-

Table 1. Comparison of Counts Obtained with *Aerobacter aerogenes* Using Conventional Plating Technics and Membrane Filters

Replicate	Nutrient agar 35 C, 48 hr — Estimated bacteria per ml	Replicate	Membrane filter — Mean cells per ocular grid	Membrane filter — Estimated bacteria per ml
1	56×10^7	1	14.5	55×10^7
2	49×10^7	2	12.7	48×10^7
3	54×10^7	3	14.0	53×10^7
4	52×10^7	4	14.0	53×10^7

lation reported, it was found necessary to use shake cultures, subject them to brief sonic dispersion and prepare gravimetric dilutions. The degree of accuracy obtained is in agreement with the data reported by Ehrlich (1955).

Natural waters containing dissolved carbon dioxide have characteristically given poor or no correlation between the numbers of bacteria known to be present in the water through direct microscopic examinations and the results obtained by plating the bacteria. We have devised several techniques for culturing the micro-organisms from such an environment with varying degrees of success. Without exception, we have always obtained the best results by the use of a medium prepared from the water under investigation and incubated in a carbon dioxide atmosphere. Most commonly, our best results were obtained with Medium No. 6.

Table 2 illustrates the much larger numbers of bacteria counted in a water containing carbon dioxide when the direct counting technique is used, as compared to plating methods. This particular water was from a geologic formation at about 5000 ft. It contained large amounts of dissolved carbon dioxide and about 120,000 ppm of total solids. The temperature of the water at the wellhead was 60 C (140 F).

Table 2. Comparison of Counts Obtained with Bacteria in Natural Waters Containing Carbon Dioxide Using Conventional Plating Techniques and Membrane Filters

Sample	Medium* 35 C, 48 hr Estimated cells per ml X 10^3						Membrane filter Estimated bacteria per ml X 10^3	
	1	2	3	4	5	6	Trial 1	Trial 2
1	0	0.04	13	0	14	52	90	97
2	0	0.23	21	0	34	74	80	84
3	0.004	0	14	0.250	17	61	140	131
4	0	0.15	19	0	24	70	124	120

*See pages 253-254 for composition of media used.

Table 3. Comparison of Counts Obtained with Bacteria in Natural Water Containing Crude Oil Using Conventional Plating Techniques and Membrane Filters

Sample	Medium* 35 C, 48 hr Estimated bacteria per ml X 10^3			Membrane filter Estimated bacteria per ml X 10^3	
	1	2	7	Trial 1	Trial 2
1	0.2	500	900	1900	1840
2	0.11	200	310	480	490
3	0.9	750	810	880	910
4	0.1	100	140	1100	1210

*See pages 253-254 for composition of media used.

Waters produced with crude oil may contain many water-soluble hydrocarbon fractions in small quantities. There are many bacteria that can utilize some or all of the fractions of hydrocarbons, and it has proved to be quite difficult to arrive at an estimate of the total numbers of bacteria present in such waters without laborious and time-consuming plating techniques involving media incorporating many different hydrocarbons. We have consistently obtained higher counts of bacteria in these waters by the use of membrane filters than was possible by plating techniques in the field.

An example of results obtained by plating and by direct membrane filter examination is seen in Table 3. Similar to the waters containing dissolved carbon dioxide, the most reliable plating technique used the source water as a base medium. Better growth almost certainly results from dissolved trace nutrients in the water. This is dramatically demonstrated when one is examining plates and one often finds a plate crowded with bacteria when 1.0 ml inoculum is used, but no growth in the dilutions. A photomicrograph illustrating the appearance of a filter prepared from produced water is seen in Fig. 1.

Discussion

Jannasch (1958) pointed out that direct microscopic examinations characteristically give higher counts of bacteria than do counts made by cultures. It is un-

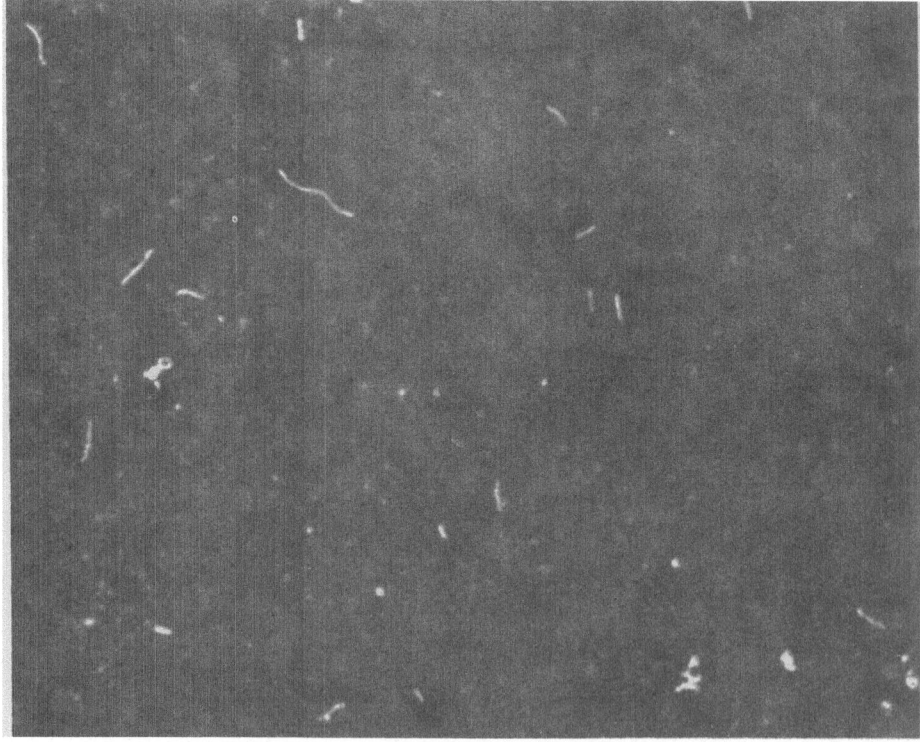

Fig. 1. Photomicrograph of membrane filter showing bacteria from oil field water. Dark phase contrast; approx. 800×.

doubtedly true that a portion of the bacteria so counted are dead, but Strugger (1959) showed by acridine—orange absorption that only a small percentage of the bacteria in a natural water population was nonviable.

Although not included in the results reported here, some work has been done using the membrane filter to detect bacteria utilizing gaseous hydrocarbons, specifically methane. Waters containing traces to moderate amounts of methane in or near producing oil formations are common. A methane-oxidizing bacterium appears very commonly in such waters and it is possible to culture this organism under methane gas, but the numbers resulting from such techniques in no way approximate the true numbers of bacteria shown to be present by direct microscopic examination.

Aside from those bacteria that will not grow under artificial culture, the greatest source of error in the traditional counting of bacteria lies in the error caused by a group or clump of bacteria reproducing as one colony. In waters containing traces of emulsified oil, such as "produced" water in the petroleum industry, this factor appears to be particularly significant, with many of the bacteria present occurring in clumps of 5 to 15 organisms.

The greatest single difficulty encountered in the direct counting of bacteria on membrane filters is the interference of detritus on the surface of the filter. This may be partially overcome by treatment with hydrochloric acid or solvents, but it is inevitable that some artifacts will be tallied incorrectly as bacteria. Ecker and Lockhart (1954) have pointed out that for an experienced observer the percentage of artifacts counted will remain relatively constant.

As originally pointed out by Jannasch (1958) and confirmed by Ecker and Lockhart (1959), the deposition of bacteria evenly on the surface of the filter is a function of volume and depth of fluid being filtered. The present study confirmed these observations. In addition, work with in-line pressure membrane filter holders has always given even distribution of microorganisms on the surface of the filter. In this situation, a dynamic system is operating, and for practical purposes an infinite depth exists. No distribution difficulties would be expected under these conditions as long as adequate flow through the filter is maintained.

It is recognized that a direct microscopic count of bacteria in oil field waters still may not show the total bacterial population. For example, Conn (1948) has discussed the occurrence of bacteria which are too small to be resolved on the surface of filters. Coccoid bodies about 1μ or less in diameter have been observed in waters containing dissolved carbon dioxide but have been regarded as artifacts and not counted in this study. It should be recognized that these may be dwarf forms and proper cultural techniques would show them to be bacterial.

In spite of some limitations of the direct counting technique, it is felt that more reliable estimates of the total numbers of bacteria present are obtained in these unusual environments than can be obtained with traditional plating techniques.

Summary

A simple technique is described for using membrane filters in the determination of the number of bacteria present in oil field waters. Good correlation has been found between conventional plate counts and the direct counting procedure when a laboratory culture of *Aerobacter aerogenes* was used, thus demonstrating the reliability of the method. The technique has been found to yield much higher counts of bacteria when natural waters containing dissolved carbon dioxide, hydrocarbon gases, or traces of crude oil are examined. It is postulated that the higher counts from natural waters are due to the inability of usual cultural techniques to provide conditions allowing growth of all or most of the microorganisms present in these natural waters. This technique also allows the individual counting of bacteria occurring in clumps that would produce only one colony when cultured.

REFERENCES

Conn. H. J. 1948 The most abundant group of bacteria in soils. Bact. Revs. **12, 3, 257.**

Ecker. R. E.. and Lockhart, W. R. 1959 A rapid membrane filter method for direct counts of microorganisms from small samples. Jour. Bact. **70, 3, 265-268.**

Ehrlich, Richard 1955 Technique for microscopic count of microorganisms directly on membrane filters. Jour. Bact. **70, 3, 265-268.**

Jannasch, H. W. 1958 Studies on planktonic bacteria by means of a direct membrane filter method. Jour. Gen. Microbiol. 18, 3, 609-620.

Porter, David B. 1959 Personal communication. Millipore Filter Corporation.

Strugger, S. 1959 Fluorescens Mikroskopie und Mikrobiologie. Hannover. M.V.H.Schaper.

PRESIDENTIAL ADDRESS·

THE PAST AND FUTURE OF THE SOCIETY FOR INDUSTRIAL MICROBIOLOGY

C. L. Porter

Dept. of Biological Sciences, Purdue University

It seems fitting on this, the tenth anniversary of the Society for Industrial Microbiology, to discuss with you where we have been and whither we are going.

Not only is this a fitting time for review and prognostications but I feel that I am pretty well qualified to discuss this history with you. Not only am I a charter member of the society, but I have served in an official capacity as treasurer, secretary, and president since its organization.

The society was organized at a meeting in New York City on December 29, 1949. During the summer of 1949, letters were mailed by a publicity committee headed by Walter N. Ezekiel, announcing the time and place of the meeting and explaining the objectives to be achieved by such an organization. I read these letters with little enthusiasm as I was a member of the Mycological Society of America and I felt that there was little need for this new society. My lack of enthusiasm was shared by John S. Karling, head of my department at Purdue, who was also a member of the Mycological Society. We had many conferences concerning this proposed organization of microbiologists and finally just before our departure for the New York meeting we decided that if we couldn't lick them we had better join them.

There were about 300 present at the organization session and as is customary with such biological phenomena, the birth was accompanied by many obvious signs of distress, groans, tears, and almost bloodshed. It seems that there were two groups interested in gaining control of this new venture. There was much shouting and gesticulating and for a while, I was certain that the police would be called to settle the disturbance. Finally, one group prevailed and a temporary organization was effected. Dr. Charles Thom was elected temporary chairman of the meeting and much to my surprise, because I had taken no active part in the debate, I was elected temporary secretary. Dr. Thom was elected president until a constitution could be written and a permanent organization effected. At Dr. Thom's insistence I remained as secretary-treasurer until elections could be held under a constitution. Only thirty-three persons signed membership cards at the close of this initial meeting. It was decided here also that the new society would put on its first program at the 1950 annual meetings of the American Institute of Biological Sciences (AIBS) in Columbus, Ohio, September 11-13. Dr. Thom appointed a committee of organization consisting of M. M. Baldwin, W. N. Ezekiel, F. G. Walter Smith, E. A. Walker, W. L. White, and C. L. Porter. This committee was to write a constitution and prepare a program for the Columbus meeting. Dr. Thom's influence in this first year of our organization cannot be overemphasized. He was one of the most outstanding microbiologists of his day and a man of international reputation. His prestige meant much to the development of this new infant. Until his death, he was a loyal booster for this society. However, having appointed an organizing committee, Dr. Thom left further details of organization to the committee. He told me as secretary that I would be responsible for calling meetings of the committee and seeing to it that they accomplished their assigned tasks.

This committee met several times during the first year, prepared a program for the Columbus meeting, and wrote a constitution which after several minor changes was adopted at the first business meeting of the new organization. This constitution has been amended and is still undergoing revision but it continues to be the framework of our governing laws.

The society has been fortunate in having a succession of able and devoted men to serve as president, the present administration not included. These presidents include Dr. Charles Thom, who served two terms, Dr. Benjamin Duggar, Dr. Kenneth Raper, Dr. James Horsfall, Dr. H. Boyd Woodruff (two terms), Dr. J. M. McGuire, and Dr. C. W. Hesseltine.

Our official meetings have always been held with AIBS, although we have held a number of additional meetings with The American Assoc. for the Advancement of Science. Our programs have always been adequate, but the earlier meetings were not outstanding. Beginning with the meeting of The Univ. of Florida in Gainesville in 1954, there was a distinct improvement in the programs, and in my opinion the Gainesville meeting and all succeeding meetings have offered programs superior to those presented by any other society belonging to AIBS. The annual banquet was first instituted at Gainesville and has since become a tradition. At the next meeting in East Lansing, 1955, we began to put out SRO signs and these have been in order for some of our sessions at every meeting since.

At the end of our first year the society registered some 75 members and by agreement, all those joining during the first year were considered charter members. Our present membership lists about 550 members.

I have related the lack of enthusiasm which I felt for this new organization in 1949. However, having worked with it in an intimate fashion for ten years, I have become perhaps its most enthusiastic booster.

At the time of organization there appeared to be a real need for this new society.

The purely practical and industrial aspects of a learned society were mostly neglected by older groups including the bacteriological, mycological, botanical, zoological, and chemical societies. It was difficult in 1949 for a purely practical paper in microbiology to secure a place on the programs existing at that time or to find opportunity for publication in the older journals.

The changed thinking which has occurred since 1949 has been brought about by our society. The Mycological Society of America has become more industrially minded and practical papers concerned with microbiological developments are now accepted by that society's journal, Mycologia. Also, the Mycological Society now arranges joint sessions with both the Botanical Society of America and the Society for Industrial Microbiology, at which papers of industrial significance are read. The greatest changes however, have developed in the thinking and attitude of the Society of American Bacteriologists. This organization began the publication of a new journal, Applied Microbiology, which is devoted entirely to the broader and more practical aspects of microbiology. Applied Microbiology has had continued encouragement from our society and until 1960 was voted the official organ of the society. Many bacteriologists now call themselves "microbiologists." At the national meetings of S.A.B. they now have an industrial section at which papers of interest to industrial microbiologists are read.

As a society, S.I.M. has little to fear from the mycologists. Their membership at present is but little larger than ours without our potential for further growth. While they now tolerate discussions of the more practical aspects of fungi, they do not encourage such discussions and give only lukewarm support to industrial mycology. The Mycological Society is concerned, and properly so, with taxonomy and morphology of the fungi and with the more theoretical aspects of fungus physiology.

The greatest threat to the existence and continued growth of the Society for Industrial Microbiology is the Society of American Bacteriologists. This is a very large group numbering several thousand members. Their dues are high and they have no special financial worries. They can afford to do many things to attract the support of industrial microbiologists which we are not able to do financially.

I have no quarrel with S.A.B. I have been a member of that society for many years but if we are to continue to exist as a distinct organization we must decide in what areas we can compete successfully with S.A.B. and how we can provide services to industries with which no other organization can compete.

It is my purpose to suggest a few ideas which may be helpful in making us, after another decade, a stronger, sturdier, and more useful society than we are at present.

MEMBERSHIP

There are a number of organizations including the Society of American Bacteriologists and the American Chemical Society which have grown so large that they have become unwieldy. The administration of these groups has lost personal contact with the membership. In order to proceed efficiently with such large numbers they must enforce rules and regulations which hamper individual initiative, incentive, and recognition. At their meetings the sessions devoted to volunteer papers are so clogged that each presentation is limited to 10 or 12 min with little opportunity following the delivery for a discussion of the more salient points. In such large societies there is little opportunity for the majority of their membership to achieve recognition in the counsels and decisions of the group. There is little chance to become personally acquainted with the officers or, to become an officer. There is about as much chance for the rank and file to express a gripe or offer a constructive suggestion as there is in the Teamsters' Union. Such a situation does not now prevail in the Society for Industrial Microbiology and I hope that we never grow so large as to be guilty of such autocratic practices.

However, at present, our group is too small to be as effective as it should be in carrying out its objectives. I have a distinct feeling that with every one working on this job we could double our present membership before 1962. A membership of 1200 would not be too unwieldy and would permit us to accomplish many things, even at present dues levels, that are now beyond our grasp.

I believe that a membership beyond 1200—1500 should be discouraged, for then, we, like other groups mentioned, would lose that personal contact which now is one of the principal attractions. When we reach this level I would favor a waiting list, adding new members as present members drop out because of death, retirement, change of occupation etc.

Our membership includes chemists, bacteriologists, mycologists, physiologists, engineers, geologists, zoologists, marine biologists and plant pathologists. The several organizations in which these scientists work include industrial and educational institutions as well as federal and state laboratories. The principal fields of interest include deterioration, antibiotics, and other products produced by microorganisms, metabolism, and control. As to occupation they may be classified as technicians, supervisors of research and production, managers, teachers and investigators concerned with pure research. Since we accept all of these diverse groups into membership we must devise a program that will be of interest to all of them. If we are to be a successful society this program must include not only the few days of the annual meeting but be of appealing interest throughout the entire year. This takes extensive and intensive planning and devotion, hard work, and personal sacrifice on the part of the society's members. This is a responsibility that we dare not shirk if we are to survive, especially in the face of the competition which is becoming more acute.

Geographically, our membership represents all sections of the country, with the Eastern Seaboard and the Midwest having the greatest membership, both regions being highly industrialized. Forty of the 50 states of the United States have one or more members. We also have members on our rolls from Canada, Mexico, Honduras, England, Italy, and Sweden. We are really an international organization.

Every scientific society is actually a service organization which, if it justifies its reasons for existence, must seek constantly to promote the professional interests and opportunities of its members.

Many industries are represented in our membership. In fact, the majority of our members are associated in some fashion with industry. We should give every possible assistance that we are capable of rendering to these industrial concerns.

A few ways suggest themselves and my list of possibilities is certainly not complete. Some industries represented in our group are very large corporations with well-stocked libraries, and with research laboratories which are better equipped than most university laboratories. Such industrial concerns are more or less self-contained and actually need little outside assistance. On the other hand, we have represented in our society, industries that are too small at present or too new to have adequate research and library facilities. They need more promotional aid, which we could provide. These newer and smaller industries need more of our help than do the larger and older ones. But whether large or small, well-established, or struggling to become established, they should be able to discover unexpected dividends from their association with the Society for Industrial Microbiology.

Discoveries made in industrial laboratories which are patented, ready for market, and no longer secret should be publicized in our news sheet. Such discoveries are stimulating to others and provide new and better ideas for industrial research in addition to informing the public of new and useful products.

It would be even more profitable if results of experiments in academic laboratories could be made more quickly available to industrial laboratories. Much academic research is published in technical journals but much more is hidden in theses which never are published except possibly in abstract. The humblest thesis for a master's degree, which does not contain sufficient positive results to justify publication, often contains ideas of technique and results which, though inconclusive, should be explored further. Even negative results, if known, would save time, effort, and expense, in continuing an investigation involving the same or similar problems.

If the salient and significant results of unpublished theses could be made available to the membership of this society we could render invaluable service. I have an idea of how this could be accomplished which I would be happy to present to our executive committee.

Many industries are equipped and prefer to train their own technical personnel in the peculiar and special methods characteristic of their laboratory and development. Other companies are required to take raw high school or college graduates and hope that their academic training has been sufficient to permit them to do adequate work. Much time is lost. Young women get married before they have become efficient laboratory workers and leave their employment. Many of the young men never do learn the required skills and leave as failures, to go into teaching or the ministry.

The suggestion, originated with Dr. E. L. Dulaney, has been made that S.I.M. take over the sponsorship of summer classes for industrial personnel to provide additional training in the study and manipulation of microorganisms, particularly fungi. These classes would provide instruction somewhat similar to that now provided at Purdue by the Microbiological Institute, now in its thirteenth year. These courses should be more closely tailored to the specific needs of industrial personnel. S.I.M. would conduct a publicity campaign to call attention to the availability of such courses. Industries in our organization could provide speakers, exhibits, and suggestions for the content of such courses. I consider the idea a practical one that would be of great use to many companies concerned with microbiological processes.

This year for the first time S.I.M. has offered industrial memberships with dues set at $50.00 per year. A number of companies have availed themselves of this opportunity. It provides additional funds for the society and gives us more economic breathing space. However, we should not look on industrial memberships as charitable gifts any more than personal memberships are so considered. Every membership which we accept, whether personal or industrial, imposes upon us an obligation for rendering service. This year, we have not promoted either personal or industrial membership to the extent of our opportunities.

Our society should never become so large as to lose the personal contact with individual members. We should encourage colleges and universities as well as federal and industrial laboratories to give us information concerning their needs for prospective employees. Also, we should assume a certain responsibility for assisting our members to find suitable employment.

PUBLICITY

A young society such as ours cannot afford to hide its light under a bushel if it wishes to become influential and to attract members. Even this year after its ten years of existence, I have received letters from people desiring more information about S.I.M. and stating that they have just heard in circuitous fashion of the existence of such an organization catering especially to the interests of industry.

Publicity should not be confined to events taking place at an annual meeting. Any event that may be newsworthy within the knowledge of any member and which in any way will advertise the activities of S.I.M. as a society should be transmitted as soon as possible to the publicity committee or to the secretary. The publicity committee is not in a position to ferret out such items, its responsibility is only to act following reception of information.

The secretary is responsible for the issuance three times a year of News Letters. Our present secretary, Dr. Brinton Miller, is especially adept in this particular enterprise. He has enthusiasm, imagination, and ability. Located as he is in the laboratories of Merck Sharp & Dohme Research Laboratories, his office is a strategic listening post for items of interest to the membership. He is in a position to evolve our present News Letter into a valuable industrial journal or book. Lack of funds is all that prevents such an evolution. There are several possibilities that might provide a more adequate journal-news letter.

This financial dilemma which now prevents the publication of a more elaborate News Letter could be solved in part by a significant increase in personal and corporate memberships. But, if this news sheet is to evolve some day into the really good microbiological journal that some of us dream about, other measures must be considered. Two come to mind: (1) advertising—a journal going out to a thousand or more members should make many types of advertising a profitable venture to companies which participate. (2) We should investigate the possibility of combining with certain existing trade journals which would be happy to be a partner in such an undertaking. I have been told that there are several such journals which would appreciate such an opportunity.

Such a journal would help us pursue successfully a number of those projects which I have mentioned but which may at the moment be beyond our grasp.

Microorganisms have great industrial potential which we are just beginning to appreciate. We must be alert and imaginative and foresee trends which will develop into new and now undreamed of areas of exploitation of microorganisms.

Not too long ago, filamentous fungi appeared important only because of their nuisance value or as a sort of intellectual hobby. Journals in that era which carried any information at all about fungi concerned themselves principally with taxonomic disputes and mycological forays. This was the squabble and picnic era of mycology. Then came the discovery of penicillin and what a difference this made in the mycological outlook. If S.I.M. had existed in the decade following Fleming's discovery and if we had had the insight to grasp the potentialities, we wouldn't be wondering today if we were going to be organized out of existence.

Two events occurring at this meeting indicate that we have an awareness of things to come. viz., the symposia on the microbiology of outer space and hallucinogenesis. Industrial laboratories with a few exceptions are followers, not leaders. Apparently, they use little imagination and are not aware that the application of pure research today is the lifeblood of tomorrow. S.I.M. can function by indicating at early stages exciting possibilities.

One area that is not new, but which should receive the attention and support of this society is taxonomy. Taxonomy is the mother and the father of all microbiological research. It is as important to know what a fungus *is* as what it *does*. We need to know natural relationships in order to select our organisms intelligently and efficiently. We should be able to produce strains of species and relate them to each other and to their near relatives. Taxonomy is deadly monotony and few care to engage in it. The taxonomist is poorly paid during his lifetime and achieves no glory. His heirs can only afford for him an unmarked gravestone after he dies.

We need more monographs on fungus groups, such as the work on Penicillia and Aspergilli published by Thom and Raper and assisted most ably by Dorothy Fennell. Incidentally, all three of these great scientists are, or have been members of S.I.M. Their books are invaluable in industrial laboratories.

The great work of our previous president, C. W. Hesseltine, on the Mucorales of Wisconsin has never been published in detail because of the lack of funds.

The Deuteromycetes, otherwise known as Fungi Imperfecti, are in an amazing state of chaos and will remain so until monographs on the various taxa are written. I don't know how many appreciate the fact, but until the confusion of this great class is removed by adequate monographs many of these fungi cannot be investigated adequately. Monographs entail much research and frequently take a lifetime to complete and few have the funds or other incentives to sacrifice a lifetime in such unrewarded work. I see no solution to this problem until industries become sufficiently enlightened to their own best interests to subsidize such work with generous grants. I believe it to be one of the opportunities of S.I.M. to urge continuously that industries give financial support to such fundamental projects.

It is for the Society for Industrial Microbiology to decide whether it wishes to become a historical statistic, to continue in mediocrity, or to go to work and achieve greatness.